Vestibular Autonomic Regulation

Edited by

Bill J. Yates
Alan D. Miller

CRC Press
Boca Raton New York London Tokyo

Vestibular Autonomic Regulation

Library of Congress Cataloging-in-Publication Data

Vestibular autonomic regulation / edited by Bill J. Yates and Alan D.
Miller.
p. cm.
Includes bibliographical references and index.
ISBN 0-8493-7668-8 (alk. paper)
1. Vestibular apparatus. 2. Autonomic nervous system. I. Yates
Bill J. II. Miller, Alan D.
QP471.V46 1996
612.8'9--dc20 95-40474
CIP

Direct all inquiries to CRC Press, Inc., 2000 Corporate Blvd., N.W., Boca Raton, Florida 33431.

International Standard Book Number 0-8493-7668-1
Library of Congress Card Number 95-40474
Printed in the United States of America 2 3 4 5 6 7 8 9 0
Printed on acid-free paper

PREFACE

Movements of the head and changes in posture produce a number of compensatory responses. Probably the least understood of these are vestibular effects on autonomic control. For the first time, this book brings together an international group of experts who examine the interaction of the vestibular system with the autonomic nervous system from basic science and clinical perspectives. Topics addressed in this book include: (1) vestibular contributions to cardiovascular and respiratory homeostasis, (2) nausea, vomiting and autonomic changes during motion sickness, (3) effects on the autonomic nervous system produced by exposure to altered gravitational environments, (4) the relationship between vestibular disorders and panic attacks, and (5) orthostatic hypotension resulting from vestibular dysfunction. The book also contains several introductory chapters (to describe the control of balance, autonomic function, and respiration) so that investigators, clinicians, and students in many disciplines can fully comprehend the specialized material presented.

Bill J. Yates
Pittsburgh, Pennsylvania

Alan D. Miller
New York, New York

THE EDITORS

Bill J. Yates, Ph.D., is Assistant Professor of Otolaryngology in the School of Medicine and Assistant Professor of Neuroscience in the Faculty of Arts and Sciences at the University of Pittsburgh (Pennsylvania). He is also a member of the Adjunct Faculty of the Rockefeller University (New York), where he was an Assistant Professor from 1990 to 1994.

Dr. Yates obtained his Ph.D. degree in 1986 from the University of Florida College of Medicine. He served as a Postdoctoral Fellow at the Rockefeller University from 1986 through 1989.

Dr. Yates is a member of the Society for Neuroscience, International Brain Research Organization, New York Academy of Sciences, American Physiological Society, American Association for the Advancement of Science, the Bárány Society, and the Association for Research in Otolaryngology. Amongst his awards are the Irma T. Hirschl Career Scientist Award.

Dr. Yates has published over 40 manuscripts and book chapters, most of which deal with vestibular influences on autonomic control. He has received continuous grant support from the National Institutes of Health since 1990 to study vestibular influences on the sympathetic nervous system, and holds a National Institutes of Health grant jointly with Dr. Miller to study vestibular effects on respiration.

Alan D. Miller, Ph. D., is Associate Professor of Neurophysiology at The Rockefeller University (New York). He also served as a Visiting Associate Professor at the Faculté des Sciences, St. Jérôme (Marseille, France) in 1990. Dr. Miller received his Ph.D. degree in Physiology from The University of Western Ontario (London, Canada) in 1980.

Dr. Miller is a Councilor of the Central Nervous System Section of the American Physiological Society. He is also a member of the International Brain Research Organization, Society for Neuroscience, and Society for the Neural Control of Movement.

Dr. Miller has a long-standing interest in the control of nausea and vomiting, as well as in the respiratory and vestibular systems. He has published over 60 research papers and reviews. His work is supported by the National Institute of Neurological Disorders and Stroke, the National Institute on Deafness and Other Communication Disorders, and private industry. He has also received research funding from the National Aeronautics and Space Administration and the National Science Foundation.

Contributors

Carey D. Balaban, Ph.D.
University of Pittsburgh
Departments of Otolaryngology and Neurobiology
Eye & Ear Institute
203 Lothrop Street
Pittsburgh, PA

Susan M. Baser, M.D.
University of Pittsburgh
Department of Neurology
Scaife Hall
Pittsburgh, PA

Robert S. K. Cheung, Ph.D.
Defence and Civil Institute Environmental Medicine
North York, Ontario
CANADA

Kazuhisa Ezure, Ph.D.
Tokyo Metropolitan Institute for Neuroscience
Department of Neurobiology
2-6 Musashidai
Fuchu, Tokyo 183
JAPAN

Joseph M. Furman, M.D., Ph.D.
University of Pittsburgh
Departments of Otolaryngology, Neurology, and Electrical Engineering
Eye & Ear Institute Bldg.
203 Lothrop Street
Pittsburgh, PA

Laurent Grélot, Ph.D.
Faculté des Sciences Saint Jérôme
Département de Physiologie et Neurophysiologie
Case 351-352, CNRS URA 1832
F-13397 Marseille cédex 20
FRANCE

Rolf G. Jacob, M.D.
University of Pittsburgh
Departments of Psychiatry and Otolaryngology
Western Psychiatric Institute and Clinic
3811 O'Hara Street
Pittsburgh, PA

James R. Lackner, Ph.D.
Brandeis University
Ashton Graybiel Spatial Orientation Laboratory
Waltham, MA

Alan D. Miller, Ph.D.
The Rockefeller University
Department of Neurophysiology
1230 York Avenue
New York, NY

Kenneth E. Money, Ph.D.
Defence and Civil Institute Environmental Medicine
North York, Ontario
CANADA

James M. Perel, Ph.D.
University of Pittsburgh
Departments of Psychiatry and Pharmacology
Western Psychiatric Institute and Clinic
3811 O'Hara Street
Pittsburgh, PA

Fred H. Previc, Ph.D.
United States Air Force Armstrong Laboratory
Crew Systems Directorate
Flight Motion Effects Branch
2504 Gillingham Drive
Brooks Air Force Base, TX

David A. Ruggiero, Ph.D.
Cornell University Medical College
Departments of Neurology and Neuroscience
411 E. 69th Street
New York, NY

Robert H. Schor, Ph.D.
University of Pittsburgh
Departments of Otolaryngology and Neurobiology
Eye & Ear Institute
203 Lothrop Street
Pittsburgh, PA

Alan F. Sved, Ph.D.
University of Pittsburgh
Department of Neuroscience
446 Crawford Hall
Pittsburgh, PA

David L. Tomko, Ph.D.
NASA Ames Research Center
Vestibular Research Facility
Moffett Field, CA

Bill J. Yates, Ph.D.
University of Pittsburgh
Departments of Otolaryngology and Neuroscience
Eye & Ear Institute
203 Lothrop Street
Pittsburgh, PA

TABLE OF CONTENTS

1

Overview of Vestibular Autonomic Regulation

Bill J. Yates and Alan D. Miller

CONTENTS

1 WHY SHOULD THE VESTIBULAR SYSTEM INFLUENCE AUTONOMIC REGULATION?

This book is the first compilation of information and theories regarding vestibular influences on autonomic control. Movement and changes in posture can profoundly affect the cardiovascular and respiratory systems. For example, when the long axis of the body is suddenly subjected to the force of gravity, as when we stand from a supine position, there is a strong tendency for blood to pool in the legs and abdomen (see Reference 5 for review). Rapid corrections need to be made in the circulation or the return of blood to the heart will decrease, cardiac output will drop, perfusion of the brain will diminish, and symptoms such as lightheadedness and even syncope (fainting) may result. Movements can also place mechanical constraints on respiratory muscles and change their resting length;[3,4] these perturbations must be quickly corrected if stable blood oxygenation is to be maintained.

A number of sensory receptors detect disturbances in blood pressure and ventilation. These include baroreceptors, spindle receptors in respiratory muscles, pulmonary stretch receptors, and peripheral and central receptors that detect blood CO_2, O_2, and pH.[1] However, reliance on "error" signals from these receptors to maintain consistent perfusion and oxygenation of the tissues is not adequate; after all, these receptors only elicit changes in autonomic activity when significant disturbances in homeostasis already exist. The fact that blood pressure and oxygenation remain reasonably stable even during rapid changes in posture suggests that adjustments probably start at the beginning of a movement that is likely to affect homeostasis.

0-8493-7668-1/96/$0.00+$.50

Among the sensory systems, the best candidate for eliciting alterations in circulation and respiration at an early point during challenging movements is the *vestibular system,* which has receptors located in the labyrinth of the inner ear. Vestibular receptors signal the direction and velocity of head movements and the static position of the head with respect to gravity. These inputs constantly provide the central nervous system with the information needed to produce correction for ongoing movements.[2]

2 PROCESSING AND EFFECTS OF VESTIBULAR SIGNALS

Work throughout this century has discerned many of the mechanisms through which the vestibular system influences motor control. A simplified view of vestibular actions on motor pathways is shown in Figure 1A. The vestibular nuclei receive inputs from vestibular receptors as well as from the body's proprioceptors, especially those in the neck. The processing of signals by the vestibular nuclei is modulated by the cerebellum. The efferent outflow of the vestibular nuclei is directed to motoneurons and pre-motor interneurons in the spinal cord; this so-called vestibulospinal pathway plays a large role in maintaining balance and stable posture. Neurons in the vestibular nuclei also have ascending connections to motoneurons in cranial nerve nuclei that innervate the extraocular muscles, as well as to nearby interneurons. Vestibular projections to the neural circuitry that controls eye position form the "vestibulo-ocular" pathway.

Although early studies demonstrated the presence of vestibular effects on respiration and circulation, research on vestibulo-respiratory and vestibulo-cardiovascular pathways is still limited. However, many recent studies have focused on these connections. As described above, and as indicated in Figure 1B, one function of vestibulo-autonomic pathways is to produce compensatory changes in breathing and circulation that are necessary during movement and changes in posture. However, this is not the only significance of vestibulo-autonomic pathways. It is well known that the vestibular system has prominent influences on the neural circuitry that mediates nausea and vomiting. For this reason, emesis is a common symptom of vestibular disease. Under normal conditions, vestibular inputs are involved in producing nausea and vomiting during motion sickness. Finally, the vestibular system has powerful connections with brainstem centers that contain monoaminergic neurons which bias the activity of many neurons in the central nervous system. Atypical or inadequate vestibular inputs to nucleus locus coeruleus, which contains norepinephrine-releasing neurons, and other brainstem pathways can be manifested as psychiatric disorders such as agoraphobia and panic attacks.

3 ORGANIZATION AND PURPOSE OF THIS BOOK

The first section of this book (Chapters 2–5) provides background on the vestibular system, the neural regulation of circulation, the neural control of breathing, and the neural basis of nausea and vomiting. The next section (Chapters 6–8) reviews data collected in animal models that (1) demonstrate the presence of vestibular influences on respiration and circulation and (2) indicate which neural pathways participate in relaying vestibular signals to neurons that regulate breathing and blood pressure. The third section (Chapters 9–12) reviews clinical findings regarding autonomic disturbances that result from vestibular dysfunction (or abnormal processing of vestibular signals). Chapter 9 discusses the literature concerning motion sickness, and presents a theory explaining why this condition may occur. Chapter 10 discusses autonomic disturbances that occur when the gravitational environment, and thus the pattern of vestibular inputs, is altered (e.g., during microgravity

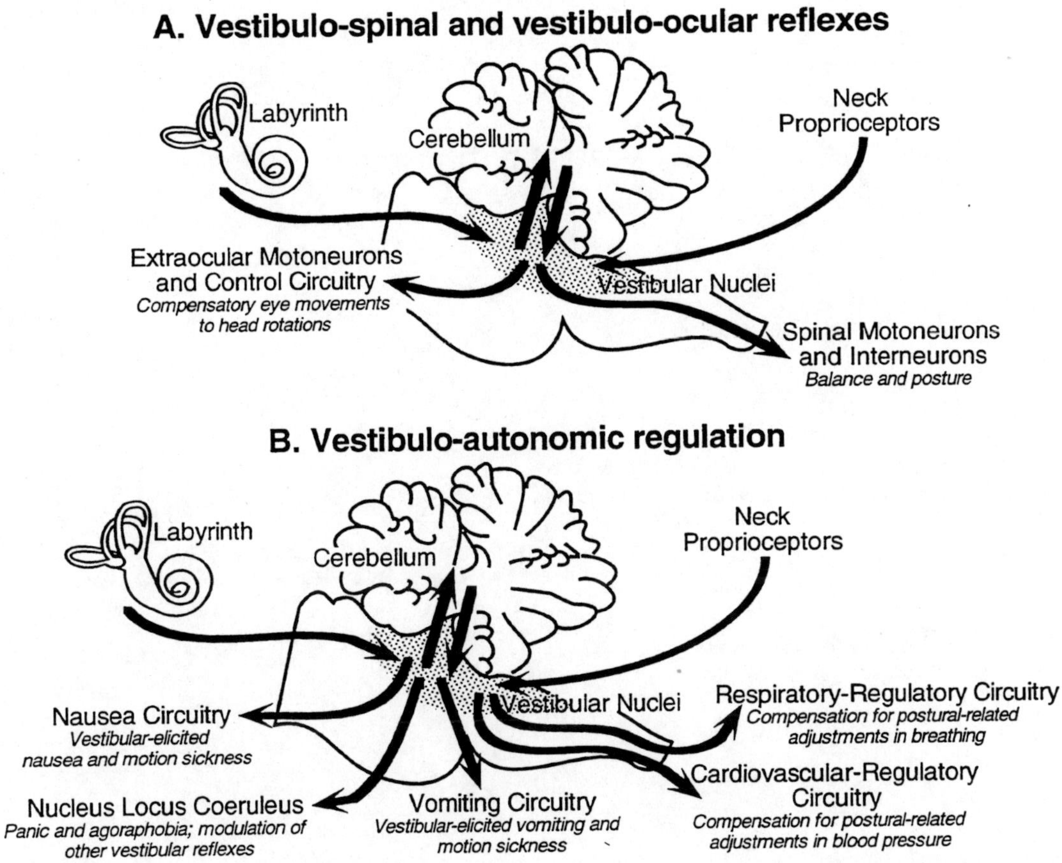

FIGURE 1
An overview of vestibular influences on motor and autonomic control. (A) Vestibulo-spinal and vestibulo-ocular pathways that maintain balance and stable eye position, respectively, during movement and changes in posture; (B) vestibular influences on autonomic control.

conditions in space or hypergravity conditions during flights in high-performance aircraft). Chapter 11 reviews recent findings that dysfunction in vestibular inputs to brainstem autonomic and monoaminergic pathways can result in some conditions that have classically been considered psychiatric disorders, including agoraphobia and panic attacks. Finally, Chapter 12 discusses posturally related hypotension (orthostatic hypotension) and possible links of the vestibular system with this condition.

REFERENCES

1. Guz, A., Regulation of respiration, *Annu. Rev. Physiol.*, 37, 303, 1975.
2. Goldberg, J. M. and Fernández, C., The vestibular system, in *Handbook of Physiology, Sec. 1: The Nervous System. Vol. III: Sensory Processes, Part 2*, Brookhart, J. M. and Mountcastle, V. B., Eds., Washington, D.C., American Physiological Society, 1984, 977.
3. Leevers, A. M. and Road, J. D., Effect of lung inflation and upright posture on diaphragmatic shortening in dogs, *Resp. Physiol.*, 85, 29, 1991.
4. Newman, S. L., Road, J. D., and Grassino, A., *In vivo* length and shortening of canine diaphragm with body postural change, *J. Appl. Physiol.*, 60, 661, 1986.
5. Yates, B. J., Vestibular influences on the sympathetic nervous system., *Brain Res. Rev.*, 17, 51, 1992.

Section I

ORGANIZATION OF VESTIBULAR AND AUTONOMIC CONTROL

2

The Vestibular System

Robert H. Schor and David L. Tomko

CONTENTS

0-8493-7668-1/96/$0.00+$.50

1 INTRODUCTION — OVERVIEW OF VESTIBULAR FUNCTION

1.1 Head Movements in Space

The vestibular system consists of movement sensors in the labyrinth of the inner ear, several brainstem nuclei, and projections to various brainstem and spinal motor centers. Vestibular neural circuits serve to detect, and compensate for, head movements in three-dimensional space. The traditional view of the vestibular system, which will be reviewed in this chapter, emphasizes two major vestibular reflex systems, an ascending vestibuloocular reflex that aids in the control of eye movements and a descending vestibulospinal system involved in limb and head postural control (reviewed in References 17 and 38). As will be developed in this volume, other pathways and reflex systems are beginning to be studied and described that relate the vestibular system to other homeostatic control systems in the body.

Head movements that take place on the surface of the Earth are influenced by the Earth's gravitational field and can be broadly divided into three classes: rotations about an Earth-vertical axis (such as shaking your head no), rotations about an Earth-horizontal axis (such as nodding your head yes), and translations that do not involve rotations (such as the forward and up/down bobbing motion of the head during walking). These classes of head movements, in turn, can be analyzed in terms of combinations of angular (rotatory) and linear (non-rotatory) accelerations and the forces they impose on the head. As we shall shortly see, rotatory stimuli are sensed by the semicircular canal system of the labyrinth, while the linear accelerations are sensed by the otolith organs.

The angular accelerations are intuitively easy to grasp; any turning of the head about any axis involves an angular acceleration. Linear acceleration stimuli, however, can arise from two sources. One is to simply impose a movement on the head (such as the example of walking). The other stimulus arises from the presence of the Earth's gravitational field, which exerts a constant (spatially) downward pull on the head. If the head orientation with respect to gravity is changed by rotating the head about some Earth-horizontal axis, the head will experience a change *in direction* of the gravitational pull of the Earth. Hence, head movements involving rotations about horizontal axes produce both angular and linear acceleratory stimuli.

1.2 Detection of Head Movements *Per Se*

Why is the detection of head movements important? The head carries a number of sensors (such as the eyes, nose, ears, whiskers) and effectors (mouth, tongue) which interact with the essentially stationary external world. During head movements, whether the result of an active head movement, a relatively passive head movement as the result of an active body movement (such as the head movements experienced during locomotion), or an unexpected movement (for example, the result of a slip or fall), the relative orientation between the head's sensors and effectors and the external world will change. One function of the vestibular system is to provide an internal reference frame to enable head-referenced sensory and motor systems to stay aligned with external targets. For example, if you turn your head to look more closely at an object of interest, the head rotation would cause a movement of the target relative to the head and its eyes, degrading the visual image. As we shall see, however, one very important vestibular reflex, the vestibuloocular reflex, serves to drive the eyes in the head to help stabilize the eyes in space during movements of the head.

1.3 Movements of the Head on the Body

Biomechanically, the head represents a significant mass at the end of a flexible neck. Active movements of the head are produced by the coordinated action of neck muscles pulling against the much larger, ground-supported mass of the rest of the body. In addition to such active head movements, the head/neck system is subjected to a number of passive forces arising from the head's mass sitting on the end of the vertebral column. One such force is gravity, which is constantly pulling the head down (usually flexing the neck). Another arises during normal body movement, as the body moves under the head; for example, if you suddenly start to move your body forward by pushing your trunk forward with your legs, the inertia of the head, which tends to keep it stationary in space, will induce a backward extension of the neck.

There are two sensory systems that serve to monitor the movement of the head. One, originating largely in neck muscle spindles, monitors the relative motion of the head on the neck. The other, the vestibular system, monitors the absolute movement of the head in space. Examining the ability of animals (including man) with labyrinth deficiencies to control their heads, one notes two major features. One is the presence of extraneous motion, often oscillatory in nature, and the difficulty in holding the head still.[27,29] The other is the inability to make accurate rapid head movements without overshooting the target position, as though the mass of the head were not being adequately controlled.[3,14]

1.4 Postural Control

The maintenance of upright posture, both during quiet stance and during locomotion, also involves input from the vestibular system. One element of such postural control, namely the sensing and control of head movement on the body, has been noted above. Vestibular input to the "antigravity" extensor muscles has been known since the work of Magnus,[27] and is in part responsible for the extensor rigidity seen in the decerebrate preparation.

In addition to helping control posture, the vestibular system can act as a sensor of changes in posture for other regulatory systems. Postural changes, such as going from lying down to a standing position, may require significant alterations of cardiovascular, respiratory, and other homeostatic control systems. Such significant changes in the orientation of the trunk and limbs will result in passive movement of the head, thus stimulating the vestibular system at the time of movement. These vestibular signals can potentially anticipate the changes in the demand on these regulatory systems, thereby allowing for rapid compensation for the altered posture. For example, the labyrinth would signal the change in posture occasioned by standing well before sensors in the carotid sinus would detect an orthostatic drop in blood pressure.

2 VESTIBULAR SYSTEM ANATOMY

2.1 Overview

The vestibular sensory apparatus is part of the membranous labyrinth, a fluid-filled structure located within the temporal bone in the inner ear. Figure 1 shows the major parts of the labyrinth, including the auditory cochlea, the three angular-acceleration-sensitive semicircular canals, and the two gravity- and linear-acceleration-sensitive otolith organs

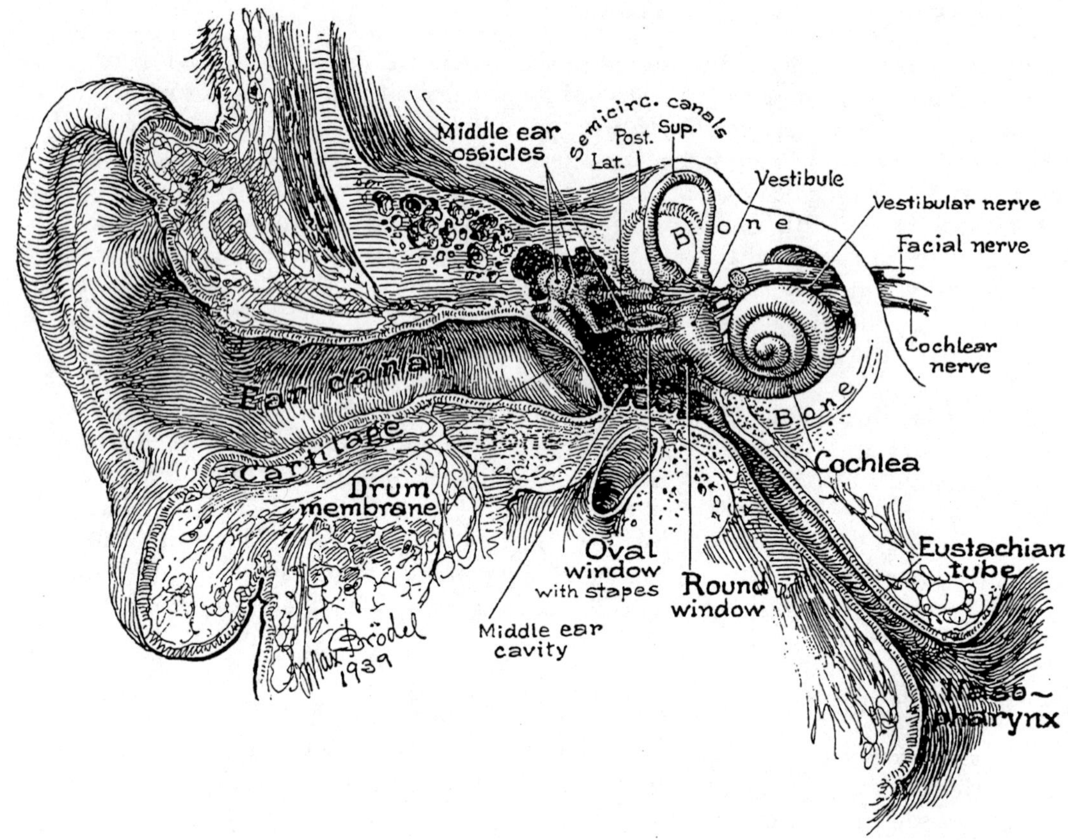

FIGURE 1
Diagram of the human ear, showing the vestibular labyrinth and cochlea within the temporal bone.[6] Note the horizontal orientation of the lateral semicircular canal, and the vertical orientation of both the anterior and posterior (vertical) semicircular canals. The two otolith organs, the utricle and saccule, are located with the central vestibule of this organ.

which are located in the common vestibule from which the semicircular canals originate. Sensory neurons from the eighth cranial nerve, the vestibulocochlear nerve, innervate hair cells within these specialized receptors. Most vestibular sensory neurons terminate in brainstem vestibular nuclei, which serve to relay information about head movements to oculomotor, postural, and other regulatory motor systems.

2.2 Vestibular Sensory Receptor Cells

Transduction in the labyrinth (the semicircular canals, the otolith organs, and the auditory cochlea) starts with hair cells, arranged in specialized epithelia within the membranous labyrinth. As shown in Figure 2, these cells bear on their surface two kinds of hairs: staggered rows of stereocilia, and a single eccentric kinocilium at the long end of stereociliar bundle. Transduction occurs by bending the hairs; as the stereocilia are bent towards the kinocilium, channels near their tip are thought to open, allowing ions to enter and depolarize the cell.[23] Thus each hair cell acts as a tiny accelerometer, maximally sensitive to acceleration applied along a single axis. This structure imparts to each cell a property called morphological polarization.

Vestibular nerve afferents form chemical synapses at the base of hair cells. These afferents have a high resting activity (50 to 100 impulses per second), which allows them

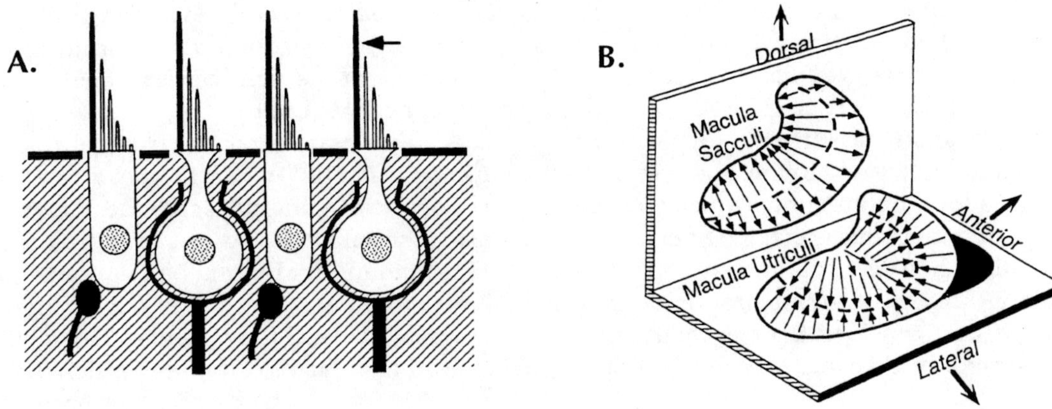

FIGURE 2
Schematic vestibular hair cells (A) bear on their surface staggered rows of stereocilia and an eccentric, longer kinocilium. As the arrow indicates, deflection of the hairs in the direction of the kinocilium leads to increased firing of the vestibular afferent fibers which either surround or make more localized synapses with the hair cell bodies. Within each semicircular canal, the morphological polarizations arising from this asymmetric arrangement of hairs are all parallel; a given head movement will excite or inhibit all of the hair cells of a particular canal. Within the otolith organs, as illustrated in B, the orientations of the hair cell polarizations basically cover either the horizontal (in the case of the utricle) or vertical (for the saccule) plane. Thus if one considers the stimulus to be the downward acceleration due to gravity, each head orientation will result in excitation of a different population of utricular and saccular afferents.

to effectively signal both excitatory (depolarizations, bending of hairs towards the kinocilium) and inhibitory (hyperpolarizations due to bending away from the kinocilium) stimuli.

2.3 Semicircular Canals

In the semicircular canals, the hair cells are located at one end of each canal in a swelling called the ampulla. Within the ampulla is a structure, the crista, which bears the hair cells on its surface; the hairs, in turn, are embedded in a gel-like structure called the cupula. The crista and cupula are arranged to block the flow of endolymph around the canal; head movements that would tend to cause fluid to flow around the ring of a particular canal result in a deformation of the cupula, and subsequent bending of the embedded hair cells.

Because of their circular construction, the canals are designed to respond to rotations, preferably about an axis perpendicular to the plane in which they lie. During normal upright stance, the lateral (or horizontal) semicircular canals will be approximately horizontal with respect to the ground, with the other two canals being vertical. The two vertical canals, besides being perpendicular to the horizontal canal, are also perpendicular to each other, and lie in planes tilted about 45° from both the frontal and sagittal planes. Their names, the superior (or anterior) and inferior (or posterior) semicircular canals, describe their location. Although this orientation of the labyrinth puts no canal in the body's sagittal or frontal plane, a bilateral symmetry emerges with each canal lying in the same plane, but mirror reflected, as a canal on the opposite side. Thus the coplanar canal pairs are the two horizontal canals, the right anterior/left posterior, and left anterior/right posterior vertical canals. Any head rotation, about any axis in space, can be broken down into equivalent rotations in the three mutually perpendicular planes of the semicircular canals.

Within each canal, the hairs are all arranged in parallel, so any particular movement excites or inhibits all of its hair cells. Fluid flowing in the canal towards the ampulla excites the horizontal canal and inhibits the vertical canal afferents. An easier rule to remember is that ipsilateral rotation excites each canal. Thus for the right horizontal canal, turning the head to the right is excitatory. It is a bit more complicated for the right vertical canals, but rotating the head in the frontal plane to the right excites them both.

The adequate stimulus for the canals is head angular acceleration (from physics, Newton's Laws of Motion, the endolymph trying to "stay still" when the head rotates exerts a force on the cupula equal to its moment of inertia times angular acceleration). One consequence of this is that long periods of rotation at a constant velocity (not a very natural situation, however) fail to stimulate the canals. Fluid motion within the narrow canal is also influenced by fluid drag and the elasticity of the cupula. The net result of these mechanical considerations is that for the normal range of head movements, the signal presented by the canal afferents is proportional to head angular velocity (see Section 3.3).

2.4 Otolith Organs

The other vestibular sensors are the otolith organs, located within the vestibule of the labyrinth. Within these two structures, the hair cells are arranged in a sheet, the macula, with the hairs embedded in a gelatinous matrix weighed down by calcium carbonate otoconia or otoliths (literally ear-stones). If the macula is tilted, or otherwise subject to an acceleration, the otoconia "roll downhill", bending the underlying hairs. As illustrated in Figure 2, when the head is erect, the macula in the utricle (or utriculus) lies roughly horizontal; the macula in the saccule (or sacculus) lies near the sagittal plane, and is therefore vertical. The orientation of the hair cells in the maculae point in all directions, unlike the parallel arrangement in the canals. Thus for any degree of tilt from an erect posture, some utricular afferents will be excited, and some will be inhibited; similarly, upward (or downward) movements will excite some saccular afferents and inhibit others. Static orientations of the head can be encoded by the pattern of otolith organ afferents. For example, with the head erect, the utricular afferents are all near their resting baseline activity, while populations of saccular afferents are maximally excited or inhibited (if a person were to do a headstand, the utricular pattern would remain the same, but the saccular pattern would reverse). Tilting the head in any direction would change the pattern of activation of the utricular afferents, exciting some and inhibiting others; in particular, a lateral tilt of the head would excite about 75% of the ipsilateral utricular afferents.

Combining canal and otolith information can provide information for all types of head movement. The canals inform the CNS about rotations in three dimensions. The otolith organs provide information about three dimensions of head accelerations, a special case of which is to provide information about head orientation with respect to gravity.

2.5 Central Vestibular Connections

2.5.1 Afferent Projections to the Vestibular Nuclei

Vestibular afferents that innervate the labyrinth form part of the eighth cranial nerve, which enters the medulla at the level of the inferior cerebellar peduncle and terminates primarily within the vestibular nuclei located in the rostral medulla on the ipsilateral side. Four major subdivisions can be distinguished on a cytoarchitectonic basis — the superior vestibular nucleus, the lateral vestibular nucleus (also called Deiters' nucleus), the medial vestibular nucleus, and the inferior vestibular nucleus (also called the descending nucleus). Vestibular afferents enter the vestibular nuclei at the level of the lateral nucleus, where they typically bifurcate, with a descending branch terminating in the medial and inferior vestibular nuclei, and an ascending branch that branches within the superior vestibular nucleus, as well as sending projections to the cerebellum (primarily to the nodulus and uvula). Two other regions that receive significant primary vestibular input are a group of cells in the medulla called the y group and the interstitial nucleus of the vestibular nerve.

Afferents from the three semicircular canals and the two otolith organs project differentially to the various vestibular nuclei, although there is considerable overlap in their projections (suggesting possible convergence of vestibular input). The following represents the major vestibular inputs to each nucleus (see References 8, 22 for more detailed descriptions). The vertical semicircular canals project heavily to the superior vestibular nucleus, while the horizontal canal projects to the medial and descending nucleus. Both the medial and descending nuclei receive utricular and saccular inputs; another region receiving significant saccular input is the small group of cells called the y group. The direct projection to the lateral vestibular nucleus appears to be relatively sparse, though the more ventral neurons do receive some input from the otolith organs, as well as from the horizontal canal. In addition to these direct monosynaptic inputs from the vestibular nerve, neurons in the vestibular nuclei also receive polysynaptic vestibular input. Thus the physiological response of neurons to natural vestibular stimulation is more widespread, and shows more convergence of vestibular input, than may be expected from the anatomy alone.

2.5.2 *Other Inputs to the Vestibular Nuclei*

There are several other inputs to the vestibular nuclei which have implications for the integrative function of this region of the brainstem. One is the connections between the vestibular nuclei and the cerebellum, treated below and in Chapter 8. Another is an extensive inhibitory commissural system that interconnects the vestibular nuclei of the two sides. As will be described below, head movements that activate vestibular neurons on one side of the head typically decrease the activity of corresponding neurons on the opposite side of the head. The inhibitory commissural system enhances this push-pull excitation/inhibition, since excitation of one vestibular nucleus leads (via the inhibitory commissure) to inhibition of the contralateral nucleus.

Vestibular neurons also receive somatic input from the neck, body, and limbs.[32] This input can be direct, typically as collaterals of spinocerebellar tracts, or more indirect relays through spinal inputs to the reticular formation or to the cerebellum. The presence of such inputs suggests that neurons in the vestibular nuclei are integrating multiple sensory inputs concerning head and body orientation. For example, as described below, many vestibulospinal neurons show a response to neck rotation that is opposite to their response to vestibular stimulation. One consequence of this organization is that such neurons show little modulation during head rotation on a stationary body, which activates both the neck and vestibular inputs.

2.5.3 *Ascending Projections to Oculomotor and Higher Centers*

The vestibular nuclei are closely connected with the oculomotor centers; many vestibular nucleus neurons receiving monosynaptic input from the labyrinth project, largely via the medial longitudinal fasciculus, to the three extraocular motor nuclei where they synapse on oculomotor neurons.[8,22] This three-neuron arc forms the basis of the vestibuloocular reflex (VOR). The general organization of these direct vestibuloocular pathways is illustrated in Figure 3. Using the principle that the purpose of the VOR is to maintain gaze stable in space, one can deduce much of the basic circuitry (although the details may vary somewhat). For example, horizontal rotation of the head requires a horizontal compensatory rotation of the eyes. This can be accomplished by an excitatory connection from labyrinth to ipsilateral vestibular nuclei to contralateral abducens nucleus

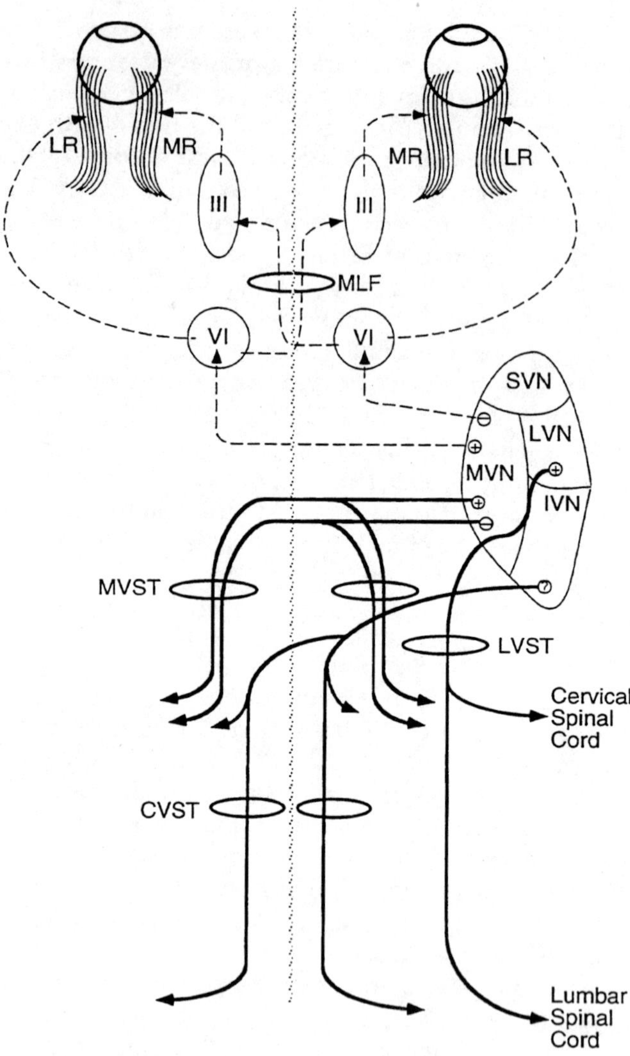

FIGURE 3
Schematic direct reflex projections from the vestibular nuclei. The upper part of the figure illustrates some of the pathways serving the horizontal vestibuloocular reflex (similar pathways connect to the other four extraocular muscles to produce the vertical and torsional VOR). Note that rotation to the right, which would excite vestibular neurons in the diagrammed right vestibular nuclei, result in excitation of the left lateral rectus motoneurons within the abducens nucleus (VI) and movement of the left eye to the left, thus keeping the eye fixed in space. Also shown is the recrossed excitation of the right medial rectus motor pool (located in the oculomotor nucleus, III) by abducens interneurons. A similar inhibitory projection to the ipsilateral abducens nucleus further contributes to this reflex. Not illustrated is the crossed inhibitory influence from the contralateral labyrinth relayed by the vestibular commissural system. The lower part of the figure illustrates the three descending vestibulospinal pathways. Note that the medial vestibulospinal tract (MVST) does not extend to the lumbar spinal cord. In addition, this tract has been shown to have both excitatory and inhibitory influences on its target motoneurons.

which innervates the contralateral lateral rectus muscle, counter-rotating the (contralateral) eye in the horizontal plane. Another excitatory pathway activates the medial rectus muscle of the ipsilateral eye, causing the two eyes to move in parallel (this particular pathway provides an "exception that proves the rule," as much of the excitatory drive comes, not directly from the vestibular nuclei, but through an internuclear relay from the contralateral abducens nucleus). Finally, there are also inhibitory projections from the vestibular nuclei to the appropriate oculomotor centers (in the horizontal example, vestibular neurons form inhibitory synapses on neurons in the ipsilateral abducens nucleus). Thus each eye muscle receives an appropriate increased or decreased drive from both labyrinths.

The nature and function of vestibular projection to the cerebral cortex is not as well known. There are several areas in the primate cortex which respond to vestibular stimulation, including neurons in area 3a (in the "neck region"), area 2v of the parietal cortex, a few neurons in parietal areas 7a and 7b, and a fairly extensive region in the parieto-insular cortex.[18] Many of these neurons respond to multiple sensory inputs, particularly movements of the visual field and somatosensory input from the neck. In the parieto-insular vestibular cortical region, for example, neurons were strongly activated by whole-body rotation, with little change in firing during sustained tilt, suggesting input from the semicircular canals, but not the otolith organs. In addition, most of the neurons in this region could also be activated both by manipulation of the neck, shoulders, and forelimbs and by rotation of a striped drum around the (stationary) animal. Where dual inputs were tested, there appeared to be no clear pattern of interaction; the responses could either cancel (as illustrated in Figure 5) or facilitate each other.[19]

2.5.4 *Vestibulospinal Tracts*

Vestibular information is directly relayed to spinal motor centers via the vestibulospinal tracts.[39] The medial vestibulospinal tract (MVST) originates primarily in the medial nucleus, with some input from the lateral nucleus, and projects bilaterally and monosynaptically to axial motoneurons of the neck and upper thoracic cord. It contains both excitatory and inhibitory neurons (note this is an exception to the general rule that descending inhibition takes place through interneurons). The lateral vestibulospinal tract (LVST), originating primarily in the lateral vestibular nucleus and the rostral part of the inferior vestibular nucleus, is a primarily ipsilateral excitatory pathway which traverses the entire cord. Its action is to facilitate limb extension, which it appears to do by making excitatory connections on spinal interneurons which, in turn, excite these muscles. In addition, the caudal part of the medial and descending nuclei gives rise to a third vestibulospinal projection, the caudal vestibulospinal tract (CVST).[30] The function of this tract, which appears to project bilaterally to all levels of the spinal cord, is still unknown.

The vestibular nuclei have extensive connections with other regions of the brainstem, especially with the underlying ponto-medullary reticular formation. Neurons in these regions give rise to the reticulospinal tracts, which descend bilaterally to all levels of the spinal cord. Reticulospinal neurons respond to both electrical and natural vestibular stimulation, and thus provide a route for vestibular information to reach motoneurons at all levels of the spinal cord. Similarly, the interstitial nucleus of Cajal, which appears to be involved in vertical and rotatory eye movements, also gives rise to a primarily ipsilateral projection to all levels of the spinal cord. It appears particularly to play a role in the control of head posture.[13]

2.5.5 *Vestibular-Cerebellum Connections*

In addition to projections up to extraocular muscles and down to the spinal cord, the vestibular system is interconnected with regions of the cerebellum.[24] Vestibular afferents, besides synapsing in the vestibular nuclei, also project to the flocculus, nodulus, and uvula of the cerebellum; these cerebellar regions are often called the vestibulocerebellum. These cerebellar regions also receive a secondary vestibular input, primarily from the inferior vestibular nucleus. The cerebellar cortex projects, in turn, back to the vestibular nuclei, both directly via axons of Purkinje cells (which inhibit their target neurons) and through projections from the fastigial nuclei. In addition to projections from the vestibulocerebellum (which appear to target all of the vestibular nuclei except the lateral vestibular nucleus), the vermis of the anterior lobe projects to the dorsal part of the lateral vestibular nucleus. These cerebellar regions additionally receive visual, oculomotor, and proprioceptive inputs, and are believed to play a role in the modification of vestibular reflex activity. These topics will be treated in further detail in Chapter 8 by Balaban.

2.5.6 Vestibular Efferents

The labyrinth receives a small bilateral input from a group of vestibular efferent neurons lying in the medulla, near the abducens nucleus. Although cochlear efferent systems are inhibitory, the vestibular efferent system appears to facilitate the activity of mammalian vestibular afferents, and perhaps to increase the sensitivity of some of these afferents to natural vestibular stimulation.[16,17]

3 VESTIBULAR PHYSIOLOGY

3.1 Natural Vestibular Stimulation

The stimuli for the vestibular system are angular and linear head accelerations. Such accelerations are conventionally described using coordinate axes fixed to the head. The naso-occipital axis passes from the back of the head out the nose. Rotation about this axis tends to raise one ear and lower the other; this motion is called roll. The interaural axis passes through the ears; rotation about this axis raises and lowers the nose, and is called pitch. The third axis is the head's dorso-ventral axis; rotation about this axis (as though saying no) is called yaw.

When considering the activation of the semicircular canal, yaw rotations, produced by either active head movements or passive rotations on a swivel chair while sitting upright, will selectively stimulate the horizontal semicircular canals. Pitch and roll stimulate the vertical canals, with pitch (as when nodding yes) producing an activation pattern that simultaneously excites, say, both anterior semicircular canals while inhibiting the posterior canals, and roll activating both vertical canals on one side of the head while inhibiting the vertical canals on the other.

Linear accelerations, which serve to stimulate the otolith organs, can be produced either by accelerating the head (for example, vertical accelerations often accompany locomotion, particularly during running) or by changing the orientation of the head with respect to gravity. These latter changes in orientation (tilt) are commonly produced by roll and pitch movements of the head, assuming the body is erect. Thus nodding the head yes would stimulate not only the vertical semicircular canals, but also the otolith organs.

3.2 Responses of Vestibular Nerve Afferents

As noted previously, vestibular afferents have a resting discharge rate of 50–100 nerve impulses per second in the absence of any motion stimulus. Head angular accelerations in the planes of the various semicircular canals will modulate the firing of afferents innervating those canals; similarly, changes in head orientation with respect to gravity, or changes in the forces due to linear acceleration of the head, will result in altered firing patterns of otolith organ afferents.

The firing pattern of otolith afferents can be understood by recalling the morphological polarization imposed on the hair cell by the arrangement of stereocilia and kinocilium;

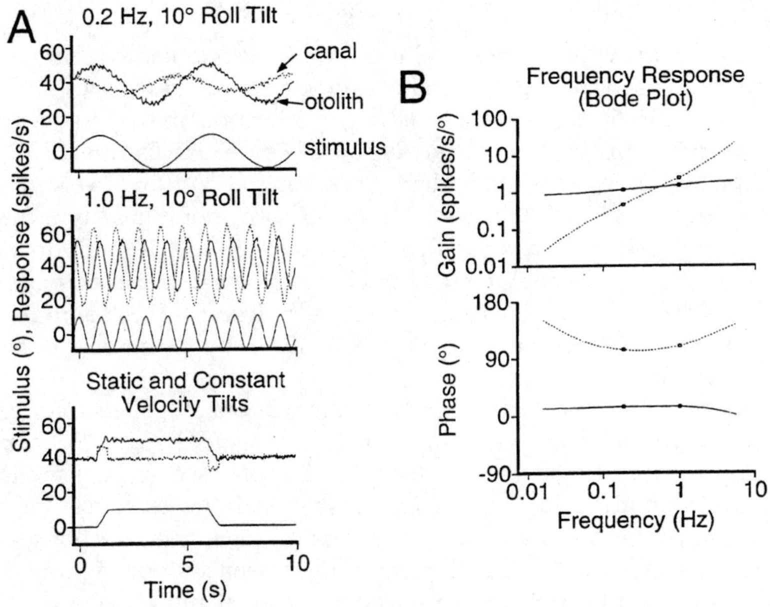

FIGURE 4
Responses of a typical otolith and canal afferent to natural tilt stimuli. Responses are modeled on squirrel monkey experimental data.[9,11] (A) The sinusoidal, or "position trapezoid," stimulus is shown on the lower traces, while the response of an otolith (solid line) and canal (dotted line) afferent are shown above. The two sinusoidal stimuli have the same amplitude of tilt, but the stimulus at 1 Hz has a higher peak velocity (63° per second) than does the 0.2-Hz stimulus (13° per second). The velocity response of the canals is especially evident on the bottom stimulus, where the canal response basically reproduces the two pulses of constant positive and negative velocity. (B) The Bode plot summarizes the frequency response of these afferents; the points on the graphs correspond to the gain and phase of the 0.2- and 1-Hz responses on the left.

bending of the hairs toward the kinocilium is excitatory. Hair cells in the utricular macula have their morphological polarizations laying in roughly the horizontal plane; these afferents will therefore be stimulated to varying degrees by roll and pitch tilts, but will be relatively unaffected by vertical accelerations.[10] Similarly, afferents innervating the sacculus will be stimulated by vertical accelerations and pitch, but not particularly by roll tilts. It is possible to deduce the orientation of the polarization vector by imposing a linear acceleration (or a tilt) in a variety of directions; when the tilt is aligned with the functional polarization vector, the response will be maximal, while tilts perpendicular to this vector will be ineffective. There are thus two main determinants to the firing pattern of utricular (and saccular) afferents, namely the direction of tilt (relative to the functional polarization vector), and the magnitude of tilt.

Responses of semicircular canal afferents also depend on the orientation of the axis of rotation with respect to the head, due to the three-dimensional geometry of the canals. Experimentally, one can use this fact to identify which canal an afferent innervates; rotation in only one of the three semicircular canal planes will produce significant modulation.[15] Within a particular canal, the morphological polarization pattern of hair cells is similar, so canal afferents of a particular canal are all excited or inhibited together. Although angular accelerations of the head are necessary to in fact modulate the firing of semicircular canal afferents, the mechanical properties of the membranous canal, the elastic properties of the crista and cupula, and the viscous drag of the endolymph within the canals cause the firing rate to lag behind the acceleration stimulus; the response right after a brief acceleration is more closely related to head angular velocity (see Figure 4).

3.3 Analysis of Vestibular Response Dynamics

Head movements provide the physical input to the vestibular system. There are two techniques commonly used to study the time-varying dynamic responses of the vestibular system. One, called time domain analysis, is to give an impulsive or a step stimulus and study the time-varying properties of the response. For example, the responses of the semicircular canal system have been studied by rotating at constant velocity for a period of time, then producing an abrupt stop (which can be considered a step in the velocity of the head, or a short pulse of deceleration). The response to such a stimulus typically has the form of a decaying exponential, which can be described in terms of the time constant of the decay. Thus the semicircular canals have been described as having a "long time constant" of 5–7 seconds (the response is usually better fit by a sum of exponentials, with additional time constants).

The other method used to study such dynamic responses, called frequency domain analysis, uses a sinusoidal stimulus, for example, sinusoidal rotation back and forth. Sinusoids are employed in signal analysis for a number of reasons. They are fairly easy to generate, they are "smooth" and "gentle" (contrasted with, for example, the acceleration pulse described above), and the stimulus is repetitive, permitting responses to be averaged. Sinusoids are especially useful when studying linear systems, where doubling the input produces a doubled output. For such systems, the output when a pure sinusoid is used as the input is a pure sinusoid of the same frequency, but possibly differing in relative amplitude (gain) and the relative timing of the peaks of the sinusoid (phase); any other response frequencies can be treated as noise (or a nonlinearity). Most biological systems, including the vestibular system, act like linear systems, particularly when given small signals (two common nonlinearities in neural systems are thresholds, or alternatively, the notion that neurons cannot fire less than 0 spikes per second, and response saturation at high levels of stimulation).

Once the responses to a range of frequencies have been obtained, the results are usually presented as a Bode plot, which plots the logarithm of the gain, and the phase vs. the logarithm of the stimulus frequency. From the form of the Bode plot, one can often deduce the underlying physical process producing the response. For example, consider the response of the semicircular canal to sinusoidal rotation. If the stimulus is considered to be angular velocity, the resulting Bode plot will show a low-frequency region characterized by a rising gain and a phase that leads angular velocity (that is, the peak of the neuron's response comes before the maximum angular velocity is attained), an intermediate region, where the gain is relatively constant and the phase is near 0 (implying the response is proportional to, and in phase with, stimulus velocity), and a high-frequency region, where the gain starts to fall and the phase begins to lag behind the stimulus. The frequencies where the gain turns the corner from rising to flat, or from flat to falling, are the inverses of the time constants that are obtained using impulsive stimuli and time domain analysis.

To illustrate the difference between canal and otolith responses in both time and frequency domains, the responses of a typical canal and otolith afferent to a variety of stimuli are shown in Figure 4A. The first two panels show the response to a 10° sinusoidal roll tilt at two frequencies, 0.2 and 1.0 Hz. Both afferents are assumed to have a resting discharge rate of 40 impulses per second. Note that the responses of the otolith afferent (solid lines) rise and fall in phase with the tilt stimulus, and that the amplitude of the response at the two frequencies is essentially constant. In contrast, the response of the canal afferent (dashed line) both grows with frequency (as does the *velocity* of the tilt stimulus) and peaks approximately a quarter cycle, or 90°, before the peak of the tilt, or, alternatively, the peak response occurs at the same time as the peak *velocity* (slope) of the tilt stimulus. The bottom panel illustrates a non-sinusoidal stimulus; a period of constant tilt, a half second of 20° per second constant velocity tilt to a new position, tilted by 10°, another period of constant tilt, and then a return to the original position. The otolith afferent response largely tracks the position of the stimulus, while the canal response basically consists of a positive, then a negative, pulse corresponding to the two periods of constant velocity movement between the two positions. This again illustrates the general principle that otolith responses can measure the position of the head with respect to gravity, while the canal responses give head velocity.

Figure 4B shows the frequency response graph, or Bode plot, for these two afferents, constructed from a series of responses to sinusoids over a wide range of frequency. The gain and phase obtained from the 0.2 and 1.0 Hz stimuli at the left of the figure are indicated by dots on the plot. The otolith afferent response (solid line) has the characteristics of a position response: a relatively constant gain across the frequency band, and a phase near 0°. The canal's velocity response is characterized by an increase with frequency, growing 10-fold for every decade of frequency, and a phase that leads position by 90°. Notice that at the lowest and highest frequencies illustrated, the canal afferent response deviates somewhat from the velocity model.

3.4 Responses within the Vestibular Nuclei

Responses of neurons in the vestibular nuclei show some variation from the responses seen in the peripheral afferents. First, responses can arise that have different spatial properties from those of the afferents. For example, one can find neurons that have a response consistent with input from a semicircular canal (i.e., response proportional to stimulus rotational velocity, no sustained discharge during static tilts), yet whose optimal stimulus plane is the roll plane, a pattern consistent with equal input from both the anterior and posterior vertical canal. Another form of convergence can arise from convergence between canal and otolith organ afferents; responses consistent with this form of convergence show sustained responses to static tilt and show position-dependent responses during very slow sinusoidal tilt; yet as the frequency of tilt increases, the response phase shifts and the gain increases proportional to the stimulus velocity. Yet another type of response is an "inhibitory" response, for example, a neuron responding to rotations in the horizontal canal plane, yet being excited during rotations towards the contralateral side. Such responses are consistent with commissural input from the contralateral vestibular nuclei.

Neurons in the vestibular nuclei also show responses consistent with convergence with other sensory and motor pathways. For example, neurons involved in the production of compensatory eye movements often show pauses in their firing during quick saccadic eye movements.[7,12] These neurons can also show modulation of their firing pattern during eye movements in the absence of head movements, indicating they receive an eye movement signal. Similarly, many vestibulospinal neurons have their firing modulated by rotations of the neck[5,26] (see Section 3.7).

Finally, there are various forms of intrinsic signal processing that take place within the central nervous system that affect the responses of central vestibular neurons. One process, called velocity storage, prolongs the decay time of the activity recorded following an abrupt rotatory stimulus from the 5–7 seconds seen in semicircular canal afferents to 15–20 seconds.[33,34] Another process can alter the dynamics of neurons receiving otolith input, producing responses that show a velocity-like increase in gain at higher frequencies.[36]

3.5 Vestibuloocular Reflexes (VOR)

The vestibuloocular reflex is a set of reflexes that tend to keep the eyes stationary in space despite movements of the head. Rotation of the head about any axis will elicit a compensatory rotation of the eye in the opposite direction of approximately equal magnitude, with the net result that the eye tends to remain fixed in space. Similarly, translations of the head can also produce compensatory eye movements to hold visual gaze fixed on a stationary target.

The basic anatomy of the vestibular pathways can explain many of the properties of the VOR. Consider, for example, yaw rotations of the head. Rotating the head to the right stimulates the right horizontal semicircular canal afferents (and inhibits those on the left). Excitation is relayed through the medial vestibular nucleus to motoneurons in the left (contralateral) abducens nucleus, which in turn innervate the left lateral rectus muscle. Thus this reflex pathway results in pulling the left eye to the left during the rightward head movement, which tends to keep the eye fixed in space. There is also an excitatory input (relayed through interneurons in the abducens nucleus) to motoneurons innervating the right medial rectus muscle, which rotates the right eye to the left. In addition, inhibitory projections from the medial vestibular nucleus to the right abducens nucleus inhibit the antagonist muscles (the right lateral rectus and left medial rectus).

The vestibular signal arising from the labyrinth receptors is modified in several ways in the vestibuloocular pathway. Signals arising from the semicircular canals undergo several stages of signal processing (including velocity storage, mentioned above) to convert the head angular acceleration/head velocity signal arising from the canal afferents into a head position signal suitable for driving the extraocular muscles. One feature of many vestibular-induced eye movements is a rhythmic resetting of the eyes, nystagmus, which breaks programmed eye movements that are physically outside the normal oculomotor range into smaller pieces, resetting the eye near the center of the orbit between each piece. Nystagmic patterns are another form of signal observed on vestibular neurons.

For the angular VOR, the amount of eye movement necessary to fixate a target while shaking or nodding the head is largely independent of the distance to the target (unless the target is very near). However, this is decidedly not true for the linear vestibuloocular reflex. For targets sufficiently far away, no eye movement at all is required to compensate for head translations, while for near targets, the amount of required eye movement goes up as the distance to the subject decreases. Thus production of the linear component of the vestibuloocular reflex requires additional processing to take account of target location in space.

The vestibuloocular reflex also provides a good illustration of the adaptive plasticity of these vestibular reflex pathways. Interactions between the vestibular and visual systems allow VORs to be "re-calibrated" to improve their performance.[28] Since the purpose of VORs is to stabilize the eyes in inertial space, errors will lead to blurring of vision. Thus the neural circuits that control the VORs receive visual information to provide feedback about how well the VOR is stabilizing the eyes in space. This feedback is then used to adjust the gain and phase of the reflex to improve its performance. One common cause for the need to re-calibrate is the wearing of spectacles that magnify or minify visual inputs. Changes in the relationship between head movements and movements of the visual world alter the requirements of the VOR for providing visual stability. In a similar fashion, changes in the relationship between head movements and forces on the labyrinth, for example during spaceflight, change the requirements for VOR performance to stabilize vision.

3.6 Control of Head Movement

A second vestibular reflex, the vestibulocollic reflex (VCR), is concerned with stabilization of the head.[31,37] As with the vestibuloocular reflex, the VCR involves a three-neuron arc, with vestibular signals relayed via the MVST and LVST to neck motoneurons. In addition, there are more indirect vestibular relays to the neck, involving the reticulospinal and interstitiospinal tracts. This reflex appears to be functionally wired much like the VOR.

Each neck muscle moves the head in a particular direction, and hence excites a particular set of canals; those canals, in turn, will generally inhibit that muscle (thereby opposing the resulting head movement). It also works in the other direction; head movements, which excite canals, activate neck muscles, which tend to cause head movements in the opposite direction. The vestibulocollic system is a bit more complex than the vestibuloocular system since (1) there are many more neck muscles to control, (2) the head has much more mass than the eyeball, and (3) the output of the reflex, head movement, is the same as its input (also head movement).

3.7 Postural Reflexes

In response to tilt or translation of the support surface, a variety of compensatory limb and trunk movements can be elicited to prevent falls and restore balance. Many of these responses were elaborated by Magnus,[27] and can be observed in the decerebrate cat. For example, roll tilt produces increased extension of the downhill limb, and decreased rigidity in the opposite limb. Some of these responses arise from vestibulospinal and reticulospinal input to limb musculature that helps maintain erect posture. In particular, one can record the responses of vestibulospinal and reticulospinal neurons in the decerebrate cat (e.g. see References 4, 25), the dynamics of which are similar to those observed in forelimb extensor muscle.[35]

When considering the vestibular input to the (body) postural control system, it is important to remember that the vestibular system does not directly sense movement of the body, but movement of the head. Consider the following two situations involving downward head pitch, namely, a stumble causing the body (and, passively, the head) to fall forward, and a downward voluntary nod of the head. In only the first case would a significant limb readjustment be appropriate. Muscle receptors in the neck give rise to a so-called tonic neck reflex, which manifests itself as an extension of the limb towards which the chin points (and a decreased drive to extensors of the opposite limb). Thus in the cat, ear-down roll of the head would lead to a vestibular reflex that would excite ipsilateral limb extensors, while the tonic neck reflex would act to inhibit these extensors, in effect canceling the vestibular signal and enabling a stable posture during the head movement. The site of this interaction is not fully known, but some of it may take place within the vestibular nuclei. Figure 5 illustrates an example of the response of a vestibulocollic neuron to three types of natural stimulation: whole-body tilt (a vestibular stimulus, solid sinusoidal line), body rotation with the head held fixed in space (activating neck, but not vestibular, receptors, dotted line), and rotation of the head on the erect, stationary body, showing the two responses essentially cancel each other.

3.8 Control of Vestibular Reflexes

The reflexes described above have as a common theme the maintenance of stability, either of the eyes or head in space (VOR, VCR) or for upright, erect body posture. These reflexes interact with each other and with other proprioceptive reflexes (such as the tonic neck reflex, discussed above) to help maintain body, head, and eyes relatively fixed in space. However, there are many situations, such as voluntary movements or orienting responses to novel stimuli, where stability is not desired, and where it may be advantageous to temporarily suppress the vestibular reflex. For example, to produce a large gaze shift, humans and animals typically produce a coordinated eye-head movement, during the initial phase of which the eye and the head both move toward the target.[2,21] During this phase of the movement, the VOR is suppressed, as it would otherwise serve to move the eye in the wrong direction (in the head). Similarly, during the head movement, the VCR

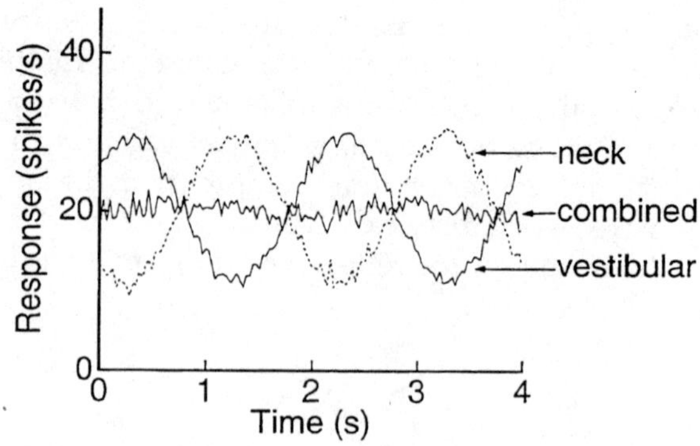

FIGURE 5
This response, modeled after responses from the decerebrate cat,[26] illustrates the response of a vestibular neuron to three forms of natural stimulus. Whole-body rotation provides a vestibular stimulus (solid line); rotation of the body, keeping the head fixed in space, selectively activates neck afferents (dotted line); rotating the head while keeping the body stationary combines the neck and vestibular inputs, and results in little modulation of the neuron.

is probably also largely suppressed (as its function would otherwise be to keep the head fixed in space). Evidence in humans, in fact, suggests that the gain of the VCR during passive body rotations in the dark is fairly low, unless conscious effort is made to track an earth-fixed imagined target.[20]

3.9 Vestibular Pathology

What happens when something goes wrong? As we discussed, the vestibular system is quite robust, having two labyrinths which interact with each other to provide some functional redundancy. In addition, other sensory cues, such as vision, neck proprioception, and other proprioceptive and somatosensory cues can provide information about movements of the body and head in space. However, disease (e.g., Meniere's Disease), aging, trauma, drugs (e.g., alcohol), or exposure to novel inertial environment (e,g., orbital space flight) can lead to alterations in the signals carried by the vestibular system.

Consider the response to a unilateral destruction of the labyrinth. Two forms of imbalance occur centrally. One is simply the loss of the tonic input from the missing labyrinth, which is reflected in a reduced firing rate in the vestibular nuclei ipsilateral to the lesion. Second, during head movements, only one labyrinth is providing a modulated input, although responses can be recorded bilaterally (due to the commissural and other intrinsic brainstem pathways). Humans to whom this happens report symptoms of vertigo, or dizziness, and they typically exhibit nystagmic eye movements consistent with illusions of self-motion toward the intact labyrinth.[1] These symptoms, however, typically resolve in days to weeks, probably due in part to compensatory mechanisms related to the observed adaptive plasticity of the vestibular system.

Under other conditions, various sensory cues from the vestibular, visual, and somatosensory systems may be in disagreement, for example, during unusual motions within a confined space (e.g., on a ship at sea, or in orbital flight). Under these conditions, a constellation of symptoms called motion sickness can arise (Chapter 9 by Money et al.). Vestibular input appears to be important (removal of the labyrinths on both sides can prevent most of these symptoms), though the nature of the vestibular involvement in this phenomenon is still not clear. Of special interest is that many of the symptoms associated with motion sickness have visceral, autonomic components.

4 SUMMARY

The vestibular system, originating in the labyrinth of the inner ear, provides the central nervous system with information about the orientation and movement of the head in space. Within the labyrinth, the semicircular canals provide information about head rotations in three dimensions; the otolith organs provide signals related to linear acceleration of the head, most particularly the orientation of the head with respect to the downward force of gravity. Vestibular information comes into four major brainstem vestibular nuclei, from whence it can go directly to motor neurons controlling eye and head movements. Thus the classical picture of vestibular reflexes centers around their maintenance of stable orientation, either of the eyes in space (the vestibuloocular reflex), the head (on the neck) in space (the vestibulocollic reflex), or upright body posture (the body in space). One purpose of this volume is to explore other, less direct means by which the vestibular system aids in serving other mechanisms of body homeostasis.

REFERENCES

1. Baloh, R. W. and Honrubia, V., *Clinical Neurophysiology of the Vestibular System,* F.A. Davis Co., Philadelphia, 1990.
2. Bizzi, E., Eye-head coordination, in *Handbook of Physiology, Sec. I: The Nervous System. Vol. II: Motor Control,* Brooks, V. B., Ed., American Physiological Society, Bethesda, MD, 1981, chap. 29.
3. Bizzi, E., Dev, P., Morasso, P., and Polit, A., Effect of load disturbances during centrally initiated movements, *J. Neurophysiol.,* 41, 542, 1978.
4. Bolton, P. S., Goto, T., Schor, R. H., Wilson, V. J., Yamagata, Y., and Yates, B. J., Response of pontomedullary reticulospinal neurons to vestibular stimuli in vertical planes. Role in vertical vestibulospinal reflexes in the decerebrate cat, *J. Neurophysiol.,* 67, 639, 1992.
5. Boyle, R. and Pompeiano, O., Convergence and interaction of neck and macular vestibular inputs of vestibulospinal neurons, *J. Neurophysiol.,* 45, 852, 1981.
6. Brödel, M., Malone, P. D., Guild, S. R., and Crowe, S. J., *Three Unpublished Drawings of the Anatomy of the Human Ear,* W.B. Saunders, Philadelphia, 1946.
7. Buettner, U. W., Büttner, U., and Henn, V., Transfer characteristics of neurons in the vestibular nuclei of the alert monkey, *J. Neurophysiol.,* 41, 1614, 1978.
8. Büttner-Ennever, J. A., Patterns of connectivity in the vestibular nuclei, in *Sensing and Controlling Motion. Vestibular and Sensorimotor Function,* Cohen, B., Tomko, D. L., and Guedry, F., Eds., New York Academy of Science, New York, 1992, 363.
9. Fernández, C. and Goldberg, J. M., Physiology of peripheral neurons innervating semicircular canals of the squirrel monkey. II. Response to sinusoidal stimulation and dynamics of peripheral vestibular system, *J. Neurophysiol.,* 34, 661, 1971.
10. Fernández, C. and Goldberg, J. M., Physiology of peripheral neurons innervating otolith organs in the squirrel monkey. II. Directional selectivity and force-response relations, *J. Neurophysiol.,* 39, 985, 1976.
11. Fernández, C. and Goldberg, J. M., Physiology of peripheral neurons innervating otolith organs in the squirrel monkey. III. Response dynamics, *J. Neurophysiol.,* 39, 996, 1976.
12. Fuchs, A. F. and Kimm, J., Unit activity in vestibular nucleus of the alert monkey during horizontal angular acceleration and eye movement, *J. Neurophysiol.,* 38, 1140, 1975.
13. Fukushima, K., The interstitial nucleus of Cajal and its role in the control of movements of head and eyes, *Prog. Neurobiol.,* 29, 107, 1987.
14. Goldberg, J. and Peterson, B. W., Reflex and mechanical contributions to head stabilization in alert cats, *J. Neurophysiol.,* 56, 857, 1986.
15. Goldberg, J. M. and Fernández, C., Physiology of peripheral neurons innervating semicircular canals of the squirrel monkey. I. Resting discharge and response to constant angular accelerations, *J. Neurophysiol.,* 34, 6638, 1971.
16. Goldberg, J. M. and Fernández, C., Efferent vestibular system in the squirrel monkey: anatomical location and influence on afferent activity, *J. Neurophysiol.,* 43, 986, 1980.

17. Goldberg, J. M. and Fernández, C., The vestibular system, in *Handbook of Physiology, Sec. I: The Nervous System. Vol. III: Sensory Processes, Part 2*, Darian-Smith, I., Ed., Am. Physiol. Soc., Bethesda, MD, 1984, chap. 21.
18. Grüsser, O.-J., Guldin, W., Harris, L., Lefèbre, J.-C., and Pause, M., Cortical representation of head-in-space movement and some psychophysical experiments on head movements, in *The Head-Neck Sensory Motor System*, Berthoz, A., Graf, W., and Vidal, P. P., Eds., Oxford University Press, New York, 1992, chap. 79.
19. Grüsser, O.-J., Pause, M., and Schreiter, U., Vestibular neurones in the parieto-insular cortex of monkeys (*Macaca fascicularis*): visual and neck receptor responses, *J. Physiol.*, 430, 559, 1990.
20. Guitton, D., Kearney, R. E., Wereley, N., and Peterson, B. W., Visual, vestibular and voluntary contributions to human head stabilization, *Exp. Brain Res.*, 64, 59, 1986.
21. Guitton, D., Munoz, D. P., and Galiana, H. L., Mechanisms of gaze control and eye-head coupling in the cat whose head is unrestrained, in *The Head-Neck Sensory Motor System*, Berthoz, A., Graf, W., and Vidal, P. P., Eds., Oxford University Press, New York, 1992, chap. 81.
22. Highstein, S. M. and McCrea, R. A., The anatomy of the vestibular nuclei, in *Neuroanatomy of the Oculomotor System*, Büttner-Ennever, J. A., Ed., Elsevier, Amsterdam, 1988, chap. 5.
23. Hudspeth, A. J., Extracellular current flow and the site of transduction in vertebrate hair cells, *J. Neurosci.*, 2, 1, 1982.
24. Ito, M., *The Cerebellum and Neural Control*, Raven Press, New York, 1984.
25. Kasper, J., Schor, R. H., and Wilson, V. J., Responses of vestibular neurons to head rotations in vertical planes. I. Response to vestibular stimulation, *J. Neurophysiol.*, 60, 1753, 1988.
26. Kasper, J., Schor, R. H., and Wilson, V. J., Responses of vestibular neurons to head rotations in vertical planes. II. Response to neck stimulation and vestibular-neck interaction, *J. Neurophysiol.*, 60, 1765, 1988.
27. Magnus, R., *Körperstellung*, Julius Springer, Berlin, 1924.
28. Melvill Jones, G., Adaptive modulation of VOR parameters by vision, in *Adaptive Mechanisms in Gaze Control*, Berthoz, A. and Melvill Jones, G., Eds., Elsevier, Amsterdam, 1985, chap. 2.
29. Money, K. E. and Scott, J. W., Functions of separate sensory receptors of nonauditory labyrinth of the cat, *Am. J. Physiol.*, 202, 1211, 1962.
30. Peterson, B. W., Maunz, R. A., and Fukushima, K., Properties of a new vestibulospinal projection, the caudal vestibulospinal tract, *Exp. Brain Res.*, 32, 287, 1978.
31. Peterson, B. W. and Richmond, F. J., *Control of Head Movement*, Oxford University Press, New York, 1988.
32. Pompeiano, O., Spinovestibular relations: anatomical and physiological aspects, in *Basic Aspects of Central Vestibular Mechanisms*, Brodal, A. and Pompeiano, O., Eds., Elsevier, Amsterdam, 1972, 263.
33. Raphan, T., Cohen, B., and Matsuo, V., A velocity-storage mechanism responsible for optokinetic nystagmus (OKN), optokinetic afternystagmus (OKAN) and vestibular nystagmus, in *Control of Gaze by Brain Stem Neurons*, Baker, R. and Berthoz, A., Eds., Elsevier/North-Holland Biomedical Press, Amsterdam, 1977, 37.
34. Robinson, D. A., Vestibular and optokinetic symbiosis: an example of explaining by modelling, in *Control of Gaze by Brain Stem Neurons*, Baker, R. and Berthoz, A., Eds., Elsevier/North-Holland Biomedical Press, Amsterdam, 1977, 49.
35. Schor, R. H. and Miller, A. D., Vestibular reflexes in neck and forelimb muscles evoked by roll tilt, *J. Neurophysiol.*, 46, 167, 1981.
36. Schor, R. H., Miller, A. D., Timerick, S. J. B., and Tomko, D. L., Responses to head tilt in cat central vestibular neurons. II. Frequency dependence of neural response vectors, *J. Neurophysiol.*, 53, 1444, 1985.
37. Wilson, V. J., Boyle, R., Fukushima, K., Rose, P. K., Shinoda, Y., Sugiuchi, Y., and Uchino, Y., The vestibulocollic reflex, *J. Vestib. Res.*, 5, 147, 1995.
38. Wilson, V. J. and Melvill Jones, G., *Mammalian Vestibular Physiology*, Plenum Press, New York, 1979.
39. Wilson, V. J. and Peterson, B. W., Vestibulospinal and reticulospinal systems, in *Handbook of Physiology. Sec. I: The Nervous System. Vol. II: Motor Control*, Brooks, V. B., Ed., American Physiological Society, Bethesda, MD, 1981, chap. 14.

3

THE AUTONOMIC NERVOUS SYSTEM STRUCTURE AND FUNCTION

Alan F. Sved and David A. Ruggiero

CONTENTS

0-8493-7668-1/96/$0.00+$.50

1 INTRODUCTION

The autonomic nervous system (ANS) innervates smooth muscle, cardiac muscle, and glands and controls their response patterns to signals that emerge in the brain and periphery. Much of the thinking about autonomic function had its origins more than half a century ago stemming from the work of Langley[66] and Cannon.[19] However, during the past 15 years, many of the classical concepts about the autonomic nervous system have been revised. The purpose of this chapter is not to provide a synopsis of autonomic function (which can be obtained in any good textbook of physiology), but rather to highlight current concepts and issues relating to the functional organization of the ANS.

The classical definition of the ANS referred to the innervation of the viscera, and it was limited to the peripheral nerves and their central site of origin. As such, the ANS consisted of the preganglionic and postganglionic neurons of the sympathetic nervous system (SNS) and parasympathetic nervous system (PNS). However, current definitions frequently include visceral afferent nerves as well as the central neural circuitry involved in mediating visceral reflexes. As such, the autonomic nervous system could be considered as including the efferent innervation of the viscera (the classical ANS) and the central neural circuitry involved in transmitting and integrating ANS inputs and outputs. Indeed, this chapter will follow that structure, while maintaining the popular definition of the ANS as being limited to the preganglionic and postganglionic neurons. Neurons of the enteric nervous system, which were included in Langley's definition of the ANS but are omitted from most current discussions of the ANS, will not be considered here (see Reference 150 for review) because their organization and functional regulation are notably different from the other components of the ANS.

2 FUNCTIONAL OVERVIEW

The ANS consists of two components, the parasympathetic "craniosacral" and the sympathetic "thoracolumbar" divisions that serve different but integrated functions. Differences in anatomical structure and chemical characteristics accounted for the initial distinction of the two divisions. Functionally, the SNS acts to generate appropriate responses to homeostatic challenges whereas the PNS allows for maintenance of function or restitution of body energy stores during rest. The early use by Hess of the terms *ergotropic* (energy mobilizing and utilizing) and *trophotropic* (growth promoting) to refer to the influences of the SNS and PNS, respectively, highlighted their functional differentiation. In this framework, it is the general consensus that the PNS acts to promote digestion and energy storage. In contrast, the SNS increases energy mobilization, cardiac output, muscle blood flow, and airway patency.

Most visceral tissues are innervated by both the SNS and the PNS, and these innervations have antagonist functions. Functionally antagonistic interactions of the SNS and PNS are not manifest solely at the level of the target tissue, but also reflect presynaptic interactions. Despite the antagonistic nature of the SNS and the PNS on visceral targets, it is more appropriate to envisage their actions as synergistic because the central neural regulation of the ANS is organized to co-regulate sympathetic and parasympathetic neural activity in opposite directions. Thus, under physiological conditions both systems are not activated simultaneously. Rather, a stimulus that increases the activity of one system reciprocally inhibits the activity of the other. For example, consider the innervation of the heart, where sympathetic innervation acts to increase cardiac output and parasympathetic innervation acts to decrease it. Since both systems are tonically active at rest, heart rate can be increased by either increasing sympathetic outflow or decreasing parasympathetic outflow. However, physiological stimuli that increase heart rate accomplish this via both mechanisms. This joint regulation of the SNS and PNS requires central neural integration of inputs to these systems, as discussed below.

Whereas most visceral effector tissues receive dual, antagonistic innervation that is regulated in tandem, others are innervated by only one division of the ANS. For example, vascular smooth muscle and sweat glands are innervated exclusively by the SNS whereas the stomach is subject predominantly to parasympathetic control. The functional significance of the single innervation pattern is consistent with the physiological control of that tissue. For example, at rest, under conditions where PNS activity would predominate, it is appropriate that a major control of tissue blood flow would be intrinsic to the vasculature (autoregulation) rather than some influence of the PNS. Furthermore, local increases or decreases in tissue blood flow should not reflect rest but rather active compensations to homeostatic challenges (action of the SNS).

Autonomic outflow is highly controlled by the central nervous system (CNS). Although Cannon characterized the SNS as discharging as a unit, the so-called "fight-or-flight" response, it is now abundantly clear that the SNS is also regulated in a more selective manner. In response to certain stimuli, highly patterned responses occur including concomitant increases or decreases in sympathetic outflow to specific target organs. As described below, sympathetic outflow is subject to exquisitely precise differential control resulting in highly patterned responses characteristic of distinct stimuli. The PNS more typically is considered as differentially regulated, depending on the target. But, as with the SNS, there are also instances when parasympathetic outflow appears to be influenced as a unit.

Mechanisms involved in central autonomic regulation and the neural networks involved in this regulation serve as the focus of the remainder of this chapter.

3 AUTONOMIC OUTFLOW

3.1 Postganglionic Neurons

The innervation of visceral effector tissues is provided by postganglionic neurons: cells that lie outside the CNS in autonomic ganglia (see Figures 1 and 2). The term *postganglionic* is somewhat of a misnomer, since these neurons comprise the ganglia. Whereas parasympathetic ganglia lie embedded in the tissue they innervate or near to it and have short postganglionic axonal projections, SNS postganglionic neurons are located in sympathetic chain (paravertebral) or prevertebral ganglia distant from their targets.

Postganglionic neurons innervating different targets have different functional properties. As would be expected based on the unique control characteristics of different targets, the reflex control of postganglionic neurons innervating these targets depends upon differential input to postganglionic neurons. (For a description of the properties of different sympathetic outflows, the reader is referred to an excellent review by Jänig and McLachlan.[54]) It is important to keep in mind that different postganglionic axons, even within a single nerve, serve different functions, are differentially regulated, and have different firing patterns. This is in sharp contrast to the classical view that the SNS functions as a unit and is independent of discrete control. Based on their responses to sustained depolarization, sympathetic postganglionic neurons have been differentiated according to whether they fire continuously at low frequency (tonic neurons) or fire a short phasic burst of action potentials initially in response to the depolarization (phasic neurons).[21,37] These firing patterns reflect different membrane conductances and perhaps

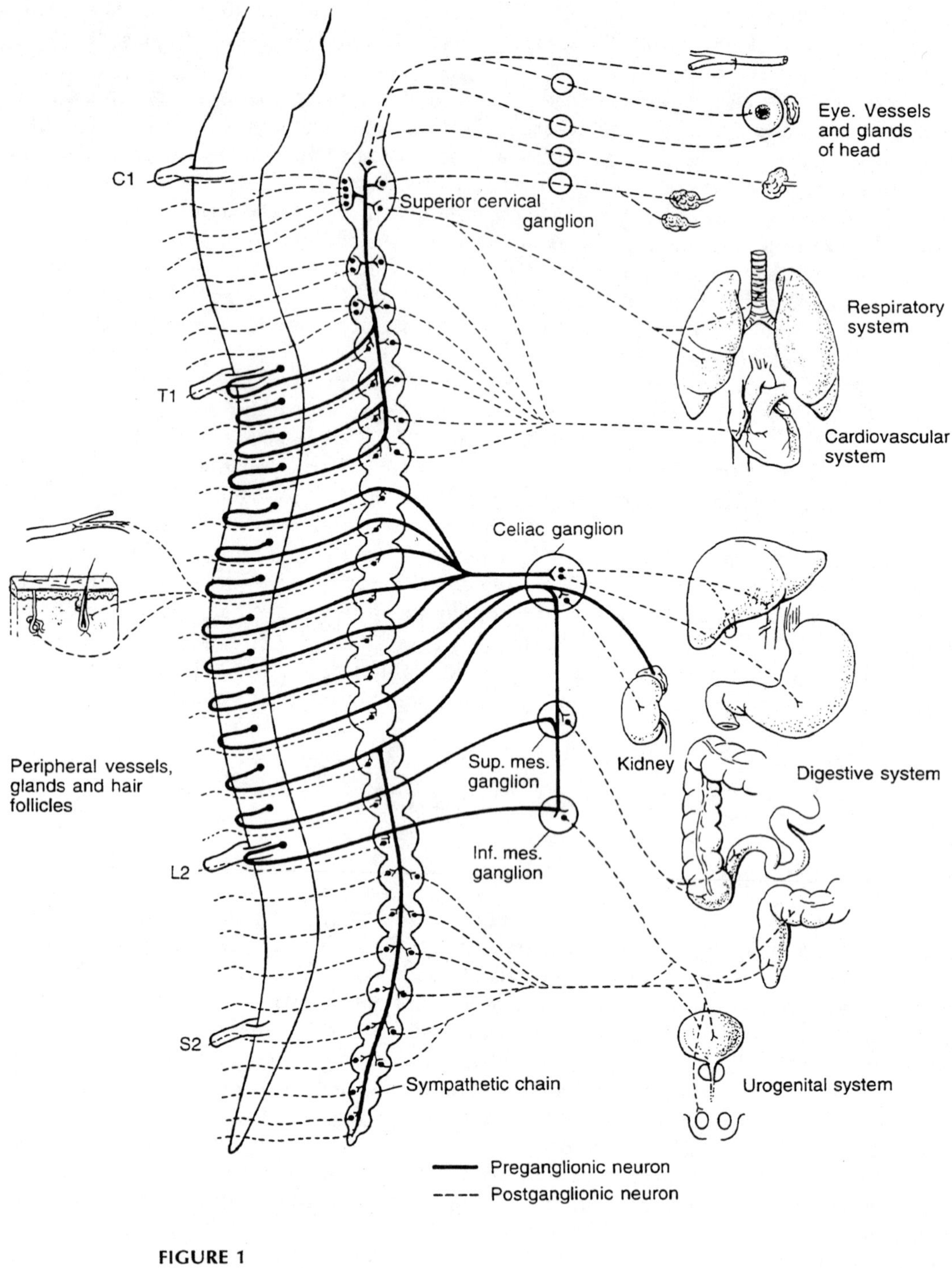

FIGURE 1
Schematic diagram of the sympathetic thoracolumbar nervous system.

different functional classes of postganglionic neurons.[21,81] In support of this idea is evidence of McLachlan and co-workers that phasic neurons in the lumbar sympathetic chain have the characteristics of vasoconstrictor neurons whereas tonic neurons in the inferior mesenteric ganglion have properties suggesting that they may regulate intestinal motility.[21,81]

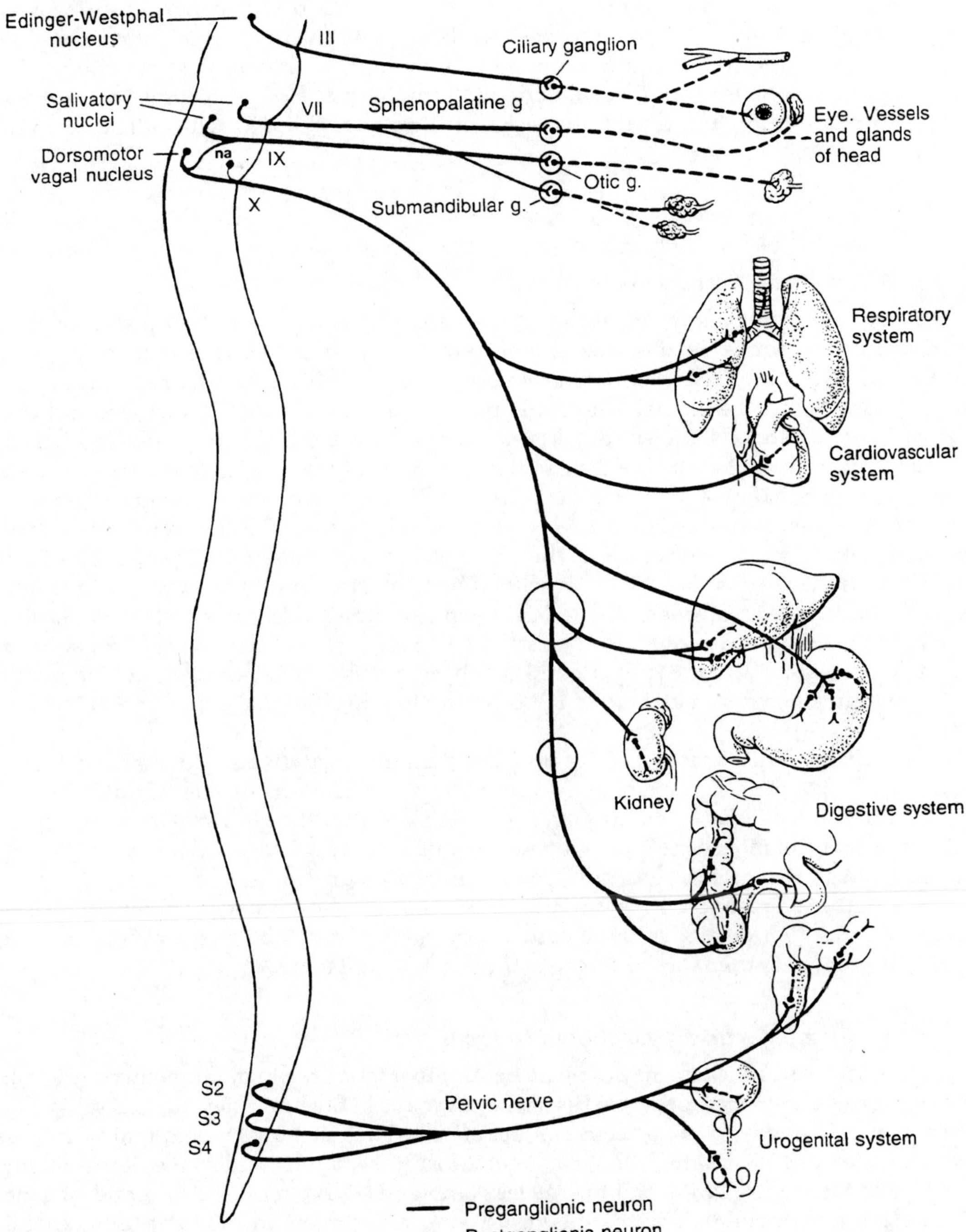

FIGURE 2
Schematic diagram of the parasympathetic craniosacral system.

Innervation of postganglionic neurons arises almost exclusively from preganglionic neurons. However, it is overly naive to envisage ganglia as merely relays. A great deal of signal processing likely occurs in ganglia, an idea supported by both convergent and divergent innervation of postganglionic neurons by preganglionic neurons.[106] Furthermore, the strength of synaptic drive of postganglionic neurons by preganglionic neurons is not uniform. Some are subject to a very powerful synaptic drive, thereby ensuring that preganglionic activity will drive the postganglionic neuron. In contrast, other neurons receive weaker inputs that require summation of multiple excitatory postsynaptic potentials.[18] As noted above, differences among the firing patterns of postganglionic neurons may also relate more to intrinsic properties of the postganglionic neuron than afferent input.

3.2 Preganglionic Neurons

3.2.1 Sympathetic Preganglionic Neurons

Sympathetic preganglionic neurons (SPNs) are found within the gray matter of thoracic and upper lumbar levels of the spinal cord (see Figure 1). Although the majority of SPNs are found within the intermediolateral cell column (IML), three other spinal regions also contain SPNs, the central autonomic nucleus, the intercalated cell group, and the lateral funicular area.[18] Different target tissues are innervated by SPNs located in distinct segments of the spinal cord. The distributions of SPNs innervating different tissues were retrogradely labeled in rat by injection of dye into different sympathetic ganglia.[135] Similar studies have been performed in cats.[50,97] Two noteworthy principles arise from this type of analysis, which are not conveyed in the schematic representation in Figure 1. First, the sympathetic innervation of specific target tissues arises from multiple levels of the spinal cord. Second, there is considerable overlap in segments innervating different target tissues. For example, SPNs that project to the rat stellate ganglion, which provides sympathetic innervation of the heart, are spread fairly evenly across five spinal segments and in many of these segments larger numbers of SPNs project to other ganglia (superior and middle cervical ganglia).

Nonetheless, this type of analysis obscures the functional diversity of neurons innervating each ganglion. For example, stellate SPNs influencing heart rate, cardiac muscle contractility, and cardiac blood flow may be differentially localized over several segments. Thus, the extent of intersegmental or intrasegmental topographical localization of different functions (e.g., vasomotor, sudomotor, or piloerector) is unclear at present.

Sympathetic preganglionic neurons display different morphological and biophysical characteristics. (For details, see review by Cabot.[18]) However, the functional correlates of these different characteristics, if they exist, are not yet apparent.

3.2.2 Parasympathetic Preganglionic Neurons

Many of the same issues may apply to parasympathetic preganglionic neurons (PPNs). However, these motoneurons form discrete cell groups in the brain stem and sacral spinal cord (Figure 2) and not a continuous column of cells. The functional topographic localization of different PPNs is thus more clear cut. Most PPNs are located in the dorsal motor nucleus of the vagus (DMN) and the nucleus ambiguus (NA) in the medulla and exit the CNS via the IXth and Xth cranial nerves to innervate targets in the neck, thorax, and abdomen. Parasympathetic innervation of the salivary and lacrimal glands and eyes amongst other viscera in the nasal and cranial cavities originate in the salivatory nuclei and Edinger-Westphal nuclei. Innervation of pelvic viscera derives from preganglionic neurons in the sacral IML.

Parasympathetic preganglionic neurons in the DMN and NA are localized in a highly topographic manner.[61,68,74] Preganglionic neurons that issue subdiaphragmatic branches of the vagus nerve lie exclusively within the DMN whereas the NA provides the major innervation of branchiomeric (striated) muscles of the neck and the thoracic viscera. Neurons within each nucleus are topographically related to their target tissue. For example, in the DMN, preganglionic neurons that innervate the pancreas lie medial to gastric efferents. In NA, neurons innervating the upper airway are rostral to those innervating the cardiomotor neurons, which lie in the ventral (external) division of the NA.[68,96]

3.3 Neurotransmitters of the ANS

The classical adrenergic and cholinergic classification of Dale provided a basic framework for chemical transmission in the ANS but is far from a complete account of autonomic neurotransmission. It has traditionally been considered that preganglionic neurons in both divisions of the ANS release acetylcholine that acts on nicotinic receptors on postganglionic neurons. Postganglionic parasympathetic neurons are also cholinergic, but the receptor at the neuroeffector junction is of the muscarinic type. Sympathetic postganglionic neurons are noradrenergic and act on α-adrenergic or β-adrenergic receptors depending upon the target tissue. However, it has long been known that this account is incomplete. For example, postganglionic sympathetic sudomotor neurons are cholinergic rather than noradrenergic, and the understanding of why these sympathetic neurons express a cholinergic phenotype has now progressed to the molecular level.[47] In addition, there are numerous instances of transmission in the ANS that are not sensitive to blockade of cholinergic and adrenergic receptors. One of the principal mediators of non-adrenergic non-cholinergic (NANC) transmission in the ANS is ATP.[15] Neuropeptides that are contained in preganglionic and postganglionic neurons also contribute to autonomic neurotransmission.[75,126] Nitric oxide or other nitrosyl-containing factors may also serve transcellular messenger roles in the ANS.[35,36,117]

4 AFFERENT SIGNALS CONTROLLING AUTONOMIC OUTFLOW

4.1 Spinal Afferents

Spinal afferents transmit visceral nociceptive and non-noxious stimuli to anatomically discrete laminae of the spinal cord. Spinal laminae that receive input from primary visceral afferents include I, V, VII and X.[25,65,137] Electrophysiological studies agree that most visceroceptive spinal neurons also receive input from somatic receptive fields.[2]

4.2 Cranial Nerve Afferents

The nucleus of the solitary tract (NTS) is the principal site of termination of primary visceral afferents of cranial nerves VII, IX and X. The NTS constitutes a mosaic of longitudinally organized cell columns,[29,59-61] with topographic localization of afferents conveying different sensory inputs (see Figures 3 and 4). Projection fields of sensory afferents to the NTS have been mapped in several species by using transganglionic transport of axonal tracers (typically WGA-HRP) injected into individual cranial nerve branches or visceral organs. Primary visceral afferents are compartmentalized in NTS and exhibit a modular organization. Two modes of first-order projection appear to exist: organ specific and integrative. Visceral afferents from individual organs preferentially terminate within one locale termed the principal terminal field. A *viscerotopic distribution pattern* suggests that individual areal subunits may be devoted to processing organ-specific information. An *integrative projection pattern* is suggested by evidence that primary visceral afferents extend into adjacent fields representing the principal targets of other organ systems. Indeed, most neurons receive converging sensory input.[84,130] Convergence of primary visceral afferents, e.g., in the medial-commissural NTS, may allow for interactions between signals arising from different visceral organs or receptor types. As with spinal visceroceptive neurons,

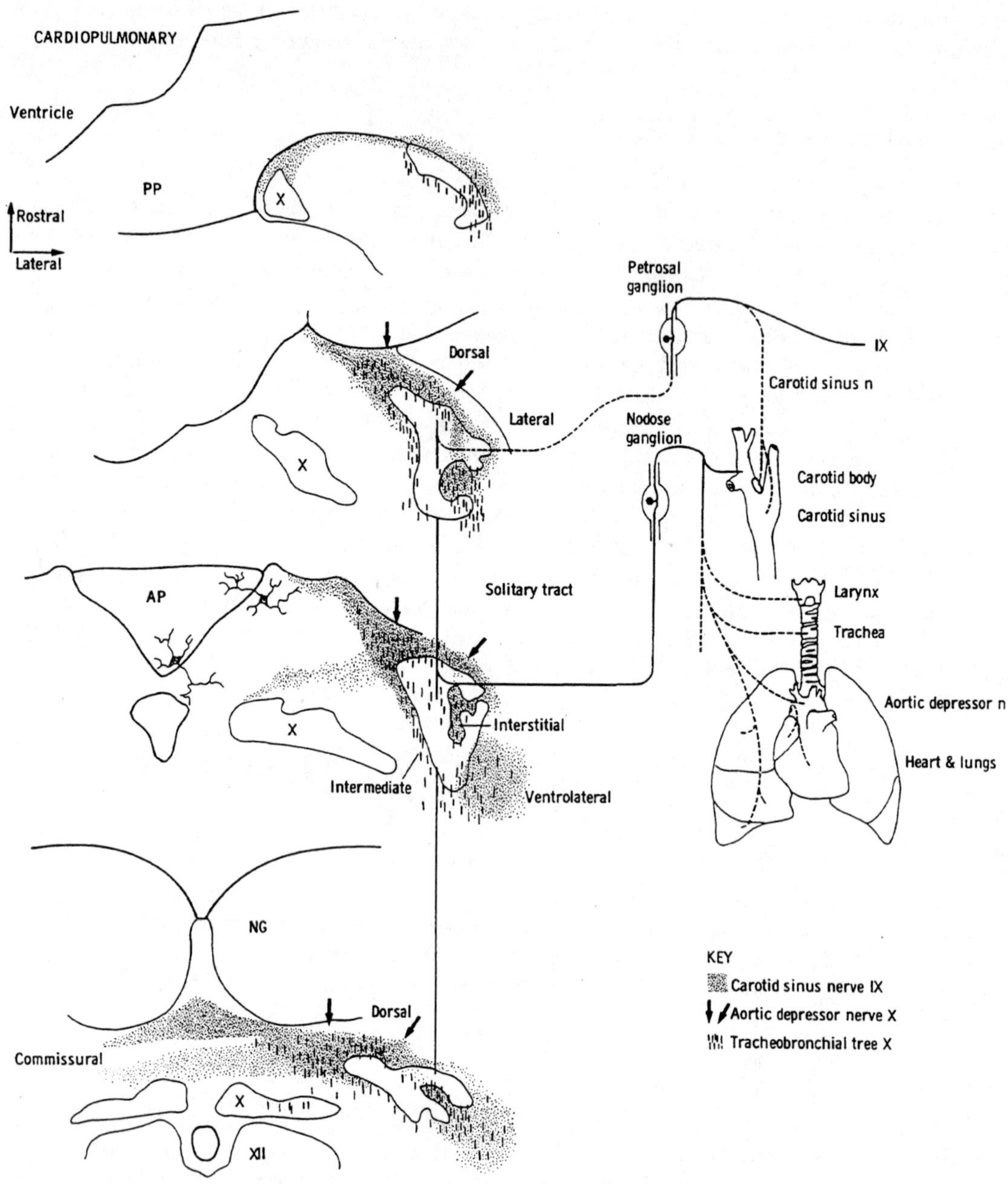

FIGURE 3

Viscerotopic organization of primary arterial baroreceptor, chemoreceptor, cardiopulmonary, and upper-airway afferents in the NTS. This schematic representation of afferent projections to the NTS is based on a synthesis of reports in the literature. Aortic and carotid sinus afferents preferentially project to dorsolateral, intermediate, and lateral commissural nuclei in cats[30,38,63] and rats.[29,42,49,52,127] Overlap between the terminal fields of these two sets of afferents is extensive. The carotid sinus afferents terminating in the commissural subnucleus may be preferentially chemoreceptive.[28,147] Laryngeal and tracheobronchial afferents preferentially project to interstitial, intermediate, and ventral nuclei.[4,61,62,103] Abbreviations: AP, area postrema; NG, nucleus gracilis; PP, nucleus prepositus; X, dorsal motor vagal nucleus; XII, hypoglossal nucleus. (From: Ruggiero, D. and Feinstein, D. in *Neurophysiological Basis of Cerebral Blood Flow Control: An Introduction*, Mraovitch, S. and Sercombe, R., Eds., John Libbey and Company, London, 1995, Figure 8b. With permission.)

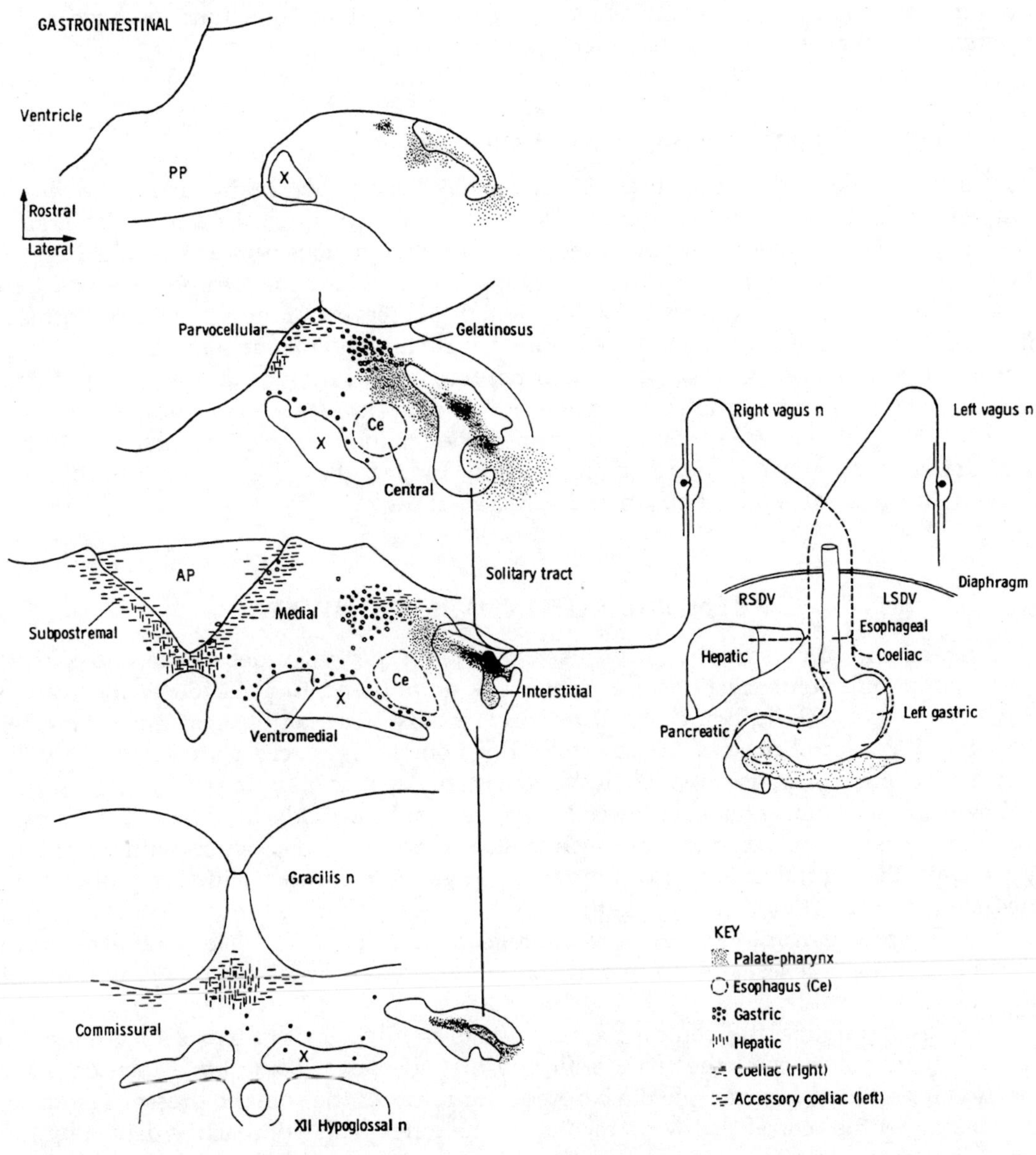

FIGURE 4
Viscerotopic organization of primary gastrointestinal afferents in the NTS. This schematic representation is compiled from reports in the literature.[4,61,70,95,129] Note the distinct projection fields of the different afferents. Abbreviations: AP, area postrema; LSDV and RSDV, left and right subdiaphragmatic vagi; PP, nucleus prepositus; X, dorsal motor vagal nucleus. (From: Ruggiero, D. and Feinstein, D. in *Neurophysiological Basis of Cerebral Blood Flow Control: An Introduction*, Mraovitch, S. and Sercombe, R., Eds., John Libbey and Company, London, 1995, Figure 8c. With permission.)

interactions of somatosensory inputs with visceral inputs also occur in the NTS.[104] Somatic input to NTS arises from secondary somatic sensory neurons in the spinal-trigeminal nucleus and spinal laminae that process visceral and noxious afferent information[83,94] and the vestibular complex.[152] These pathways provide substrates for somato-visceral integration in the NTS.

In addition to visceral and sensory inputs, the NTS receives afferents from a wide variety of brain regions, including hypothalamus, amygdala, and insular cortex.[118] Thus, it is important to consider the NTS as a site of integration of multiple, functionally distinct inputs, rather than a sensory relay nucleus.

4.3 Non-neural Inputs: The Circumventricular Organs

Periventricular areas of the brain lacking a functional blood brain barrier are potentially sensitive to circulating substances that reflect the physiological status of the organism. Specialized neurovascular structures without tight junctions between capillary endothelial cells are termed circumventricular organs. These include the area postrema (AP) in the dorsomedial medulla adjacent to the NTS, and three diencephalic regions: the subfornical organ, the organum vasculosum of the lamina terminalis, and the median eminence. Each of these areas contains receptors for various peptide hormones; for example, the AP has a high concentration of receptors for angiotensin[45] and vasopressin.[105] Furthermore, the AP, subfornical organ, and organum vasculosum of the lamina terminalis are interconnected with brain areas involved in autonomic control.[57] Indeed, the AP even receives primary visceral afferents carried in the IXth and Xth cranial nerves.[30,102]

5 CENTRAL AUTONOMIC PATHWAYS

Visceral and somatic inputs drive visceral reflex responses. Indeed, it can be considered that *all* autonomic responses are reflexive, as opposed to consciously motivated. However, not all of the afferent stimuli are truly sensory. For example, fear is not typically considered a visceral or somatic sensation, but nonetheless elicits characteristic conditioned autonomic responses that could be considered reflexive. In this context emotional or imagined sensation arising from cortical sites would provide the afferent limb of a conditioned reflex arc. Autonomic adjustments also occur in conjunction with voluntary movement; the central command to move a muscle is integrated with sympathetically mediated increased blood flow to that muscle.[39,87] Thus, in considering the central neural pathways involved in autonomic control, circuits must exist to mediate local reflexes, more global patterned reflexes, and the integration of visceral responses with cognitive stimuli and behavioral events.

Studies examining the central neural circuits involved with control of autonomic outflow have approached the issue with one of two biases. Some have focused on a functional response, for example the baroreceptor reflex or the somatic-pressor response, and examined the central pathways mediating the reflex. This approach is driven by the functional response being examined. Alternatively, other studies have been driven by a neuroanatomical focus; for example, extending the circuitry from the preganglionic neurons back further into the CNS. These studies are designed to explore the origins of neurons that innervate different subsets of preganglionic neurons. Despite the different approaches, the ultimate goal is the same: To understand the functional neural networks controlling the viscera. Eventually, the results of these two approaches will need to be integrated into a gestalt. The data are just now becoming complete enough to allow for this type of synthesis.

5.1 Projections of Secondary Visceral Afferents

5.1.1 Secondary Spinal Afferents

As noted above (Section 4.1), visceral afferents terminate primarily in laminae I, V, VII, and X. These laminae transmit by short intraspinal reflex circuits to regions containing autonomic preganglionic neurons,[17,32] to mediate local spinal visceral reflexes. Visceroceptive neurons may also transmit rostrally to cervical spinal neurons[3] or supraspinal centers[16,24,133] that subserve integrative reflex actions on the ANS.

5.1.2 Projections from the NTS

Sensory signals arriving in NTS may influence autonomic activity by relays to sites containing preganglionic neurons. This occurs over circuitry confined to the dorsomedial medulla and via two projection pathways: (1) a dorsal periventricular tract and (2) a lateral transtegmental fiber system[94,107] (Figure 5). Although evidence for this organization has been obtained largely in rats, similar projections of the NTS occur in other species, including cat.[72]

Visceral afferents reach the ANS via topographically organized projections from the NTS to preganglionic motoneurons or more circuitously via autonomic relay nuclei. Solitariospinal axons arise from the caudal-ventral NTS and descend in the lateral spinal funiculus to terminate in the upper thoracic IML. Direct or polysynaptic relays to IML are predicted by additional projections to laminae, e.g., I and VII, that harbor autonomic interneurons or dendrites of SPNs.[17,32] The NTS also projects to phrenic and intercostal respiratory motor nuclei in the mid-cervical and thoracic ventral horn. Solitariospinal pathways form short visceral reflex arcs through which signals from cardiorespiratory and chemoreceptors may influence, perhaps simultaneously, autonomic outflow and breathing.[93] Solitario-bulbar fibers terminate within nuclei that, in turn, descend to innervate the sympathetic IML: the rostral ventrolateral medulla, pontine A5 cell group, caudal raphe nuclei, Kölliker-Fuse nucleus, and hypothalamus (see Section 5.1).

Efforts are underway to elucidate spatial relationships between organ-specific or potentially convergent viscerosensory relay neurons in NTS and their projection fields. Insights into these organizational patterns have been gained by examining NTS projections within medulla and spinal cord. Topographically organized projections from the general visceral afferent division of NTS were mapped to the medullary reticular formation, nucleus ambiguus and spinal gray.[115] The NTS projects to cell columns in the lateral tegmentum and neighboring reticular formation that harbor (1) propriobulbar cells that subserve reflex coordination of muscles that control movements of the face, jaw and tongue, diaphragm and accessory muscles of respiration;[51,119,142] (2) general visceral efferents; and (3) reticular relay neurons, e.g., pre-parabrachial,[64] pre-hypothalamic,[124,146] or pre-spinal.[78,110]

Intranuclear (associational) networks are formed by local circuit neurons in the NTS. Primary visceral afferents and local interneurons in NTS terminate in the dorsal motor nucleus.[108,112] These circuits constitute short (monosynaptic and disynaptic) visceral reflex arcs.[109]

NTS efferents project caudally and rostrally within the periventricular gray (Figure 5). A descending limb travels caudally in proximity to the central canal to terminate in spinal lamina X. Terminal input to lamina I arises from fibers arching laterally to NTS and caudally to the NTS-projection field in the spinal trigeminal nucleus pars caudalis.[93] Although solitario-spinal fibers traveling in the periventricular gray may not directly innervate sympathetic preganglionic neurons (see below), they are still in a position to influence autonomic response. Ascending periventricular projections from the NTS terminate in the periaqueductal gray[11] and periventricular regions in the hypothalamus and preoptic area[143] and midline ("periventricular") thalamus.[99,100] Of these projection sites, only the hypothalamic paraventricular nucleus is capable of direct influence over autonomic preganglionic neurons; other interactions must be more indirect.

The transtegmental projections of the NTS radiate ventrally and diagonally across the intermediate reticular zone and ventral tegmental field (Figures 5 and 6). (The terminal

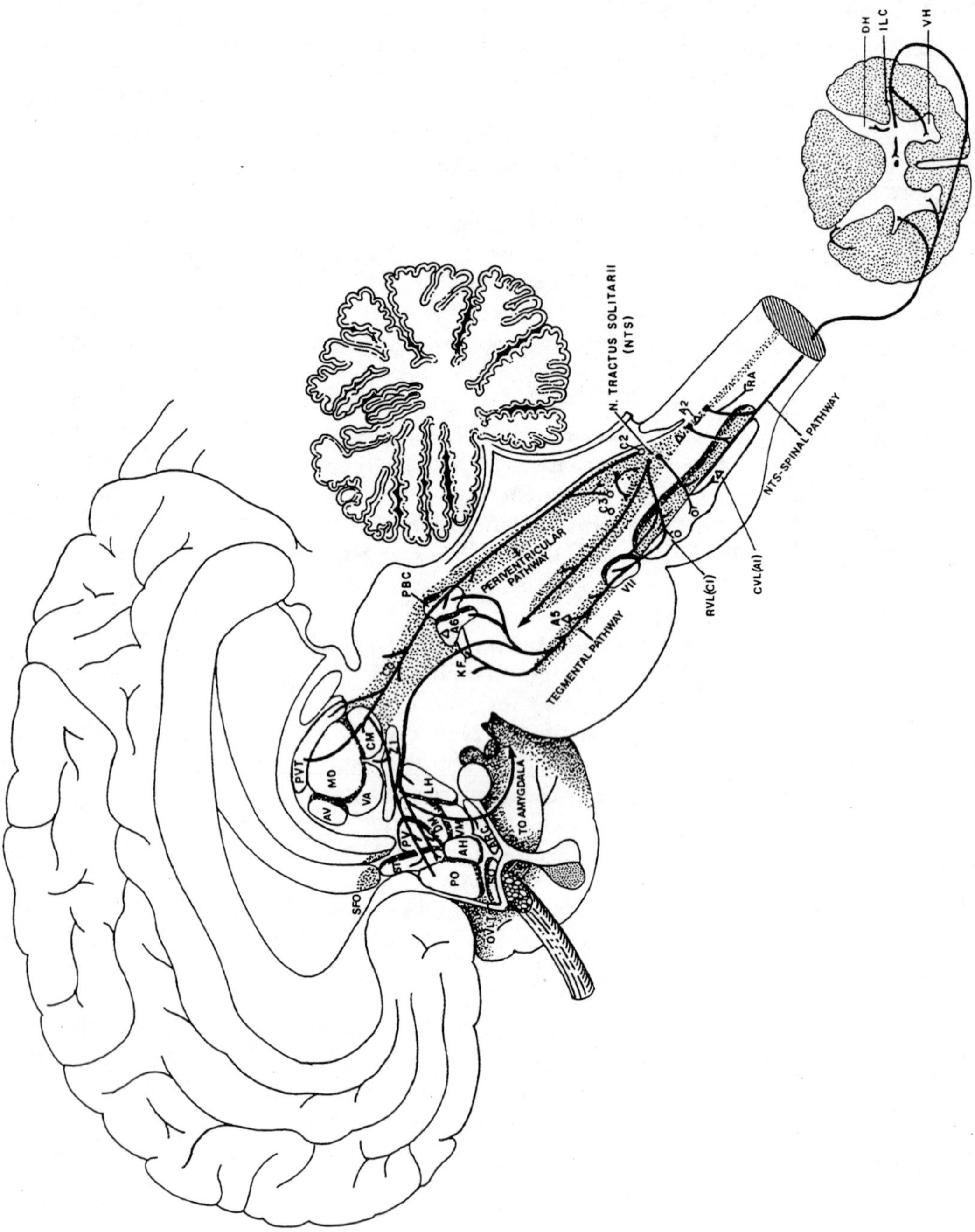
DH
ILC
VH
N. TRACTUS SOLITARII
(NTS)
NTS-SPINAL PATHWAY
CVL(A1)
RVL(C1)
PERIVENTRICULAR PATHWAY
TEGMENTAL PATHWAY
PBC
KF
A6
A5
VII
C2
TO AMYGDALA
PVT
MD
CM
VA
AV
LH
PV
PO
SO
SFO
OVLT

FIGURE 5 (Facing page)
Visceral afferent relay systems emanating from NTS. As described in the text (Section 5.1.2), the NTS projects to other autonomic centers via two pathways: a dorsal periventricular pathway and a tegmental pathway. Abbreviations: AH, anterior hypothalamic nucleus; ARC, arcuate nucleus; AV, anteroventral thalamic nucleus; CM, nucleus centre-median thalami; CVL, caudal ventrolateral medulla; DM, dorsomedial hypothalamic nucleus; KF, Kölliker-Fuse nucleus; LH, lateral hypothalamic area; MD, mediodorsal thalamic nucleus; NTS, nucleus tractus solitarii; OVLT, organum vasculosum lamina terminalis; PBC, parabrachial complex; PO, preoptic area; PV, paraventricular hypothalamic nucleus; PVT, nucleus paraventricularis thalami; RA, retroambigual area; RVL, rostral ventrolateral medulla; SFO, subfornical organ; SO, supraoptic hypothalamic nucleus; ST, stria terminalis; VA, ventral anterior thalamic nucleus; VM, ventromedial hypothalamic nucleus; ZI, zona incerta. (From: Ruggiero, D. and Feinstein, D. in *Neurophysiological Basis of Cerebral Blood Flow Control: An Introduction*, Mraovitch, S. and Serocombe, R., Eds., John Libbey and Company, London, 1995, Figure 9. With permission.)

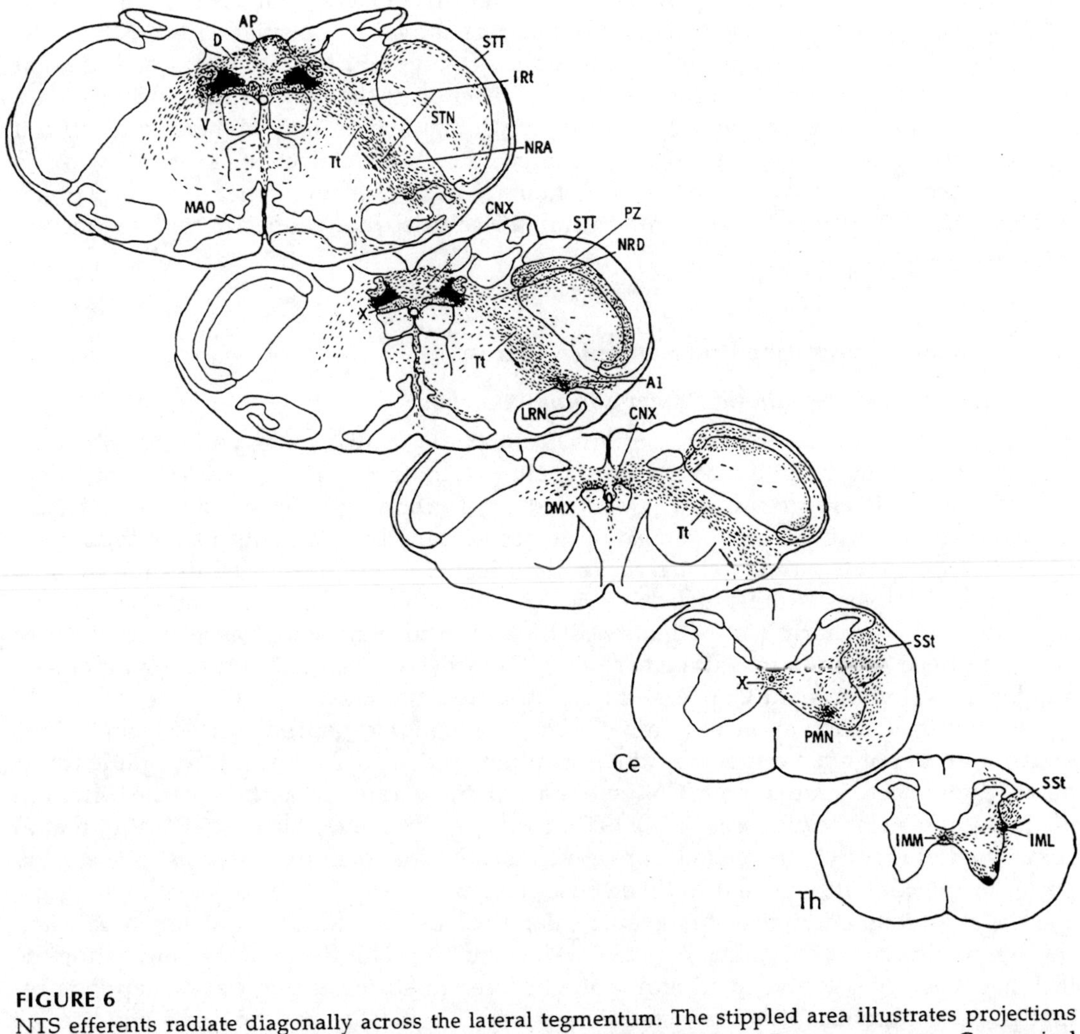

FIGURE 6
NTS efferents radiate diagonally across the lateral tegmentum. The stippled area illustrates projections from the NTS based on an anterograde tracing study.[93] Abbreviations: AP, area postrema; Ce, cervical spinal cord; CNX, commissural vagal nucleus; DMX, dorsal motor vagal nucleus; IML, intermediolateral cell column; IMM, intermediomedial cell column; IRt, intermediate reticular zone; LRN, lateral reticular nucleus; MAO, medial accessory olivary nucleus; NRA, nucleus retroambiguus; NRD, nucleus reticularis dorsalis, PMN, phrenic motor nucleus; PZ, pars zonale, SST, solitariospinal tract; STN and STT, spinal trigeminal nucleus and tract; Th, thoracic spinal cord; Tt, transtegmental tract; X, lamina 10 of Rexed.

fields of this projection will be discussed in the context of functional circuits in Sections 5.3.1 and 5.3.2.) These radiations define a transtegmental fiber system that forms a circumscribed field in the lateral reticular formation.[93] Axons issue perpendicularly to this field and extend bidirectionally along planes that roughly parallel the neuraxis (Figures 5 and 6).

5.1.3 Projections from the Circumventricular Organs

Neurons in the AP receive information related to visceral function from factors circulating in blood and neurons via direct projections from the IXth and Xth cranial nerves and projections from the adjacent NTS. Two major efferents arise from the AP. The AP projects to the medial and commissural subnuclei of the NTS, subnuclei that receive chemoreceptor, baroreceptor, and gastrointestinal afferents.[128] The AP also projects to the parabrachial nucleus, as does the NTS. The parabrachial nucleus has extensive interconnections with other central autonomic nuclei (e.g., hypothalamus, amygdala, insular cortex).[43] Furthermore, neurons in the parabrachial complex are sensitive to a variety of visceral afferent sensations[48] and autonomic responses can be elicited from this nucleus.[26] These data suggest that the AP may transmit information integrated from peripheral and central chemoreceptors and viscera. The AP/NTS circuit may mediate actions exerted by toxins or other agents on cardiovascular[41] or alimentary[14,85] reflex function. The subfornical organ and organum vasculosum of the lamina terminalis are also interconnected with autonomic centers.[57]

5.2 Pathways Innervating Preganglionic Neurons

5.2.1 Inputs to Sympathetic Preganglionic Neurons

As noted above, sympathetic outflow from the CNS originates with preganglionic neurons located in the intermediate gray matter of thoracic and lumbar levels of spinal cord. These SPNs are prominently controlled by descending supraspinal inputs, as described below. However, they are also influenced locally by spinal inputs, providing the substrate for spinal sympathetic reflexes (see Section 5.1). Thus, whereas high spinal transection abolishes supraspinal drive, it does not totally eliminate sympathetic nerve activity.[82,140,149] In quadriplegic patients with high cervical spinal cord transections a variety of sympathetic reflexes can still be evoked.[76] Nonetheless, in intact subjects sympathetic regulation is predominantly governed by supraspinal structures.

Supraspinal innervation of sympathetic preganglionic neurons arises from a very limited number of brain regions. Studies examining the retrograde and anterograde transport of dyes injected into discrete CNS sites have demonstrated projections to the IML from the following sites (see Reference 31): NTS, rostral ventrolateral medulla (RVLM), rostral ventromedial medulla, the caudal raphe nuclei (obscuris, pallidus, magnus), the A5 cell group in the caudal pons, the Kölliker-Fuse nucleus, the midbrain periaqueductal gray, specific hypothalamic nuclei (paraventricular nucleus, lateral hypothalamic area), and specific cortical areas (infralimbic, prefrontal) (Figure 7). This list of areas innervating the IML has been mostly confirmed and refined by recent studies using transsynaptic retrograde neuroanatomical tracing techniques in rat. Specifically, injection of a neurotropic virus into a peripheral tissue or sympathetic ganglion results in uptake and retrograde transport of the virus. Within the cell body of the neuron the virions replicate and travel retrogradely across the synapse, whereupon the process is repeated.[20,55] In this manner, multi-synaptic networks can be traced transsynaptically. A variety of sympathetic ganglia and sympathetically-innervated tissues have been studied in rat using this methodology including the kidney, adrenal medulla, stellate ganglion, celiac ganglion, and superior cervical ganglion.[56,125,134,136] The results demonstrate that there are five brain regions that seem to innervate at least a subpopulation of preganglionic neurons projecting to each tissue or ganglion: the RVLM, the rostral ventromedial medulla, the caudal raphe nuclei (obscuris, pallidus, and magnus), the A5 noradrenergic cell group in the caudal pons, and the parvicellular region of the hypothalamic paraventricular nucleus.

In addition to the five areas that are labeled from all sympathetic ganglia studied to date, other brain areas are uniquely labeled following injection of virus into certain ganglia. For example, virus injections into the superior cervical ganglion infected neurons in the mesencephalic central gray and lateral hypothalamic area. However, it is important to note that these neurons may represent the next higher order, i.e., tertiary relays in the neural network. Nonetheless, such differences in innervation pattern may underlie components contributing to differential regulation of sympathetic nerves. Certain areas, such as the NTS, medial prefrontal cortex, and infralimbic cortices[6,53,93] although not identified in viral-transport studies, have been reported to innervate the IML. It is important to consider that these projections may innervate interneurons in the IML rather than preganglionic neurons and represent the next higher order neuron. Alternatively, pools of neurons not susceptible to infection by the neurotropic virus[20] could result in certain neurons innervating SPNs not being labeled by this technique.

With the exception of the A5 noradrenergic cells, each region that innervates SPNs is heterogeneous in terms of neuronal projections and neurotransmitter phenotype. In each cell group, it is a minority of neurons that innervate SPNs and each area has extensive projections to other autonomic brain nuclei. However, it should be noted that the spinally projecting neurons in each area are also neurochemically heterogeneous. Although many raphe-spinal neurons contain serotonin (often co-localized with the neuropeptides substance P and thyrotropin releasing hormone) other neurons in this region are not serotonergic. Within the RVLM, approximately half of the spinally-projecting neurons contain the enzyme phenylethanolamine-*N*-methyltransferase that catalyzes the conversion of norepinephrine to epinephrine;[79,139,146] these neurons comprise the C1 cell group. Within the circumscribed vasopressor region of the RVLM, an even higher proportion (72%) of RVLM-spinal neurons belongs to the C1 cell group.[114] Despite their potential to synthesize epinephrine, these neurons do not appear to utilize epinephrine as a neurotransmitter at their terminals.[138] The neurotransmitter phenotypes of the remainder of the neurons are heterogeneous, but pharmacology studies indicate that the functional neurotransmitter at most RVLM-spinal synapses may be an excitatory amino acid.[10,86,90] The neuropeptide oxytocin is present in approximately 10–20% of spinally projecting neurons from the paraventricular nucleus.[123] Different neuropeptides are present in small percentages of other spinally projecting neurons in this nucleus and the neurotransmitter phenotype of the bulk of this projection remains to be characterized.

Although the five regions noted above seem to innervate preganglionic neurons projecting to each ganglion, the percentage of neurons from each brain region and their neurochemical content differ relative to the ganglion and target site examined.[56,125,134,136] For example, the locations of labeled neurons in the paraventricular nucleus of the hypothalamus are topographically related to the injected target. This observation is also reflected by the percent of viral-labeled neurons in the paraventricular nucleus of the hypothalamus expressing the neuropeptide vasopressin. Differences are also apparent at medullary levels. Following stellate ganglia injections most neurons infected with the virus are found in the rostral ventromedial medulla whereas with injections into the kidney labeled neurons predominate in the RVLM. The constancy of these five areas might reflect their integrated roles in the fight-or-flight pattern of global sympathetic discharge. However, this is pure speculation and as regards the RVLM seems not to be the case (see below). Conversely, the differential localization and neurotransmitter content of neurons innervating different populations of SPNs are likely related to the differential control exerted by the sympathetic nervous system over different ganglia and target organs.

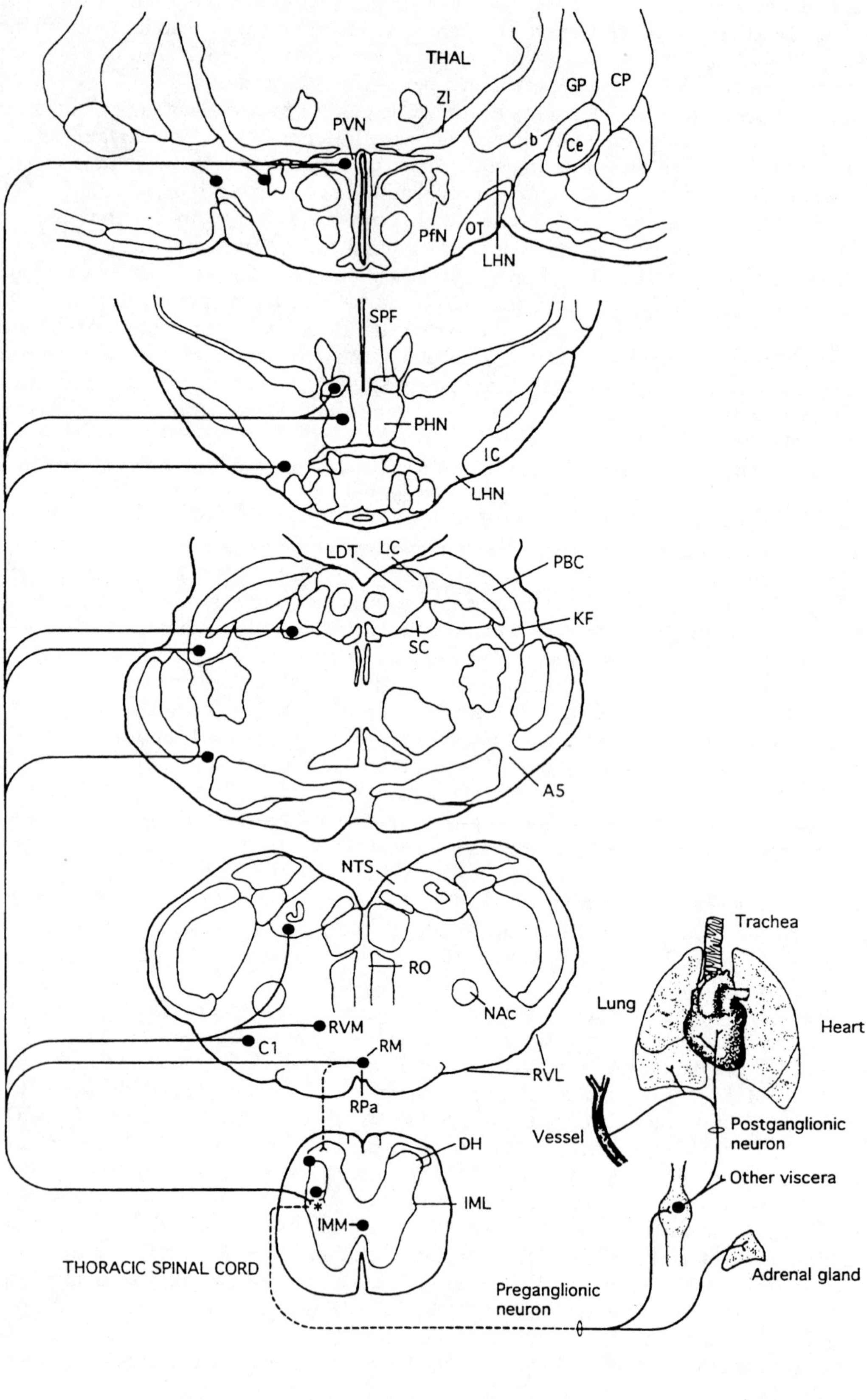
THAL
GP
CP
ZI
PVN
b
Ce
PfN
OT
LHN
SPF
PHN
IC
LHN
LDT
LC
PBC
KF
SC
A5
NTS
RO
NAc
RVM
C1
RM
RVL
RPa
DH
IML
IMM
THORACIC SPINAL CORD
Trachea
Lung
Heart
Vessel
Postganglionic neuron
Other viscera
Adrenal gland
Preganglionic neuron

FIGURE 7 (Facing page)
Schematic of CNS regions innervating IML. As described in the text (Section 5.2.1), the IML is innervated by distinct cell groups in the brain and spinal cord. Abbreviations: b, nucleus basalis; CP, caudate-putamen; C1, C1 cell group; Ce, central amygdaloid nucleus; DH, dorsal horn; GP, globus pallidus; IC, internal capsule; IML, intermediolateral cell column; IMM, intermediomedial cell column; KF, Kölliker-Fuse nucleus; LC, locus ceruleus; LDT, laterodorsal tegmental nucleus; LHN, lateral hypothalamic nucleus; NAc, nucleus ambiguus (compact formation); NTS, nucleus tractus solitarii; OT, optic tract; PBC, parabrachial complex; PfN, perifornical hypothalamic nucleus; PHN, posterior hypothalamic nucleus; PVN, paraventricular hypothalamic nucleus; RM, nucleus raphe magnus; RMe, nucleus raphe medianus; RO, nucleus raphe obscurus; RPa, nucleus raphe pallidus; RVL, rostral ventrolateral medulla; SC, subceruleus; SPF, subparafascicular nucleus; THAL, thalamus; ZI, zona incerta.

Given evidence that these five brain regions innervate SPNs projecting to all sympathetically-innervated tissues, the question arises as to whether a single neuron might collateralize to innervate different tissue-specific populations of preganglionic neurons. For example, might a single RVLM neuron branch innervate SPNs providing projections to the kidney, heart, and adrenal medulla? The absence of single brain neurons retrogradely labeled from multiple levels of the spinal cord led Tucker and Saper[145] to conclude that few if any neurons innervate different subsets of preganglionic neurons. This is in contrast to reports that neurons in the caudal raphe and RVLM can be antidromically activated from multiple levels of the IML.[8,91] Furthermore, a large percent of A5 neurons is retrogradely labeled by injection of virus into each visceral organ explored thus far,[56,125,134,136] indicating that single A5 neurons innervate preganglionic neurons that subserve different functions. A similar argument can be made for spinally projecting C1 neurons.

It should be noted that the above description of neurons providing innervation of SPNs is based largely on studies conducted in rats. As noted below, differences exist between the rat and cat, the most common animal model used in studies of vestibular physiology. Nonetheless, the feline RVLM (referred to as the subretrofacial nucleus), caudal raphe nuclei, A5 neurons, and the hypothalamic paraventricular nucleus innervate the IML. The cardiovascular-active zone of the rostral ventromedial medulla defined in rat[1] has not been studied in cat. However, the relative importance and control of these cell groups likely differ between these mammals. One approach that has been used in cats to identify neurons antecedent to SPNs, has been to identify brain neurons whose activity correlates with sympathetic outflow rhythms.[44] Spinally projecting neurons in the RVLM and caudal raphe nuclei appear to contribute to these rhythms, and these rhythms are imparted on those neurons largely by innervation provided by a circumscribed dorsal region of the lateral tegmental field (LTF).[9,44,155] Although the LTF appears to play a major role in the regulation of sympathetic outflow in cats, if a functionally analogous region exists in rats it has remained elusive. Whereas the lateral tegmentum is a general term referring to the lateral third of the pontomedullary reticular formation, the feline LTF has come to refer to a specific functional region in the dorsal aspect of the lateral tegmentum.

The functions of the five supraspinal areas that innervate SPNs at all levels of the thoracic cord have not been thoroughly examined. Certainly the most well studied of these areas is the RVLM due to its role in cardiovascular control. Discrete bilateral inhibition of this region produces a marked decrease in blood pressure, a decrease equivalent to that elicited by total blockade of sympathetic vasomotor outflow consequent to spinal cord transection or ganglionic blockade.[111] Numerous studies now support the notion that the RVLM is the principal site involved in the tonic and reflex control of sympathetic vasomotor tone and the maintenance of blood pressure.[34,46] Even so, differential regulation of sympathetic vasomotor outflow may also be subserved by different subsets of neurons in this region. Pre-sympathetic vasomotor neurons in this region are highly topographically organized,[34,80] at least in cats. However, this topographical organization does not appear to exist in rat based on the findings that preferential responses of different sympathetic nerves can be elicited from sites that appear intermixed in the RVLM.[12] This lack of topographical organization of the rat RVLM is also supported by the seemingly similar locations of RVLM neurons innervating different levels of the IML.[145] Nonetheless, the relatively small size of the rat RVLM may obscure a topographical organization.

Independent of species, several distinct cell populations within the RVLM defined by electrophysiological or neurochemical criteria seem to be involved in cardiovascular regulation.[34,46] Neurons in the RVLM that project to spinal SPNs receive a variety of inputs, including cardiovascular, visceral, and somatic including spinal and vestibular.[40,46,153] Signals generated by each of their receptor organs including muscle, skin, and the vestibular labyrinth modify cardiovascular responses.

Although the RVLM plays a major role in cardiovascular regulation, its role in controlling sympathetic outflow to other visceral tissues is more limited, at least in cats. McAllen demonstrated that stimulation of neurons in the cat RVLM by local injections of excitatory amino acids evokes large increases in sympathetic vasomotor tone, but not pupillary dilation, sudomotor responses, or marked suppression of gastric contractions.[77,80] However, the situation in the rat is quite different. Glutamate injections into the RVLM, at sites that elicit vasomotor responses, also elicit pupillary dilation.[156] Interestingly, injections of muscimol into the caudal ventrolateral medulla, which produce a marked increase in blood pressure by increasing the activity of RVLM-spinal neurons (see below), do not alter pupil size.[156] Thus, although stimulation of the RVLM in the rat increases sympathetic outflow to the vasculature and the iris, these outflows are apparently differentially regulated.

Increased sympathetic outflow brought about by decreased baroreceptor afferent input is accompanied by increased electrical activity of RVLM-spinal neurons.[46] Similarly, spinal somatic afferent stimulation-induced increases in sympathetic vasomotor tone are accompanied by increased activity of RVLM neurons.[92] Studies using the evoked expression of the immediate early gene c-fos as an index of increased neuronal activity[89] have corroborated these observations. RVLM-spinal neurons, many of which belong to the C1 catecholaminergic cell group, express the immediate-early gene c-fos in response to decreased baroreceptor input-induced increase in sympathetic vasomotor tone in cats[79] and rats.[139] In contrast, insulin or 2-deoxyglucose-induced increases in sympathoadrenal activity are not accompanied by an induction of c-fos in the RVLM.[157] Thus, despite the widespread innervation of the IML by RVLM neurons and the prominent role of this area in cardiovascular control, an overall picture of how cells in the area are involved in autonomic integration has been slow to develop. Furthermore, the emerging picture appears clouded by marked species differences in this region between rat and cat.

The roles of the rostral ventromedial medulla, caudal raphe nuclei, A5 area, and paraventricular nucleus of the hypothalamus in regulation of the sympathetic nervous system are less clear. Each region has been implicated in cardiovascular regulation control mechanisms.[33] However, with the exception of the paraventricular nucleus of the hypothalamus, little consideration has been given to their actions in other aspects of sympathetic neural function. The hypothalamic paraventricular nucleus seems to be involved in the control of a variety of sympathetic outflows, and possibly serves as a primary site of integration and coordination of autonomic responses.

5.2.2 Inputs to Parasympathetic Preganglionic Neurons

The primary input to parasympathetic preganglionic neurons arises from primary or secondary visceral sensory afferents involved in reflex neural control of the target tissue. Thus, gastric vagal preganglionic neurons receive direct first order afferent input from the stomach[108] as well as input from neurons in the NTS that receive gastric afferents.[109] Other afferents to the DMN include hypothalamic nuclei, the central nucleus of the amygdala, mesencephalic central gray, parabrachial complex, A5 cell group, and raphe obscuris and pallidus.[74] A similarly long list of afferents to the DMN was drawn from viral-transport studies of the parasympathetic projection to the pancreas.[73] Evidence that brain regions innervating SPNs in the spinal cord also project to the DMN, predicts that these afferents may arise as collaterals to sympathetic and parasympathetic motoneurons.

5.3 Functional Autonomic Circuits

5.3.1 Baroreceptor Reflex

The baroreceptor reflex serves as a good illustration of a functional autonomic circuit. The baroreceptor reflex is mediated via a circuit from the NTS, the site of termination of baroreceptor afferents, to vasomotor and cardiomotor preganglionic sympathetic neurons in the spinal cord and cardiomotor preganglionic parasympathetic neurons in the medulla. These circuits were elucidated through a combination of neuroanatomical mapping studies, electrophysiological recording, and stimulation and lesion studies. It is now clear that the minimal circuitry necessary to elicit baroreceptor reflex mediated changes in sympathetic outflow consists of the following components: an excitatory solitario-reticular projection from the NTS to a group of interneurons in the caudal ventrolateral medulla, an inhibitory "intrareticular" projection from the caudal ventrolateral medulla to a circumscribed vasopressor region in the RVLM, and an excitatory "reticulospinal" projection from the RVLM that descends to innervate sympathetic preganglionic neurons. In the framework of NTS efferents described above, the projection from NTS to the caudal ventrolateral medulla and the intrareticular projection to the RVLM are components of the transtegmental fiber system. A neuroanatomically defined projection from the baroreceptive region of the NTS directly to the vasopressor area of the RVLM, does not appear to be sufficient to mediate the baroreceptor reflex. The reflex circuit mediating baroreceptor control of heart rate is less clear. Specifically, whether it is mediated via a direct projection from NTS to cardiovagal preganglionic neurons or obligate interneurons in the ventrolateral medulla is uncertain.

At each locus involved in this visceral reflex pathway, modulation of the response is possible, and likely serves as one substrate by which responses are integrated and patterned. For example, during stress the baroreceptor reflex is inhibited thereby facilitating the stress-evoked cardiovascular changes. This modulation appears to occur primarily at the level of the NTS and RVLM. In addition, the prominent reciprocal innervation of the NTS with each of its targets suggests the existence of feedback autonomic circuits.[94]

5.3.2 Spinal Somatic-Autonomic Reflex Circuits

Somatic-autonomic reflex circuits highlight the multiple levels of parallel processing of signals that influence autonomic responses. Spinal neurons exert effects on sympathetic nerve discharge and perhaps the parasympathetic nervous system through local propriospinal and supraspinal connections with the ANS. Myriad circuits exist for a range of electrophysiologically identified short and long latency responses of sympathetic and adrenomedullary visceromotor neurons to peripheral afferent stimulation.[5,88,122]

Spinal-Autonomic Reflex Arcs: The Early Components — Short intraspinal (elementary) reflexes are mediated by myelinated and unmyelinated afferents and characterized by an early component based on recordings of sensory-evoked changes in sympathetic nerve discharge. Local nociceptive group IV reflexes, for example, provoke regional, unilateral visceromotor responses[5] that in spinal-injured patients are used to monitor potentially dangerous signals below the level of analgesia.[120] Anatomical substrates are predicted by short axonal projections from lamina I of the spinal dorsal horn[32] to the sympathetic cell column. Autonomic interneurons also reside in laminae V and VII and were labeled transneuronally from the superior cervical ganglion.[17] Longer, more complex reflex pathways may also exist. These are suggested by the effects exerted by neurons in the upper cervical spinal cord on abdominal sympathetic nerve activity,[141] and inferential evidence of long, propriospinal projections from the spinal enlargement to thoracic segments that harbor sympathetic preganglionic neurons.[148]

Supraspinal Autonomic Reflex Arcs: The Late Components — Spinal projections ascend via a medial periventricular fiber system and a lateral transtegmental fiber system and terminate in the NTS and along viscerosensory output pathways.[144,154] Terminal fields of spinal afferents to the NTS are concentrated at levels caudal to obex and overlap sites that receive first-order afferents from the upper alimentary tract, tracheobronchial tree, and the carotid body (see Figures 3 and 4). In contrast, spinal afferents avoid projection targets of primary baroreceptor afferents, the subdiaphragmatic vagus, and gustatory nerves (i.e., the rostral nuclei). A potential role of a spinal-afferent input to NTS in cardiorespiratory reflex function is supported by the following evidence: (1) the cells or origin derive from laminae I and X that transmit nociceptive stimuli[83] and respond to noxious stimulation of the peritoneum and colorectal distension,[67] and (2) neurons within these loci in NTS descend via the solitariospinal tract to innervate phrenic and intercostal motor neurons and the thoracic IML.[93]

Spinal Afferents to Brainstem Tegmentum — Spinal afferents also terminate within regions that relay visceral afferents from the NTS.[107,112] Evidence of convergence predicts that these neurons may integrate signals that influence autonomic outflow. These regions include the lateral reticular formation,[133] pontine parabrachial nuclei,[24] central gray,[71] and lateral hypothalamus.[16]

5.3.3 Behavioral and Integrative Circuits

Another level of circuit complexity is apparent when considering circuits involved in the higher order integration of visceral and behavioral responses. For example, the characteristic autonomic response to cold includes cutaneous vasoconstriction, piloerection, and energy mobilization. However, in a behaving animal the autonomic response pattern is integrated (adjusted) with behaviors appropriate to the stimulus — huddling and moving to a warmer location. Similarly, fear and rage are characterized by integrated autonomic, endocrine, and behavioral responses. The cerebral cortex has long been recognized for its role in mediating integrated somatic, visceral and electrocorticographic response patterns characteristic of organized behaviors.[58] The role of specific cortical regions, most prominently the posterior insular area, in central autonomic regulation is apparent in the site-specific effects of chemical or electrical stimulation in altering blood pressure and heart rate[23,116] and in provoking cardiac arrhythmias.[98] The insular cortex is extensively connected with the medial prefrontal cortex, the medial thalamus, the amygdala and neurons distributed along their output pathways. An important insight into the pathways mediating cortically evoked changes in autonomic outflow was gained by the discovery that the lateral hypothalamus serves as a mandatory relay for the insular-pressor response.[22] This finding ties with evidence that neurons in this area descend to terminate within sensory (NTS), internuncial (lateral tegmentum), and motor (DMN and IML) limbs of the autonomic reflex arc.[113,118,121]

The central nucleus of the amygdala, an outlet critical for emotional motor expression receives convergent input predominantly from areas of cerebral cortex endowed with complex integrative capacity and non-discriminative thalamo-amygdala relay neurons that form a critical afferent link in the emotional conditioning circuit. The central nucleus transmits, by way of the lateral hypothalamus and central gray, the autonomic (cardiovascular) and skeletomotor (freezing) components of the conditioned-fear response.[69] Projections of the amygdala and substantia innominata include a periventricular nucleus in the midline thalamus, "nucleus paraventricularis".[101] This nucleus, which is activated during stress,[27] projects to medial and lateral areas of prefrontal cortex and a site in the ventral striatum, the nucleus accumbens, involved in motivated behaviors.[13] Collaterals to the midline thalamus and NTS were recently found to arise from viscerolimbic areas of forebrain (e.g., insular cortex), diffuse projection systems (e.g., locus coeruleus and raphe), and neurons intercalated along ascending visceral relay systems.[99] Collaterals might distribute signals in parallel to nuclei involved in generating behavior and autonomic adjustments and, perhaps, serve to coordinate components of emotional expression.

5.3.4 Vestibular-Autonomic Reflexes

Given this organization of the autonomic nervous system, how might vestibular inputs influence autonomic function? (This topic is also examined by Yates in Chapter 6 of this volume.) Although it is assumed that primary vestibular afferents do not terminate on neurons in the NTS or form axodendritic synapses on processes extending outside the border zone of the viscerosensory projection field, this issue has not been critically evaluated. To date, the collective evidence indicates that vestibular spinal neurons do not innervate spinal sympathetic or sacral-parasympathetic visceromotor nuclei. The most direct input of vestibular information into autonomic circuits is a direct pathway from the medial and inferior vestibular nuclei to the NTS and dorsal motor nucleus in cat,[152] rabbit,[7] and rat.[158] The sites innervated by the medial vestibular nucleus overlap with neurons that receive somatic afferents from the spinal trigeminal nucleus caudalis and spinal cord[83] as well as visceral afferents from chemoreceptors, pulmonary receptors, and upper alimentary tract.[4,38,62] Vestibular inputs to NTS also overlap with regions that project to the ventrolateral medulla,[112] and terminate on spinal-autonomic and respiratory motor nuclei.[93] These circuits may mediate vestibular influences on sympathetic outflow and breathing, respectively. Such a picture is consistent with the effects exerted by natural[131] or electrical[153] vestibular afferent stimulation on the electrophysiological activity of neurons in the RVLM. Furthermore, the RVLM is required for vestibulo-sympathetic reflexes.[151,153] However, the NTS is not essential for vestibulo-sympathetic reflexes, since they were not attenuated by extensive lesions of the NTS.[132] Likely alternative circuits involve projections from the vestibular nuclei to the LTF or caudal ventrolateral medulla (see Yates, Chapter 6). The cerebellum may provide another route via which vestibular inputs may influence autonomic function (see Balaban, Chapter 8).

6 SUMMARY

The ANS provides integrated control over the viscera and patterns of autonomic outflow that occur under many physiological conditions. The efferent components of autonomic control, the sympathetic and parasympathetic motoneurons, are regulated by afferent sensory signals arising from the viscera as well as non-visceral tissues. The central neural networks that integrate these sensory signals and modify visceromotor output are complex. Mono- and di-synaptic reflex circuits, organized in the brainstem and spinal cord, are influenced by more elaborate pathways including forebrain circuits integrating behavioral and visceral responses. At each of these levels of processing, integration is provided necessary for differential autonomic responses to signals generated in the internal and external environments and brain.

REFERENCES

1. Aicher, S. A., Reis, D. J., Ruggiero, D. A., and Milner, T. A., Anatomical characterization of a novel reticulospinal vasodepressor area in the rat medulla oblongata, *Neuroscience*, 60, 761, 1994.
2. Akeyson, E. W. and Schramm, L. P., Processing of splanchnic and somatic input in thoracic spinal cord of the rat, *Am. J. Physiol.*, 266, R257, 1994.
3. Akeyson, E. W. and Schramm, L. P., Splanchnic and somatic afferent convergence on cervical spinal neurons of the rat, *Am. J. Physiol.*, 266, R268, 1994.
4. Altschuler, S. M., Bao, X., Bieger, D., Hopkins, D. A., and Miselis, R. R., Viscerotopic representation of the upper alimentary tract in the rat: sensory ganglia and nuclei of the solitary and spinal trigeminal tracts, *J. Comp. Neurol.*, 283, 248, 1989.
5. Araki, T., Ito, K., Kurosawa, M., and Sato, A., Responses of adrenal sympathetic nerve activity and catecholamine secretion to cutaneous stimulation in anesthetized rats, *Neuroscience*, 12, 289, 1984.
6. Bacon, S. J. and Smith, A. D., A monosynpatic pathway from an identified vasomotor centre in the medial prefrontal cortex to an autonomic area in the thoracic spinal cord, *Neuroscience*, 54, 719, 1993.
7. Balaban, C. D. and Beryozkin, G., Vestibular nucleus projections to nucleus tractus solitarius and the dorsal motor nucleus of the vagus nerve: potential substrates for vestibulo-autonomic interactions, *Exp. Brain Res.*, 98, 200, 1994.
8. Barman, S. M. and Gebber, G. L., Axonal projection patterns of ventrolateral medullospinal sympathoexcitatory neurons, *J. Neurophysiol.*, 53, 1551, 1985.
9. Barman, S. M. and Gebber, G. L., Lateral tegmental field neurons play a permissive role in governing the 10-Hz rhythm in sympathetic nerve discharge, *Am. J. Physiol.*, 265, R1006, 1993.
10. Bazil, M. K. and Gordon, F. J., Sympathoexcitation from the rostral ventolateral medulla is mediated by spinal NMDA receptors, *Brain Res. Bull.*, 31, 273, 1993.
11. Beitz, A. J., The organization of afferent projections to the midbrain periaqueductal gray of the rat, *Neuroscience*, 7, 133, 1982.
12. Beluli, D. J. and Weaver, L. C., Differential control of renal and splenic nerves without medullary topography, *Am. J. Physiol.*, 260, H1072, 1991.
13. Berendse, H. W. and Groenewegen, H. J., Restricted termination fields of the midline and intralaminar thalamic nuclei in the rat, *Neuroscience*, 42, 73, 1991.
14. Borison, H. L., Area postrema: chemoreceptor circumventricular organ of the medulla oblongata, *Prog. Neurobiol.*, 32, 351, 1989.
15. Burnstock, G., Co-transmission, *Arch. Int. Pharmacodyn.*, 304, 7, 1990.
16. Burstein, R., Cliffer, K. D., and Giesler, G. J., Direct somatosensory projections from the spinal cord to the hypothalamus and telencephalon, *J. Neuroscience*, 7, 4159, 1987.
17. Cabot, J., Alessi, V., Carroll, J., and Ligorio, M., Spinal cord lamina V and lamina VII interneuronal projections to sympathetic preganglionic neurons, *J. Comp. Neurol.*, 347, 515, 1994.
18. Cabot, J. B., Sympathetic preganglionic neurons: cytoarchitecture, ultrastructure, and biophysical properties, in *Central Regulation of Autonomic Functions*, Loewy, A. D. and Spyer, K. M., Eds., Oxford University Press, New York, 1990, 44.
19. Cannon, W. B., *The Wisdom of the Body*, W.W. Norton and Co., New York, 1939.
20. Card, J. P., Rinaman, L., Lynn, R. B., Lee, B.-H., Meade, R. P., Miselis, R. R., and Enquist, L. W., Pseudorabies virus infection of the rat central nervous system: ultrastructural characterization of viral replication, transport, and pathogensis, *J. Neuroscience*, 13, 2515, 1993.
21. Cassell, J. F., Clark, A. L., and McLachlan, E. M., Characteristics of phasic and tonic sympathetic ganglion cells of the guinea pig, *J. Physiol.*, 372, 457, 1986.
22. Cechetto, D. F. and Chen, S. J., Subcortical sites mediating sympathetic responses from the insular cortex in the rat, *Am. J. Physiol.*, 258, R245, 1990.
23. Cechetto, D. F. and Saper, C. B., Role of the cerebral cortex in autonomic function, in *Central Regulation of Autonomic Functions*, Loewy, A. D. and Spyer, K. M., Eds., Oxford University Press, New York, 1990, 208.
24. Cechetto, D. F., Standaert, D. G., and Saper, C. B., Spinal and trigeminal dorsal horn projections to the parabrachial nucleus in the rat, *J. Comp. Neurol.*, 240, 153, 1985.
25. Cervero, F. and Connell, L. A., Distribution of somatic and visceral primary afferent fibres within the thoracic spinal cord of the cat, *J. Comp. Neurol.*, 230, 88, 1984.
26. Chamberlin, N. L. and Saper, C. B., Topographic organization of cardiovascular responses to electrical and glutamate microstimulation of the parabrachial nucleus in the rat, *J. Comp. Neurol.*, 326, 245, 1992.
27. Chastrette, N., Pfaff, D. W., and Gibbs, R. B., Effects of daytime and nighttime stress on Fos-like immunoreactivity in the paraventricular nucleus of the hypothalamus, the habenula, and the posterior paraventricular nucleus of the thalamus, *Brain Res.*, 563, 339, 1991.

28. Chitravanshi, V. C., Kachroo, A., and Sapru, H. N., A midline area in the nucleus commissuralis of NTS mediates the phrenic nerve responses to carotid chemoreceptor stimulation, *Brain Res.*, 662, 127, 1994.
29. Ciriello, J., Hochstenbach, S. L., and Roder, S., Central projections of baroreceptor and chemoreceptor afferent fibers in the rat, in *Nucleus of the Solitary Tract*, Barraco, I. R. A., Ed., CRC Press, Boca Raton, FL, 1994, 35.
30. Ciriello, J., Hrycyshyn, A. W., and Calaresu, F. R., Glossopharyngeal and vagal afferent projections to the brain stem of the cat: a horseradish peroxidase study, *J. Autonom. Nerv. Syst.*, 4, 63, 1981.
31. Coote, J. H., The organization of cardiovascular neurons in the spinal cord, *Rev. Physiol. Biochem. Pharmacol.*, 110, 148, 1988.
32. Craig, B. A. D., Propriospinal input to thoracolumbar sympathetic nuclei from cervical and lumbar lamina I neurons in the cat and the monkey, *J. Comp. Neurol.*, 331, 517, 1993.
33. Dampney, R. A. L., Functional organization of central pathways regulating the cardiovascular system, *Physiol. Rev.*, 74, 323, 1994.
34. Dampney, R. A. L., The subretrofacial vasomotor nucleus: anatomical, chemical and pharmacological properties and role in cardiovascular regulation, *Prog. Neurobiol.*, 42, 197, 1994.
35. Davisson, R. L., Johnson, A. K., and Lewis, S. J., Acetylcholine (Ach)-induced vasodilation may involve the release of pre-formed stores of nitrosyl factors (NOFs) from the endothelium in the conscious rat, *FASEB J.*, 8, A531, 1994.
36. Davisson, R. L., Shaffer, R. A., Murphy, S. P., Boutelle, S. L. P., and Lewis, S. J., Immunohistochemical evidence that post-ganglionic sympathetic neurons innervating rat hindlimb vasculature contain nitric oxide synthase, *Circulation*, 88, I-425, 1993.
37. Decktor, D. L. and Weems, W. A., An intracellular characterization of neurones and neural connexions within the left coeliac ganglion of cats, *J. Physiol.*, 341, 197, 1983.
38. Donoghue, S., Felder, R. B., Jordan, D., and Spyer, K. M., The central projections of carotid baroreceptors and chemoreceptors in the cat: A neurophysiological study, *J. Physiol.*, 347, 397, 1984.
39. Eldridge, F. L., Millhorn, D. E., Kiley, J. P., and Waldrop, T. G., Stimulation by central command of locomotion, respiration, and circulation during exercise, *Respir. Physiol.*, 59, 313, 1985.
40. Ermirio, R., Ruggeri, P., Molinari, C., and Weaver, L. C., Somatic and visceral inputs to neurons of the rostral ventrolateral medulla, *Am. J. Physiol.*, 265, R35, 1993.
41. Ferrario, C. M., Barnes, K. L., Diz, D. I., Block, C. H., and Averill, D. B., Role of area postrema pressor mechanisms in the regulation of arterial pressure, *Can. J. Physiol. Pharmacol.*, 65, 1591, 1987.
42. Finley, J. C. W. and Katz, D. M., The central organization of carotid body afferent projections to the brainstem of the rat, *Brain Res.*, 572, 108, 1992.
43. Fulwiler, C. E. and Saper, C. B., Subnuclear organization of the efferent projections of the parabrachial nucleus in the rat, *Brain Res. Rev.*, 7, 229, 1984.
44. Gebber, G. L., Central determinants of sympathetic nerve discharge, in *Central Regulation of Autonomic Functions*, Loewy, A. D. and Spyer, K. M., Eds., Oxford University Press, New York, 1990, 126.
45. Gehlert, D. R., Speth, R. C., and Wamsley, J. K., Distribution of [125I] angiotensin II binding sites in the rat brain: a quantitative autoradiographic study, *Neuroscience*, 18, 837, 1986.
46. Guyenet, P. G., Role of the ventral medulla oblongata in blood pressure regulation, in *Central Regulation of Autonomic Functions*, Loewy, A. D. and Spyer, K. M., Eds., Oxford University Press, New York, 1990, 145.
47. Habecker, B. A. and Landis, S. C., Noradrenergic regulation of cholinergic differentiation, *Science*, 264, 1602, 1994.
48. Hermann, G. E. and Rogers, R. C., Convergence of vagal and gustatory afferents within the parabrachial nucleus of the rat, *J. Autonom. Nerv. Syst.*, 13, 1, 1985.
49. Higgins, G. A., Hoffman, G. E., Wray, S., and Schwaber, J. S., Distribution of neurotensin-immunoreactivity within baroreceptive portions of the nucleus of the tractus solitarius and the dorsal vagal nucleus of the rat, *J. Comp. Neurol.*, 226, 155, 1984.
50. Holets, V. and Elde, R., Sympathoadrenal preganglionic neurons: their distribution and relationship to chemically-coded fibers in the kitten intermediolateral cell column, *J. Autonom. Nerv. Syst.*, 7, 149, 1983.
51. Holstege, G., Kuypers, H. G. J. M., and Dekker, J. J., The organization of bulbar fiber connections to the trigeminal, facial and hypoglossal motor nuclei. II. An autoradiographic study in cat, *Brain*, 100, 265, 1977.
52. Housley, G. D., Martin-Body, R. L., Dawson, N. J., and Sinclair, J. D., Brain stem projections of the glossopharyngeal nerve and its carotid sinus branch, *Neuroscience*, 22, 237, 1987.
53. Hurley, K. M., Herbert, H., Moga, M. M., and Saper, C. B., Efferent projections from the infralimbic cortex of the rat, *J. Comp. Neurol.*, 308, 249, 1991.

54. Jänig, W. and McLachlan, E. M., Characteristics of function-specific pathways in the sympathetic nervous system, *Trends Neuroscience*, 15, 475, 1992.
55. Jansen, A. S. P., Farewell, C. G., and Loewy, A. D., Specificity of pseudorabies virus as a retrograde marker of sympathetic preganglionic neurons: implications for transneuronal labeling studies, *Brain Res.*, 617, 103, 1993.
56. Jansen, A. S. P., Wessendorf, M. W., and Loewy, A. D., Transneuronal labeling of CNS neuropeptide and monoamine neurons after pseudorabies virus injections into the stellate ganglion, *Brain Res.*, in press.
57. Johnson, A. K. and Loewy, A. D., Circumventricular organs and their role in visceral functions, in *Central Regulation of Autonomic Functions*, Loewy, A. D. and Spyer, K. M., Eds., Oxford University Press, New York, 1990, 247.
58. Kaada, B. R., Somatomotor, autonomic and electrocorticographic responses to electrical stimulation of "rhinencephalic" and other forebrain structures in primates, cat and dog, *Acta Physiol. Scand.*, 24 (Suppl. 83), 1, 1951.
59. Kalia, M., Fuxe, K., Hökfelt, T., Johansson, O., Lang, R., Ganten, D., Cuello, C. and Terenius, L., Distribution of neuropeptide immunoreactive nerve terminals within subnuclei of the nucleus of the tractus solitarius of the rat, *J. Comp. Neurol.*, 222, 409, 1984.
60. Kalia, M. and Mesulam, M. M., Brain stem projections of sensory and motor components of the vagus complex in the cat: the cervical vagus and nodose ganglion, *J. Comp. Neurol.*, 193, 435, 1980.
61. Kalia, M. and Mesulam, M. M., Brain stem projections of sensory and motor components of the vagus complex in the cat. II. Laryngeal, tracheobronchial, pulmonary, cardiac, and gastrointestinal branches, *J. Comp. Neurol.*, 193, 467, 1980.
62. Kalia, M. and Richter, D., Morphology of physiologically identified slowly adapting lung stretch receptor afferents stained with intra-axonal horseradish peroxidase in the nucleus of the tractus solitarius in the cat. I. A light microscopic study, *J. Comp. Neurol.*, 241, 503, 1985.
63. Kalia, M., Location of aortic and carotid baroreceptor and chemoreceptor primary afferents in the brain stem, in *Central Nervous System Mechanisms in Hypertension*, Buckley, J. P. and Ferrario, C. M., Eds., Raven Press, New York, 1981, 9.
64. King, G. W., Topology of ascending brainstem projections to nucleus parabrachialis in the cat, *J. Comp. Neurol.*, 191, 615, 1980.
65. Kuo, D. C. and De Groat, W. C., Primary afferent projections of the major splanchnic nerve to the spinal cord and gracile nucleus of the cat, *J. Comp. Neurol.*, 231, 421, 1985.
66. Langley, J. N., *The Autonomic Nervous System*, Heffer, Cambridge, 1921.
67. Lanteri-Minet, M., Isnardon, P., de Pommery, J., and Menetrey, D., Spinal and hindbrain structures involved in visceroception and visceronociception as revealed by the expression of Fos, Jun and Krox-24 proteins, *Neuroscience*, 55, 737, 1993.
68. Laughton, W. B. and Powley, T. L., Localization of efferent function in the dorsal motor nucleus of the vagus, *Am. J. Physiol.*, 252, R13, 1987.
69. LeDoux, J. E., Emotion, in *Handbook of Physiology, Sec. 1: The Nervous System. Vol. V: Higher Functions of the Brain, Part 1*, Plum, F., Ed., American Physiological Society, Bethesda, MD, 1987, 419.
70. Leslie, R. A., Gwyn, D. G., and Hopkins, D. A., The central distribution of the cervical vagus nerve and gastric afferent and efferent projections in the rat, *Brain Res. Bull.*, 8, 37, 1982.
71. Liu, R. P. C., Laminar origins of spinal projection neurons to the periaqueductal gray of the rat, *Brain Res.*, 264, 118, 1983.
72. Loewy, A. D. and Burton, H., Nuclei of the solitary tract: efferent projections to the lower brain stem and spinal cord of the cat, *J. Comp. Neurol.*, 181, 421, 1978.
73. Loewy, A. D. and Haxhiu, M. A., CNS cell groups projecting to pancreatic parasympathetic preganglionic neurons, *Brain Res.*, 620, 323, 1993.
74. Loewy, A. D. and Spyer, K. M., Vagal preganglionic neurons, in *Central Regulation of Autonomic Functions*, Loewy, A. D. and Spyer, K. M., Eds., Oxford University Press, New York, 1990, 68.
75. Lundberg, J. M. and Hökfelt, T., Multiple co-existence of peptides and classical transmitters in peripheral autonomic and sensory neurons — functional and pharmacological implications, *Prog. Brain Res.*, 68, 241, 1986.
76. Mathias, C. J. and Frankel, H. L., Cardiovascular control in spinal man, *Annu. Rev. Physiol.*, 50, 577, 1988.
77. McAllen, R. M., Action and specificity of ventral medullary vasopressor neurones in the cat, *Neuroscience*, 18, 51, 1986.
78. McAllen, R. M., Identification and properties of sub-retrofacial bulbospinal neurones: a descending cardiovascular pathway in the cat, *J. Autonom. Nerv. Syst.*, 17, 151, 1986.
79. McAllen, R. M., Badoer, E., Shafton, A. D., Oldfield, B. J., and McKinley, M. J., Hemorrhage induces c-fos immunoreactivity in spinally projecting neurons of cat subretrofacial nucleus, *Brain Res.*, 575, 329, 1992.

80. McAllen, R. M., May, C. N., and Shafton, A. D., Functional anatomy of sympathetic premotor cell groups in the medulla, *Clin. Exp. Hypertens.*, 17, 209, 1995.
81. McLachlan, E. M. and Meckler, R. L., Characteristics of synaptic input to three classes of sympathetic neurone in the coeliac ganglion of the guinea-pig, *J. Physiol.*, 415, 109, 1989.
82. Meckler, R. L. and Weaver, L. C., Splenic, renal, and cardiac nerves have unequal dependence upon tonic supraspinal inputs, *Brain Res.*, 338, 123, 1985.
83. Menetrey, D. and Basbaum, A. I., Spinal and trigeminal projections to the nucleus of the solitary tract: a possible substrate for somatovisceral and viscerovisceral reflex activation, *J. Comp. Neurol.*, 255, 439, 1987.
84. Mifflin, S. W. and Felder, R. B., Synaptic mechanisms regulating cardiovascular afferent inputs to solitary tract nucleus, *Am. J. Physiol.*, 259, H653, 1990.
85. Miller, A. D. and Ruggiero, D. A., Emetic reflex arc revealed by expression of the immediate-early gene c-fos in the cat, *J. Neuroscience*, 14, 871, 1994.
86. Mills, E. H., Minson, J. B., Pilowsky, P. M., and Chalmers, J. P., *N*-methyl-D-aspartate receptors in the spinal cord mediate the pressor responses to stimulation of the rostral ventrolateral medulla in the rat, *Clin. Exp. Physiol. Pharmacol.*, 15, 147, 1988.
87. Mitchell, J. H., Neural control of the circulation during exercise, *Med. Sci. Sports Exercise*, 22, 141, 1990.
88. Mitchell, J. H., Kaufman, M. P., and Iwamoto, G. A., The exercise pressor reflex: its cardiovascular effects, afferent mechanisms and central pathways, *Annu. Rev. Physiol.*, 45, 229, 1983.
89. Morgan, J. K. and Curran, T., Stimulus-transcription coupling in the nervous system: involvement of the inducible proto-oncogenes fos and jun, *Annu. Rev. Neuroscience*, 14, 421, 1991.
90. Morrison, S. F., Ernsberger, P., Milner, T. A., Callaway, J., Gong, A., and Reis, D. J., A glutamate mechanism in the intermediolateral nucleus mediates sympathoexcitatory responses to stimulation of the rostral ventrolateral medulla, *Prog. Brain Res.*, 81, 159, 1989.
91. Morrison, S. F. and Gebber, G. L., Axonal branching patterns and funicular trajectories of raphespinal sympathoinhibitory neurons, *J. Neurophysiol.*, 53, 759, 1985.
92. Morrison, S. F. and Reis, D. J., Reticulospinal vasomotor neurons in RVL mediate the somatosympathetic reflex, *Am. J. Physiol.*, 256, R1084, 1989.
93. Mtui, E. P., Anwar, M., Gomez, R., Reis, D. J., and Ruggiero, D. A., Projections from the nucleus tractus solitarii to the spinal cord, *J. Comp. Neurol.*, 337, 231, 1993.
94. Mtui, E. P., Anwar, M., Reis, D. J., and Ruggiero, D. A., Medullary visceral circuits: local afferents to nucleus tractus solitarii synthesize catecholamines and project to thoracic spinal cord, *J. Comp. Neurol.*, 351, 5, 1995.
95. Norgren, R. and Smith, G. P., Central distribution of subdiaphragmatic vagal branches in the rat, *J. Comp. Neurol.*, 273, 207, 1988.
96. Nosaka, S., Yamamoto, T., and Yasunaga, K., Localization of vagal cardioinhibitory preganglionic neurons within rat brain stem, *J. Comp. Neurol.*, 186, 79, 1979.
97. Oldfiled, B. L., and McLachlan, E. M., An analysis of the sympathetic preganglionic neurons projecting from the upper thoracic spinal roots of the cat, *J. Comp. Neurol.*, 196, 329, 1981.
98. Oppenheimer, S. M., Cechetto, D. F., and Hachinski, V. C., Insular stimulation, cardiac arrhythmogenesis and stroke, *Stroke*, 21, 174, 1990.
99. Otake, K., Reis, D. J., and Ruggiero, D. A., Afferents to the midline thalamus issue collaterals to the nucleus tractus solitarii: an anatomical basis for thalamic and visceral reflex integration, *J. Neuroscience*, 14, 5694, 1994.
100. Otake, K. and Ruggiero, D. A., Monoamines and nitric oxide are imployed by afferents engaged in midline thalamic regulation, *J. Neuroscience*, 15, 1891, 1995.
101. Otake, K., Ruggiero, D. A., and Nakamura, Y., Adrenergic innervation of forebrain neurons that project to the paraventricular thalamic nucleus in the rat, *Brain Res.*, in press.
102. Panneton, W. M. and Loewy, A. D., Projections of the carotid sinus nerve to the nucleus of the solitary tract in the cat, 191, 239, 1980.
103. Patrickson, J. W., Smith, T. E., and Zhou, S.-S., Afferent projections of the superior and recurrent laryngeal nerves, *Brain Res.*, 539, 169, 1991.
104. Person, R. J., Somatic and vagal afferent convergence on solitary tract neurons in cat: electrophysiological characteristics, *Neuroscience*, 30, 283, 1989.
105. Phillips, P. A., Abrahams, J. M., Kelley, J., Paxinos, G., Grzonka, Z., Mendelsohn, F. A. O., and Johnston, C. I., Localization of vasopressin binding sites in rat brain by *in vitro* autoradiography using radioiodinated V1 receptor antagonist, *Neuroscience*, 27, 749, 1988.
106. Purves, D., Rubin, E., Snider, W. D., and Lichtman, J., Relation of animal size to convergence, divergence, and neuronal number in peripheral sympathetic pathways., *J. Neuroscience*, 6, 158, 1986.

107. Ricardo, J. A. and Koh, E. T., Anatomical evidence of direct projections from the nucleus of the solitary tract to the hypothalamus, amygdala, and other forebrain structures in the rat, *Brain Res.*, 153, 1, 1978.
108. Rinaman, L., Card, J. P., Schwaber, J. S., and Miselis, R. R., Ultrastructural demonstration of a gastric monosynaptic vagal circuit in the nucleus of the solitary tract in rat, *J. Neuroscience*, 9, 1985, 1989.
109. Rogers, R. C., McTigue, D. M., and Hermann, G. E., Vagovagal reflex control of digestion: afferent modulation by neural and "endoneurocrine" factors, *Am. J. Physiol.*, 268, G1, 1995.
110. Ross, C. A., Ruggiero, D. A., Joh, T. H., Park, D. H., and Reis, D. J., Rostral ventrolateral medulla: selective projections to the thoracic autonomic cell column from the region containing C1 adrenaline neurons, *J. Comp. Neurol.*, 228, 165, 1984.
111. Ross, C. A., Ruggiero, D. A., Park, D. H., Joh, T. H., Sved, A. F., Fernandez-Pardal, J., Saavedra, J. M., and Reis, D. J., Tonic vasomotor control by the rostral ventrolateral medulla: effect of electrical or chemical stimulation of the area containing C1 adrenaline neurons on arterial pressure, heart rate and plasma catecholamines and vasopressin, *J. Neuroscience*, 4, 474, 1984.
112. Ross, C. A., Ruggiero, D. A., and Reis, D. J., Projections from the nucleus tractus solitarii to the rostral ventrolateral medulla, *J. Comp. Neurol.*, 242, 511, 1985.
113. Ruggiero, D. A., Cravo, S., Arango, V., and Reis, D. J., Central control of the circulation by the rostral ventrolateral reticular nucleus: anatomical substrates, *Prog. Brain Res.*, 81, 49, 1989.
114. Ruggiero, D. A., Cravo, S. L., Golanov, E. V., Gomez, R., Anwar, M., and Reis, D. J., Adrenergic and non-adrenergic spinal projections of a cardiovascular-active pressor area of medulla oblongata: quantitative topographic analysis, *Brain Res.*, 663, 107, 1994.
115. Ruggiero, D. A., Gomez, R. E., Cravo, S. L., Mtui, E., Anwar, M., and Reis, D. J., The rostral ventrolateral medulla: Anatomical substrates of cardiopulmonary integration, in *Cardiorespiratory and Motor Coordination*, Koepchen, H.-P. and Huopaniemi, T., Eds., Springer-Verlag, Berlin, 1991, 89.
116. Ruggiero, D. A., Mraovitch, S., Granata, A. R., Anwar, M., and Reis, D. J., A role of insular cortex in cardiovascular function, *J. Comp. Neurol.*, 257, 189, 1987.
117. Ruggiero, D. A., Mtui, E. P., Otake, K., and Anwar, M., Central and primary visceral afferents to nucleus tractus solitarii may generate nitric oxide as a membrane permeant neuronal messenger, *J. Comp. Neurol.*, in press.
118. Ruggiero, D. A., Pickel, V. M., Milner, T. A., Anwar, M., Otake, K., Mtui, E. P., and Park, D., Viscerosensory processing in nucleus tractus solitarii: structural and neurochemical substrates, in *Nucleus of the Solitary Tract*, Barraco, I. R. A., Ed., CRC Press, Boca Raton, FL, 1994, 3.
119. Ruggiero, D. A., Ross, C. A., Kumada, M., and Reis, D. J., Reevaluation of projections from the mesencephalic trigeminal nucleus to the medulla and spinal cord: new projections. A combined retrograde and anterograde horseradish peroxidase study, *J. Comp. Neurol.*, 206, 278, 1982.
120. Saper, C. B. and DeMarchena, O., Somatosympathetic reflex unilateral sweating and pupillary dilation in paraplegic man, *Annu. Neurol.*, 19, 389, 1986.
121. Saper, C. B., Loewy, A. D., Swanson, L. W., and Cowan, W. M., Direct hypothalamo-autonomic connections, *Brain Res.*, 117, 305, 1976.
122. Sato, A. and Schmidt, R. F., Somatosympathetic reflexes: afferent fibers, central pathways, and discharge characteristics, *Physiol. Rev.*, 53, 916, 1973.
123. Sawchenko, P. E. and Swanson, L. W., Immunohistochemical identification of neurons in the paraventricular nucleus of the hypothalamus that project to the medulla or to the spinal cord in the rat, *J. Comp. Neurol.*, 250, 260, 1982.
124. Sawchenko, P. E. and Swanson, L. W., The organization of noradrenergic pathways from the brainstem to the paraventricular and supraoptic nuclei in the rat, *Brain Res. Rev.*, 4, 275, 1982.
125. Schramm, L. P., Strack, A. M., Platt, K. B., and Loewy, A. D., Peripheral and central pathways regulating the kidney: a study using pseudorabies virus, *Brain Res.*, 616, 251, 1993.
126. Schultzberg, J. M. and Lindh, B., Transmitters and peptides in autonomic ganglia, in *Handbook of Chemical Neuroanatomy, Vol. 6: The Peripheral Nervous System*, Bjorklund, A., Hökfelt, T., and Owman, C., Eds., Elsevier, Amsterdam, 1988, 297.
127. Seiders, E. P. and Stuesse, S. L., A horseradish peroxidase investigation of carotid sinus nerve components in the rat, *Neurosci. Lett.*, 46, 13, 1984.
128. Shapiro, R. E. and Miselis, R. R., The central neural connections of th area postrema of the rat, *J. Comp. Neurol.*, 234, 344, 1985.
129. Shapiro, R. E. and Miselis, R. R., The central organization of the vagus nerve innervating the stomach of the rat, *J. Comp. Neurol.*, 238, 473, 1985.
130. Spyer, K. M., The central nervous organization of reflex circulatory control, in *Central Regulations of Autonomic Functions*, Loewy, A. D. and Spyer, K. M., Eds., Oxford University Press, New York, 1990, 168.

131. Spyer, K. M., Ghelarducci, B., and Pompeiano, O., Responses of brain stem reticular neurons to tilting, *Brain Res.*, 56, 321, 1973.
132. Steinbacher, B. C., Jr. and Yates, B. J., Brainstem interneurons necessary for vestibular influences on sympathetic outflow, *Soc. Neurosci. Abstr.*, 21, 892, 1995.
133. Stornetta, R., Morrison, S. L., Ruggiero, D. A., and Reis, D. J., Neurons in the rostral ventrolateral medulla mediate the somatic pressor reflex, *Am. J. Physiol.*, 256, R448, 1989.
134. Strack, A. M., Sawyer, W. B., Hughes, J. H., Platt, K. B., and Loewy, A. D., A general pattern of CNS innervation of the sympathetic outflow demonstrated by transneuronal pseudorabies viral infections, *Brain Res.*, 491, 156, 1989.
135. Strack, A. M., Sawyer, W. B., Marubio, L. M., and Loewy, A. D., Spinal origin of sympathetic preganglionic neurons in the rat, *Brain Res.*, 455, 187, 1988.
136. Strack, A. M., Sawyer, W. B., Platt, K. B., and Loewy, A. D., CNS cell groups regulating the sympathetic outflow to adrenal gland as revealed by transneuronal cell body labeling with pseudorabies virus, *Brain Res.*, 491, 274, 1989.
137. Sugiura, Y., Terui, N., and Hosoya, Y., Difference in distribution of central terminals between visceral and somatic unmyelinated (C) primary afferent fiber, *J. Neurophysiol.*, 62, 834, 1989.
138. Sved, A. F., PNMT-containing catecholaminergic neurons are not necessarily adrenergic, *Brain Res.*, 481, 113, 1989.
139. Sved, A. F., Mancini, D. L., Graham, J. C., Schreihofer, A. M., and Hoffman, G. E., PNMT-containing neurons of the C1 cell group express c-fos in response to changes in baroreceptor input, *Am. J. Physiol.*, 266, R361, 1994.
140. Taylor, R. F. and Schramm, L. P., Differential effects of spinal transection on sympathetic nerve activities in rats, *Am. J. Physiol.*, 253, R611, 1987.
141. Taylor, R. F. and Schramm, L. P., Spinally mediated inhibition of abdominal and lumbar sympathetic activities, *Am. J. Physiol.*, 254, R655, 1988.
142. Ter Horst, G. J., Copray, J. C. V. M., Liem, R. S. B., and Van Willigen, J. D., Projections from the rostral parvocellular reticular formation to pontine and medullary nuclei in the rat: involvement in autonomic regulation and orofacial motor control, *Neuroscience*, 40, 735, 1991.
143. Ter Horst, G. J., de Boer, P., Luiten, P. G. M., and Van Willigen, J. D., Ascending projections from the solitary tract nucleus to the hypothalamus. A Phaseolus Vulgaris lectin tracing study in the rat, *Neuroscience*, 31, 785, 1989.
144. Torvik, A., Afferent connections to the sensory trigeminal nuclei, the nucleus of the solitary tract and adjacent structures. An experimental study in the rat, *J. Comp. Neurol.*, 106, 51, 1956.
145. Tucker, D. C. and Saper, C. B., Specificity of spinal projections from hypothalamic and brainstem areas which innervate sympathetic preganglionic neurons, *Brain Res.*, 360, 159, 1985.
146. Tucker, D. C., Saper, C. B., Ruggiero, D. A., and Reis, D. J., Organization of central adrenergic pathways. I. Relationships of ventrolateral medullary projections to the hypothalamus and spinal cord, *J. Comp. Neurol.*, 259, 591, 1987.
147. Vardan, A., Kachroo, A., and Sapru, H. N., Excitatory amino acid receptors in commissural nucleus of the NTS mediate carotid chemoreceptor responses, *Am. J. Physiol.*, 264, R41, 1993.
148. Verburgh, C. A. and Kuypers, H. G. J. M., Branching neurons in the cervical spinal cord: a retrograde fluorescent double-labeling study in the rat, *Exp. Brain Res.*, 68, 565, 1987.
149. Weaver, L. C. and Stein, R. D., Effects of spinal cord transection of sympathetic discharge in decerebrate-unanesthetized cats, *Am. J. Physiol.*, 257, R1506, 1989.
150. Wood, J. D., Neural basis of enteric motor function, in *Nerves and the Gastrointestinal Tract*, Singer, M. V. and Goebell, H., Eds., MTP Press, Lancaster, UK, 1989, 227.
151. Yates, B. J., Vestibular influences on the sympathetic nervous system, *Brain Res. Rev.*, 17, 51, 1992.
152. Yates, B. J., Grélot, L., Kerman, I. A., Balaban, C. D., Jakus, J., and Miller, A. D., Organization of vestibular inputs to nucleus tractus solitarius and adjacent structures in cat brain stem, *Am. J. Physiol.*, 267, R974, 1994.
153. Yates, B. J., Yamagata, Y., and Bolton, P. S., The ventrolateral medulla of the cat mediates vestibulosympathetic reflexes, *Brain Res.*, 552, 265, 1991.
154. Zemlin, F. P., Leonard, C. M., Kow, L. M., and Pfaff, D. W., Ascending tracts of the lateral columns of the rat spinal cord. A study using the silver impregnation and horseradish techniques, *Exp. Neurol.*, 62, 298, 1978.
155. Zhong, S., Huang, Z.-S., Gebber, G. L., and Barman, S. M., Role of the brain stem in generating the 2- to 6-Hz oscillation in sympathetic nerve discharge, *Am. J. Physiol.*, 265, R1026, 1993.
156. Sved, A. F., Schreihofer, A. M., Ito, S., and Curtis, J. T., unpublished observations.
157. Sved, A. F., unpublished.
158. Ruggiero, D. A., unpublished.

4

Respiratory Control

Kazuhisa Ezure

CONTENTS

0-8493-7668-1/96/$0.00+$.50

LIST OF ABBREVIATIONS

AUG	Augmenting (firing pattern)
bI/E	Burst I-E phase-spanning (neuron)
BÖT	Bötzinger complex
CON	Constant (firing pattern)
DEC	Decrementing (firing pattern)
DRG	Dorsal respiratory group
E	Expiratory
I	Inspiratory
I/E	I-E phase-spanning (neuron)
NTS	Nucleus of the solitary tract
PHR	Phrenic
PIA	Post-inspiratory activity
RVLM	Rostral ventrolateral medulla
STA	Spike-triggered averaging (method)
UCIN	Upper cervical inspiratory neuron
VRG	Ventral respiratory group

1 OVERVIEW OF RESPIRATORY CONTROL

Respiration is one of the most basic activities of life in both animals and plants to maintain necessary oxygen in their tissues. In mammals, this purpose is fulfilled by ventilating the lungs with air and circulating blood between the lungs and tissues. In a narrow sense, the term *respiration* is normally, as in this chapter, used for this ventilatory activity. The ventilation of the lungs is characterized by its automatic rhythmicity and controlled by the nervous system. Since abnormalities in the respiratory nervous system, either central or peripheral, cause a number of life-threatening diseases (dyspnea), and also since the basic question of how the automatic rhythmicity of respiration is generated has attracted much attention of physiologists, respiration is one of the most important and extensively studied fields in physiology. Since the 1970s, electrophysiological and neuroanatomical methods at the level of single central neurons have greatly developed, and our knowledge about the respiratory system has accumulated. The purpose of this chapter is to describe the portions of the nervous system that control respiration based on our recent knowledge, with special emphasis on the genesis of the respiratory rhythm.

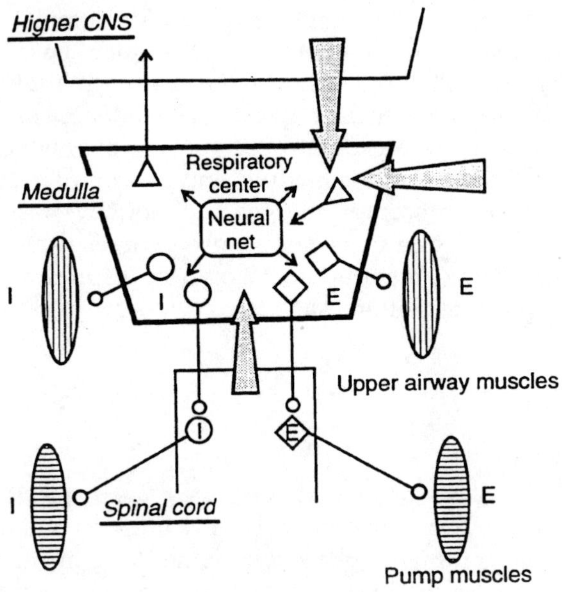

FIGURE 1
Overview of respiratory control. The respiratory center is located in the medulla, is composed of various types of respiration-related neurons, generates the automatic respiratory rhythm, and elicits spatiotemporally organized contraction of the respiratory muscles by integrating peripheral and central information. I: inspiratory neuron or muscle, E: expiratory neuron or muscle.

1.1 The Respiratory Center

In the central and peripheral nervous system, there are various types of neurons that fire in synchrony with respiration. Such neurons are called respiration-related neurons, or simply, respiratory neurons. Respiratory neurons include a wide variety of neurons (e.g., motoneurons, sensory neurons, interneurons essential for rhythm generation, higher-center neurons which monitor respiratory activity, autonomic neurons with respiratory rhythm, etc.). The mode of rhythmicity exhibited by respiratory neurons is also varied. Some respiratory neurons have clear and stable rhythm coupled strongly with respiration, whereas some have weak rhythm which is revealed only after a careful observation over a number of cycles of respiration. Respiratory neurons thus defined are distributed all over the central nervous system from the cerebrum to the spinal cord. Of these, the best known and well-studied respiratory neurons are those in the medulla, where the neural mechanisms primarily responsible for respiratory control reside.

The medullary area which plays a major role in respiratory control is called the respiratory center (Figure 1). The respiratory center largely corresponds to the medullary area where respiratory neurons with strong and clear rhythm are concentrated. However, the respiratory center is not an anatomically defined but rather a functionally defined entity. It is impossible to delineate the exact location of the respiratory center in the medulla. The basic roles of the respiratory center are to make automatic rhythmicity of respiration and to generate spatio-temporally organized contraction of a variety of respiration-related muscles. The respiratory center receives chemical and mechanical information about the results of ongoing ventilatory activity from both inside and outside of the medulla. The respiratory center integrates this information to generate respiratory motor outputs in such a way as to achieve the most efficient gas exchanges (Figure 1).

In the respiratory center, a variety of characteristic respiratory neurons exist. The core of the respiratory center is formed by a number of types of neurons which play fundamental roles in generation of the automatic rhythmicity of respiration. In addition, there are neurons which transmit the respiratory activity to the spinal cord, motoneurons which innervate respiration-related muscles, neurons which transmit information to higher central nervous system, and neurons which relay external feedback information to the respiratory center. Furthermore, there are chemoreceptor neurons, which have not yet been identified but are believed to exist in the medulla and to tonically activate the respiratory neurons. Although these chemoreceptor neurons may not have respiratory rhythm, they might have the privilege of belonging to a special group of respiratory neurons since they make an essential contribution to respiration. All of these respiratory neurons, as a whole, constitute the respiratory center (Figure 1).

1.2 Respiratory Activity

1.2.1 Respiratory Movements

Respiratory activity consists of two basic phases: an inspiratory (I) phase and an expiratory (E) phase. Respiratory movements having these two phases are produced by respiratory muscles which contract in phase with either the I or the E phase. Respiratory muscles which contract during the I phase are called I muscles and those which contract during the E phase are called E muscles. Respiratory muscles are further classified into two groups based on their functions during respiration. One is a group of pump muscles; these are the diaphragm, intercostal, and abdominal muscles. The contraction of pump muscles ventilates the lungs by producing alternating air flows inward to and outward from the lungs. The other is a group of upper airway muscles that controls air flows by opening or closing the airway of the oropharynx and the larynx. Close cooperation of both groups of muscles is essential for respiration; failure in this cooperation may occlude the airway and elicit dyspnea.

Motoneurons that drive pump muscles and upper airway muscles are located in the spinal cord and the medulla, respectively (Figure 1). I and E muscles are included in both pump and upper airway muscles. Corresponding motoneurons fire in respective phases and therefore are divided into I and E neurons (Figure 1). I and E motoneurons of the spinal cord are driven by I and E bulbospinal neurons of the respiratory center, respectively. On the other hand, upper airway motoneurons in the medulla are driven by propriobulbar respiratory neurons or by collaterals of the bulbospinal neurons of the respiratory center (Figure 1).

1.2.2 Firing Patterns

Most respiratory neurons fire during either the I or the E phase, as mentioned above. In both I and E neural populations, three basic types of firing patterns are observed. These are augmenting, constant, and decrementing types.[15] Augmenting type I (E) neurons increase their firing in a ramp-like fashion and attain their peak firing frequency near the end of the I (E) phase. Firings of constant-type I (E) neurons start at the onset of the I (E) phase and are fairly constant throughout the I (E) phase. Firings of decrementing-type neurons start at the onset of the I (E) phase and quickly attain peak firing frequency, which then declines gradually. In this chapter, these respiratory neurons are referred to, respectively, as I-AUG, E-AUG, I-CON, E-CON, I-DEC, and E-DEC neurons. Figure 2 shows five of these firing patterns.

There are variations in the above mentioned basic patterns, and also there are other firing patterns. For instance, the onsets of firing of I-AUG neurons in the I phase are various. Some I-AUG neurons start firing relatively late in the I phase, whereas others start firing at or even earlier than the start of the I phase. The I-AUG neurons that fire very late

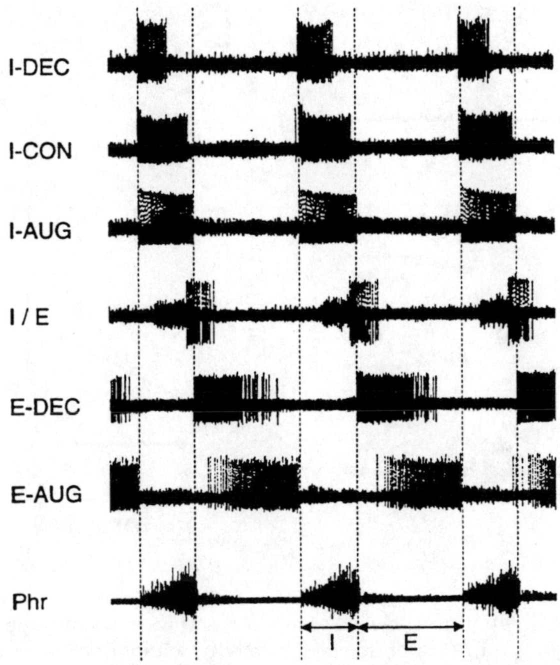

FIGURE 2
Firing patterns of respiratory neurons. I-DEC, I-CON, I-AUG, I/E, E-DEC, E-AUG: decrementing I, constant I, augmenting I, I-E phase spanning, decrementing E, and augmenting E type, respectively. Phr: phrenic nerve discharge as a monitor of respiratory activity. Firings of six neurons from different experiments are arranged in one panel as if these were recorded simultaneously, by manipulating the time base of individual neurons.

in the I phase, (e.g., at around the time of transition from the I to E phase) are called late-I neurons.[71,78,91] In an inverse way, I-DEC neurons are sometimes called early-I neurons, since their firing sometimes stops early in the I phase.[15,71,78,91] The situation is the same for E neurons. The onset of firing of E-AUG neurons is various, and E-DEC neurons stop firing at various times during the E phase. Some neurons are impossible to classify as either I or E. Of these, neurons that fire spanning both phases are called phase-spanning neurons, such as I-E phase spanning or E-I phase spanning neurons.

The term *post-I neuron* is frequently used in the literature. This term was originally introduced to discriminate neurons that fire in the post-I phase (see next section) from E-DEC neurons.[91] Recently, this term has often been used with the same meaning as E-DEC neuron.

1.2.3 Phases of the Respiratory Cycle

As a first approximation, the I and E phases correspond roughly to the active and inactive periods of the I muscles, respectively (Figures 2 and 3). When we analyze the firing of respiratory neurons, a more precise definition of the phase is required, with the phrenic nerve discharge being the standard monitor of the respiratory cycle. The start of the I phase or the end of the E phase corresponds to the time of the start of phrenic burst, at which time inward airflow starts. The start of the E phase or the end of the I phase corresponds to the time of peak phrenic activity, at which time airflow changes from inward to outward away from the lungs. It is virtually impossible to determine unequivocally the exact peak of phrenic nerve activity, although the start of phrenic nerve activity is normally clear. Therefore, ambiguity exists with respect to the time of the transition from the I to the E

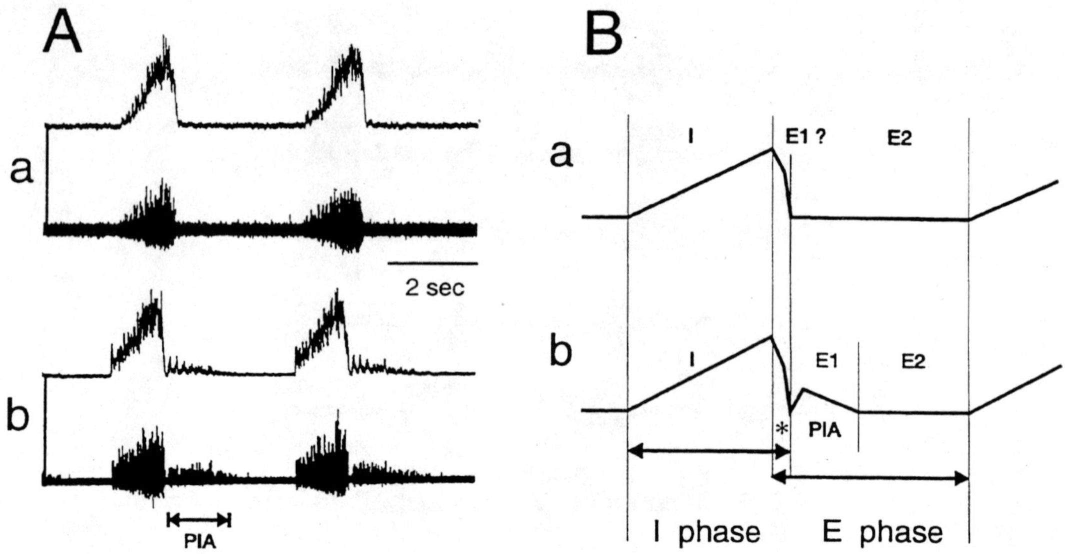

FIGURE 3
Definition of phases of the respiratory cycle. (A) Rectified and integrated (upper traces) and raw (lower traces) phrenic nerve activity; Aa, Ab: phrenic nerve activity without and with post-inspiratory activity (PIA); (B) schematic representation of rectified and integrated phrenic nerve activity. The transition period from the I to E phase is shown by an asterisk.

phase. Probably it is not useful to try to determine a certain point in the time domain as the transition point. The transition may be going on during the period spanning the time of peak phrenic activity and the time when the rapid decrease in phrenic discharge occurs.[71]

After the I phase there is a period during which residual phrenic activity is observed (Figure 3). This activity is called post-inspiratory activity (PIA) or post-inspiratory activity.[91,102] This PIA is sometimes isolated by a silence period (or a trough) from the preceding I activity (Figure 3Ab). This period is called the post-I phase or E1 phase (of the E phase), and the remaining E phase with no phrenic activity is called the E2 phase. The post-I phase is important for effective gas exchange within the lungs. During this period an unwanted too-sudden exhaust of air from the lungs is prevented both by preserving I muscle activity and by activating constrictor muscles (E muscles) of the upper airway.

Frequently, the cycle of respiration is considered to be composed of three distinct phases (I, E1 or post-I, and E2) instead of two phases (I and E).[91] Prominent PIA is observed when respiratory activity is high, such as in unanesthetized decerebrate preparations, preparations with reduced levels of anesthetics, or preparations with high CO_2 levels (Figures 3Ab and Bb). In many cases, however, the PIA and therefore the E1 or post-I phase are difficult to discern. In moderately anesthetized animals with moderate CO_2 level, the phrenic activity declines after the peak without a silent period which separates the post-I phase from the I phase (Figures 3Aa and Ba). In such cases, the division of the E phase as well as the determination of the start of the E phase are rather arbitrary. Earlier and later parts of the E phase are roughly called the E1 and the E2 phases, respectively.

1.3 Medullary Respiratory Neurons

1.3.1 Locations

The respiratory neurons have been studied largely in cats and to a lesser extent in rats, although the earliest experiment of unit recording was made in dogs,[36] and studies in rats

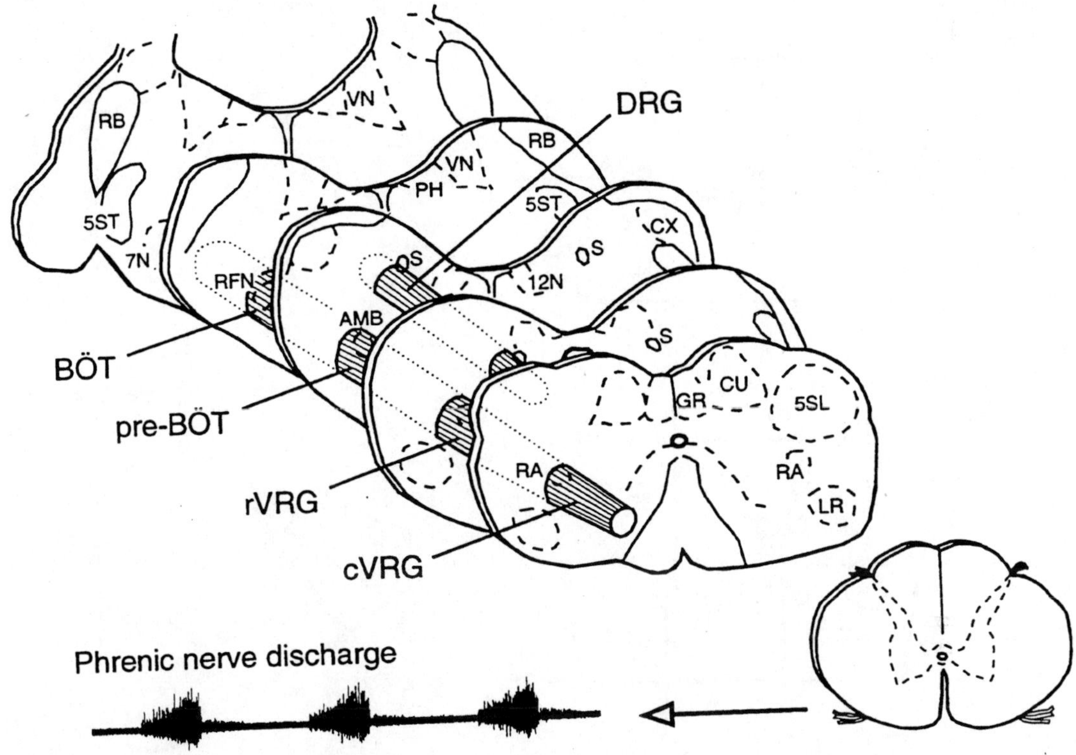

FIGURE 4
Location of respiratory groups in the medulla. BÖT: Bötzinger complex, pre-BÖT: pre-Bötzinger complex, DRG: dorsal respiratory group, rVRG and cVRG: rostral and caudal ventral respiratory group. AMB: nucleus ambiguus, CU: cuneate nucleus, CX: external cuneate nucleus, DX: dorsal motor nucleus of the vagus, GR: gracile nucleus, LR: lateral reticular nucleus, PH: nucleus prepositus hypoglossi, RFN: retrofacial nucleus, RA: nucleus retroambigualis, RB: restiform body, S: solitary tract, VN: vestibular nuclei, 5SL: laminar trigeminal nucleus, 5ST: spinal trigeminal tract, 7N: facial nucleus, 12N: hypoglossal nucleus.

are going to be more popular than before. The present description is based on cat data; however, the distribution of respiratory neurons seems basically similar among different species with minor differences.[20,30] Respiratory neurons with clear and strong respiratory rhythm are concentrated in two medullary regions: dorsomedial and ventrolateral medullary regions (Figure 4). The dorsomedial region largely corresponds to the ventrolateral nucleus of the solitary tract (NTS) and is called the dorsal respiratory group (DRG).[11] The ventrolateral region is functionally subdivided into the ventral respiratory group (VRG)[11] and the Bötzinger complex (BÖT).[71] The VRG extends in close association with the ambiguus nucleus and the retroambigual nucleus. The BÖT lies rostrally to the VRG and is situated in close association with the retrofacial nucleus (Figure 4). In this chapter, the BÖT and VRG are sometimes collectively referred to as the BÖT/VRG. The VRG is normally subdivided into rostral (rVRG) and caudal divisions (cVRG) at around the level of the obex, since the populations of respiratory neurons are different.[44] The transitional zone between the BÖT and the rostral VRG has recently started to be called the pre-Bötzinger complex (pre-BÖT). Although the term *pre-BÖT* is originally based on observations in neonatal rats,[98] it also seems applicable in cats.[17]

Respiratory neurons are also distributed, though sparsely, in other medullary areas, such as various cranial motor nuclei, the lateral reticular nucleus,[36] raphe nucleus,[49,59] and the reticular formation. The details of respiratory neurons in these areas largely remain to be studied. Closely related structures to medullary respiration-related areas exist in the pons and the spinal cord. In the pons, respiratory neurons are concentrated in the

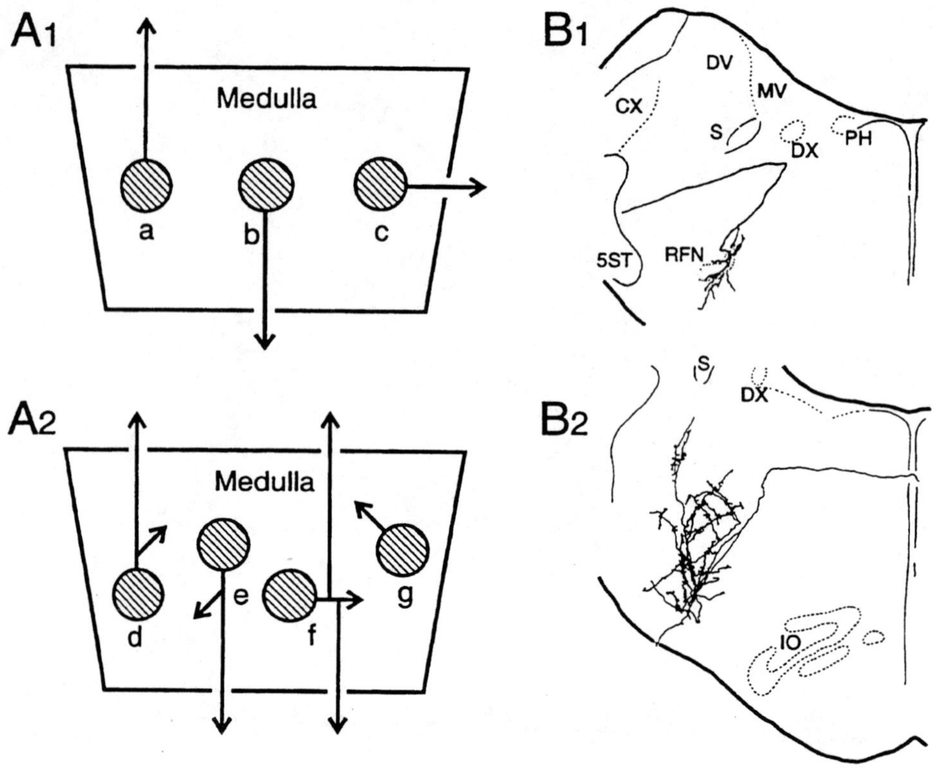

FIGURE 5
Patterns of axonal projections. A group of neurons without (A1) and with (A2) axonal projections within the medulla. B1, B2: reconstructions of neurons stained by intracellular injection of horseradish peroxidase (HRP). Vagal motoneurons with I activity in the area of the BÖT (B1) and I-CON neurons with extensive collaterals within the medulla in the pre-BÖT (B2). (Modified from Otake, K., Sasaki, H., Ezure, K., and Manabe, M., *J. Comp. Neurol.*, 302, 485, 1990.)

area of the parabrachial nucleus. The pontine area is called the pontine respiratory group. It used to be called the pneumotaxic center,[67] since that area was thought to play a critical role in respiratory off-switch mechanisms. The pontine respiratory group is situated next to or overlapping with areas of autonomic control. Throughout the spinal cord, a variety of interneurons as well as motoneurons are widely distributed. Respiratory neurons of the upper cervical cord are distributed in an area which continues caudally from the caudal VRG of the medulla. The motor output from phrenic motoneurons, which are located in the lower cervical cord, is used as the standard monitor of the respiratory rhythm (Figure 4).

1.3.2 Axonal Projections

Medullary respiratory neurons include motoneurons, bulbospinal neurons, and propriobulbar neurons (Figure 5), and show various patterns of axonal projections. Their axonal projections have been studied with neuroanatomical and electrophysiological methods: retrograde or anterograde transport of tracers, staining of individual neurons with intracellular injection of tracers, or antidromic mapping with electrical stimulation. From the point of respiratory rhythmogenesis, the fundamental concern about medullary respiratory neurons is whether their axons project to the medulla or not, since the respiratory rhythm is generated within the medulla. Neurons whose axons do not project within the medulla (Figure 5A1) cannot contribute to the medullary genesis of the respiratory rhythm, whereas neurons having axons that project within the medulla (Figure 5A2) have the potential to contribute to rhythm genesis. Axons of vagal and glossopharyngeal motoneurons leave the medulla without axon collaterals in the medulla and innervate upper airway muscles (Figure 5A1c). These motoneurons are situated in the BÖT/VRG area, and the majority of them show strong respiratory rhythm but do not participate in rhythmogenesis. Some bulbospinal neurons in the VRG and DRG have no medullary collaterals and function simply to transmit the medullary-produced rhythm to the spinal cord (Figure 5A1b).

Some bulbospinal neurons of the DRG, VRG and BÖT, especially the majority of the VRG and BÖT neurons, have axon collaterals which project to the medulla (Figure 5A2ef). Some neurons have ascending axons with medullary collaterals as well (Figure 5A2d). In the case of propriobulbar neurons, it is virtually impossible to determine whether the axons, examined either electrophysiologically or morphologically, stay strictly within the medulla (Figure 5A2g) or have other axons extending beyond the medulla. Therefore, the term *propriobulbar* is used frequently in a wider sense if neurons with medullary collaterals are not activated antidromically from the spinal cord or the cranial nerves. A number of such propriobulbar neurons have been identified, and most of them play important roles in respiration. Figure 5B shows examples of two types of axonal projections: vagal motoneurons with I activity (B1) and I-CON neurons with extensive collaterals within the medulla (B2).

1.3.3 Respiratory Groups and Identified Neurons

Since the activity of medullary respiratory neurons was first recorded from the medulla,[36] a number of interesting neurons have been identified by physiological and anatomical studies. At present we have quite good information on the medullary respiratory neurons, about their locations, firing patterns, axonal projections, input organizations, synaptic interactions, and also functions.

The DRG contains almost exclusively I neurons with a few E neurons. The majority of I neurons are I-AUG neurons: a small number of non-augmenting I neurons are intermingled with these I-AUG neurons.[82] Baumgarten and Kanzow[7] recognized the existence of two types of I neurons on the basis of their response to lung inflation. Ib (or Rb) neurons are excited by lung inflation indicating the existence of excitatory inputs from lung stretch receptors (see Section 3), but Ia (or Ra) neurons are not affected by lung inflation. The majority of the I-AUG neurons, both Ia and Ib neurons, project to the spinal cord. The DRG and adjacent NTS also contain another type of respiratory neuron, the so-called P (pump) cell, which receives excitatory inputs exclusively from lung stretch afferents.[10,19,20] The P cells do not receive central respiratory drive and therefore do not participate in rhythmogenesis. It is unknown how medullary respiratory neurons receive and use information conveyed by the P cells.[19]

In the caudal VRG, which is the functional subdivision of the VRG caudal to the obex and corresponds to the area of the nucleus retroambigualis, there is a concentration of E neurons,[44] in particular bulbospinal E-AUG neurons (see Figure 6). In the rostral VRG, which lies in close association with the ambiguus nucleus, predominantly I neurons exist.[44] These I neurons are of I-AUG type and project to the spinal cord. Importantly, many of them have medullary collaterals.[28,70]

The BÖT has been defined functionally in the vicinity of the retrofacial nucleus as the area where a newly identified group of E-AUG neurons are located. In other words, the BÖT itself is defined as the distribution of E-AUG neurons. In the BÖT area there are E-DEC neurons as well, having a population that is small as compared to that of the E-AUG

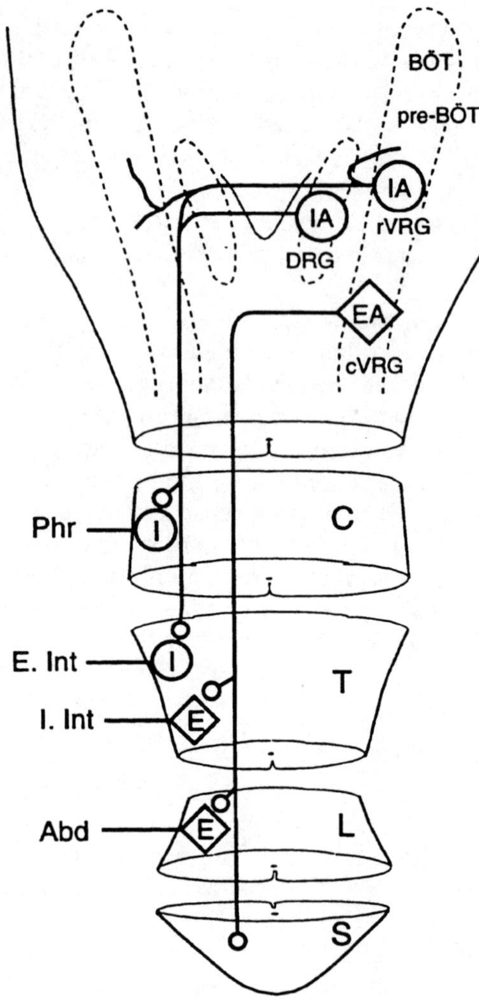

FIGURE 6
Two bulbospinal paths which convey central respiratory drive to the spinal cord. IA: I-AUG neurons of the DRG and rostral VRG. EA: E-AUG neurons of the caudal VRG. Phr, E.Int, I.Int, Abd: phrenic, external intercostal, internal intercostal, abdominal nerve, respectively. C, T, L, S: cervical, thoracic, lumbar, and sacral spinal segments.

neurons. Both E-AUG and E-DEC neurons project extensively to the respiration-related area of the medulla.[12,84] Many E-AUG neurons, but not E-DEC neurons, have spinal axons.

A variety of types of respiratory neurons are found throughout the BÖT and the VRG. I-DEC neurons are distributed in the VRG intermingled with I-AUG neurons of the rostral VRG and E-AUG neurons of the caudal VRG.[6] I-DEC neurons also exist in the BÖT. I-CON neurons are distributed in the same area as I-DEC neurons.[29] Although I-DEC and I-CON neurons are scattered throughout the BÖT/VRG, they are densely distributed in the area from the caudal part of the BÖT to the rostral part of the rostral VRG,[29] which may correspond to the pre-BÖT area. I-DEC and I-CON neurons are propriobulbar neurons with extensive medullary projections,[29,84] although a small population of I-DEC and I-CON neurons project to the spinal cord.[29] In the area dorsomedial to the area of I-DEC and I-CON neurons, there are neurons that fire around the transition from the I to E phase. These neurons, termed bI/E (burst I-E phase-spanning) neurons (see Figure 2, for firing) project to the areas of medullary respiratory neurons and may contribute to the I to E phase transition.[79] In the ambiguus and retrofacial nuclei, there are E and I motoneurons intermingled with other neurons of the BÖT/VRG.

2 MOTOR CONTROL BY THE RESPIRATORY CENTER

2.1 Respiratory Muscles

One of the characteristic features of respiratory movements is that the voluntary musculature is used for involuntary movements of respiration with automatic rhythmicity. The muscles of respiration are also involved in many other voluntary and involuntary activities. Therefore, respiratory movements are modified by or interact with other activities such as vocalization, swallowing, coughing, sneezing, vomiting, emotion, locomotion, and body position. Medullary respiratory neurons are also not independent from these activities but respond either passively or actively to these activities. It is important to keep these facts in mind, although the focus of this chapter is on respiratory movements.

2.1.1 Pump Muscles

The diaphragm is the main respiratory muscle playing a major role in respiratory pumping. Its contraction increases the volume of the thoracic cavity and causes inward air flow to the lungs. The diaphragm is innervated by the phrenic nerves whose somata are located in the lower cervical cord. The diaphragmatic activity represents the I phase of the respiratory cycle; thus during physiological experiments in animals phrenic nerve activity is used as a standard monitor of the respiratory cycle. The diaphragm is composed of several regions: the crural region which immediately surrounds the esophagus, and the sternal and costal regions. Although all these regions contract similarly during respiration, the crural region works differently from the other regions during swallowing or vomiting.[42]

Intercostal muscles work as pump muscles by widening or narrowing the thoracic cavity. The external and internal intercostal muscles contract during the I and the E phase and therefore are I and E muscles, respectively. The intercartilagenous muscles are I muscles, whereas the triangularis sterni muscles are E muscles. These muscles are innervated by the intercostal nerves whose motoneurons are located in the thoracic cord.

There are four major abdominal muscles, the external oblique, the internal oblique, the transverse abdominis and the rectus abdominis. All of these contract during the E phase. Contraction of these muscles increases intra-abdominal pressure and thereby narrows the thoracic cavity to generate outward airflow from the lungs. It is generally accepted that the abdominal muscles are not so active during quiet breathing but work effectively in forced or voluntary expiration. These muscles are importantly involved in several activities which increase abdominal pressure such as vomiting, coughing, urination, defecation, singing, or vocalization. These muscles are innervated from motoneurons extending from the mid-thoracic to upper lumbar cord.

2.1.2 Upper Airway Muscles

During normal breathing, in short, the upper airways are widened in the I phase and narrowed in the E phase by the contraction of upper airway muscles which work in close linkage with pump muscles. This close cooperation with pump muscles is controlled by the respiratory center.

The glossopharyngeal nerve shows I activity,[40] reflecting the activity of the motoneurons innervating the stylopharyngeal muscle. The vagus nerve innervating the following muscles shows both I and E activities.[2,25,39] Pharyngeal constrictor muscles (superior, middle, and inferior) are of the E type. Intrinsic laryngeal muscles show strongly modulated respiratory rhythm. Of these, the cricoarytaenoideus lateralis shows I activity and the cricothyreoideus muscle shows weak I activity. Other intrinsic laryngeal muscles, the cricoarytaenoideus, arytaenoideus, and thyreoarytaenoideus, are of the E type. Of these, the thyreoarytaenoideus shows typical E-DEC activity.

Motoneurons of these upper airway muscles are located in the ventrolateral medulla, in and around the retrofacial and the ambiguus nuclei intermingled with other types of respiratory neurons. The BÖT and the VRG, but not the DRG, include these motoneurons. In general, the firing of upper airway motoneurons is different from other medullary respiratory neurons in that AUG, DEC, and CON patterns are not stable but changeable depending on the state.[26] This may reflect the fact that these muscles are not only involved in respiration but in other related activities and receive complex controls from various sources.

Various muscles which control the oropharyngeal cavity and tongue also work during respiration. Therefore many motoneurons in the hypoglossal, facial, and trigeminal nuclei have respiratory rhythm. Of the extrinsic muscles of the tongue, the genioglossus muscle, the main protruser of the tongue, has clear I activity and plays an important role in the maintenance of airway patency.[2,66] The loss of its activity is well correlated with airway occlusion during sleep.[90] However, there are no firmly established views as to how the other extrinsic tongue muscles work during respiration.[66] Unit recording from the hypoglossal nucleus shows that hypoglossal motoneurons are largely of the I type with some of the E type included.[104]

2.2 Bulbospinal Neurons

Outputs from the medullary respiratory center to the spinal cord are conveyed by two major paths (Figure 6). One is the path from I neurons of the DRG and the VRG to spinal I motoneurons of the phrenic and external intercostal nerves. This brings excitatory drive to the I neurons. The other is the path of excitatory drive from E neurons of the caudal VRG to spinal E motoneurons of the internal intercostal and abdominal nerves. It is generally agreed that these excitatory drives may also be mediated through polysynaptic pathways which include some spinal interneurons (see below). In addition to these excitatory paths, direct inhibitory inputs to phrenic motoneurons originate from E neurons of the BÖT.

2.2.1 Inspiratory Neurons

Many I neurons of the VRG and the majority of I neurons of the DRG project to the spinal cord. A contralateral projection is prominent in cats, but bilateral projections are prominent in rats.[64] These neurons show I-AUG activity and their spinal axons travel down to the thoracic cord and project to the areas of both phrenic and intercostal motoneurons. The spike-triggered averaging method (STA, see Figure 9) showed monosynaptic connections from DRG I neurons (both Ia and Ib neurons) to phrenic motoneurons,[24,33,63,77] from DRG I neurons to intercostal motoneurons,[24] and rare connections from VRG I neurons to phrenic motoneurons.[33] However, Merrill and Lipski failed to detect monosynaptic connections from VRG I neurons to intercostal motoneurons by the STA method and emphasized the importance of interneurons for conveying central respiratory drive to intercostal motoneurons.[73] Since there are several studies which report monosynaptic connections between VRG I neurons and intercostal motoneurons using the technique of cross-correlation between the activity of VRG neurons and intercostal nerves,[18] further studies are needed to solve this controversy. A few I neurons of the VRG and DRG project as far caudal as the lumbar cord.[76] However, the functional significance of the lumbar projection is not known.

Many bulbospinal I-AUG neurons of the VRG and some of the DRG have axon collaterals in the medulla.[64,82,95] Excitatory monosynaptic connections from these collaterals to medullary I neurons have been shown.[28] The medullary collaterals also project to the hypoglossal nucleus and the nucleus of the solitary tract.[64]

2.2.2 *Expiratory Neurons*

E motoneurons in the spinal cord are driven by E neurons of the caudal VRG with characteristic E-AUG activity. The majority of E-AUG neurons of the caudal VRG project to the thoracic and the upper lumbar cord. Monosynaptic connections between E-AUG neurons to intercostal motoneurons have been proved.[55,73] However, these monosynaptic connections do not seem frequent, and the importance of pathways mediated by interneurons has been emphasized.[73] Much evidence from electrophysiological[70] and morphological[6] studies shows that E-AUG neurons of the caudal VRG have no axon collaterals in the medulla. In contrast to I-AUG neurons of the VRG and DRG, therefore, these E-AUG neurons are pure relay neurons of central respiratory drive to the spinal cord and do not contribute to rhythm generation.

Extensive antidromic mapping study by Sasaki and his colleagues has shown that many E neurons extend their axons down to the lower lumbar or sacral segments and distribute collaterals in L5-L7.[96] Some E neurons have extensive collaterals in the nucleus of Onuf and adjacent structures.[97] These results suggest that E neurons are involved not only in respiratory pumping but also in some respiratory modulated motor activities of the lower lumbar segments. It is possible that the sacral projections may prevent incontinence by causing contraction of external anal and urethral sphincter muscles during increased intra-abdominal pressure.[97] Innervation to the parasympathetic preganglionic neurons of the sacral cord is also possible.[97]

2.3 Respiratory Integration in the Spinal Cord

In the spinal cord, besides motoneurons of pump muscles, various types of respiratory neurons have been found. There are segmental interneurons, propriospinal neurons with long descending axons, neurons with ascending axons, and neurons with autonomic functions. Their firing patterns are also various.

2.3.1 *Upper Cervical Inspiratory Neurons*

Aoki and colleagues recorded inspiratory activity from neurons located in the upper cervical cord (C1-C2) in the cat.[3] These neurons, later termed upper cervical inspiratory neurons (UCIN), are also found in the rat and have features similar to those in the cat.[61] UCINs are located within the lateral part of the intermediate gray matter and sometimes in the adjacent white matter. Their firing pattern is of I-AUG type (Figure 7), which may reflect the firing pattern of bulbospinal I-AUG neurons of the DRG and VRG which project to UCINs.[46,47]

Since the identification of the UCIN a number of studies using antidromic stimulation and intracellular staining have been performed to identify their descending axons and projections. The axons of UCINs descend ipsilaterally, sometimes down to the lumbar cord (Figure 7).[48,76] Extensive collateral arborizations have been found mainly in the region of

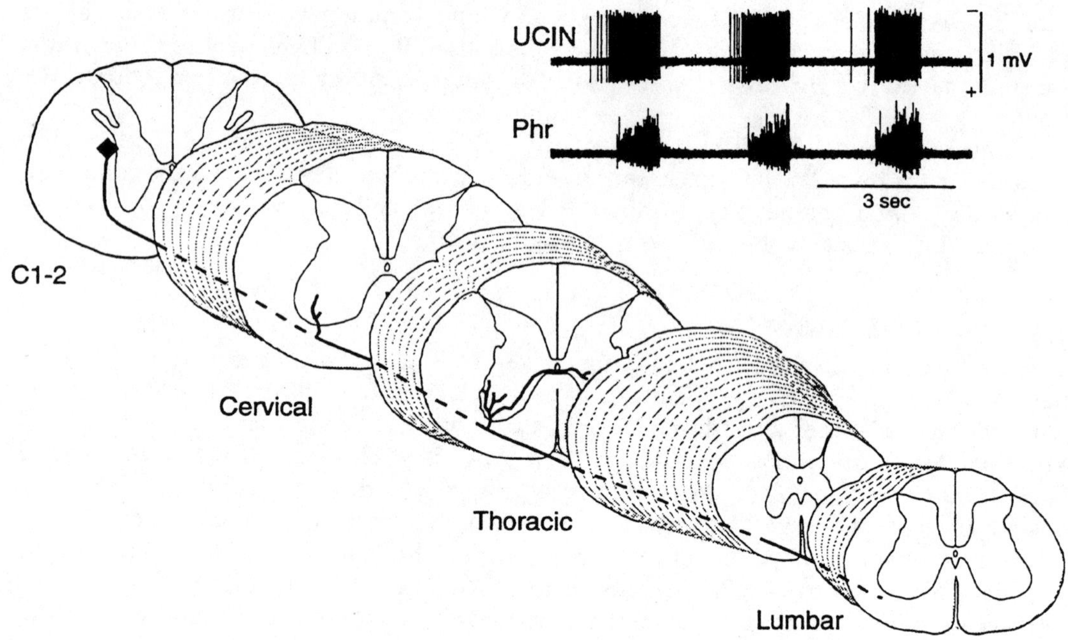

FIGURE 7
Upper cervical inspiratory neuron (UCIN). UCINs have a characteristic I-AUG discharge pattern and exhibit extensive branching in the vicinity of intercostal motoneurons of the thoracic cord but poor branching in the vicinity of phrenic motoneurons of the lower cervical cord.

the intercostal motoneurons in the thoracic cord.[23,48,60] Axons often project to the contralateral side.[23,60] In the cervical cord, only short collaterals with poor arborizations have been observed in the vicinity of the phrenic nucleus.[60,61] However, cross-correlation or spike-triggered averaging experiments have not succeeded in finding evidence for monosynaptic connections between UCINs and either intercostal (both I and E neurons) or phrenic motoneurons.[23,48,60,61] Therefore, although the exact synaptic targets are unknown, UCINs presumably project to interneurons in the areas of phrenic and intercostal motoneurons. At present the functional significance of UCINs remains totally elusive in spite of extensive studies.

2.3.2 *Thoracic and Lower Cervical Interneurons*

In the thoracic cord, Sumi[99] was the first to address the question of whether or not respiratory interneurons exist and how they take part in the control of respiration. He established the existence of both I and E interneurons intermingled with thoracic motoneurons, and concluded that a large number of interneurons participate in the regulation of respiratory motor activity in the spinal cord. Recently Kirkwood and his colleagues have done extensive studies on thoracic interneurons,[54] and suggested that there are almost 10 times more respiratory thoracic interneurons as respiratory bulbospinal neurons. One possible role of interneurons is to mediate the central excitatory drive conveyed by bulbospinal neurons to intercostal motoneurons. Another possible role is to inhibit intercostal motoneurons during the off-phase of their central respiratory drive.[54]

In search of the phrenic nucleus in the cervical cord, Baumgarten et al.,[8] found respiratory interneurons that were not antidromically activated by phrenic nerve stimulation. They found both I and E interneurons in the area of phrenic motoneurons. Modern studies have shown a more precise view of interneurons which are located in the vicinity of the phrenic motor nucleus.[9,22,41,85] Since these interneurons are considered to provide inputs to phrenic motoneurons, I and E interneurons are supposed to be excitatory and inhibitory, respectively.[9,41,85] In fact, E interneurons behave consistently with the idea that E neurons convey inhibition to phrenic motoneurons during various behavior such as vomiting and coughing.[41] Monosynaptic inhibitory connections from E interneurons to phrenic motoneurons have recently been identified.[22] I interneurons could relay central excitatory drive to phrenic motoneurons though they may not be simple relay neurons.

These lower cervical and thoracic interneurons may function to integrate various segmental and central information to elicit the most appropriate pattern of activity on spinal motoneurons critical for accomplishing complex behavioral patterns.[41] Considering the existence of a variety of interneurons, the cervical and thoracic respiratory neurons may form a neural network which is comparable to those more intensively studied in the respiratory center in the medulla. Under special circumstances, this spinal network may generate a respiration-like output.[3,4,14,103] Similar interneurons might also exist in the lower thoracic and upper lumbar cord where abdominal muscles are located; however, data are not available yet.

2.3.3 *Other Respiratory Neurons*

In the thoracic cord, Hirai and co-workers[100] showed the existence of two types of respiratory neurons which project to the cerebellum. One type is located in Clarke's column, which is the origin of the dorsal spinocerebellar tract (DSCT). These neurons respond passively to movements of the chest wall, both during expansion and retraction. They receive inputs possibly from muscle proprioceptors. The other type of respiratory neuron is located in laminae VII and VIII. These neurons receive central respiratory drive inputs of both the I or E type, as well as peripheral inputs from the chest wall. The latter neuron group corresponds to the ventral spinocerebellar tract (VSCT) neurons of limb movements. Their function in respiratory control and relation to central respiratory neurons remain to be studied.

Sympathetic preganglionic neurons in the thoracic and lumbar cord also have respiratory rhythm of various patterns.[38,87,89] (See Section 5.)

3 INPUTS TO THE RESPIRATORY CENTER

The respiratory center needs feedback information about the results of on-going respiratory movements in order to generate the appropriate respiratory activity. Mechanoreceptors, which are located peripherally, and chemoreceptors, which are located both peripherally and centrally, play critical roles in respiratory control. Although phasic feedback inputs are not necessary for medullary generation of the respiratory rhythm, tonic excitation from chemoreceptors is essential.

3.1 Mechanoreceptors

Receptors in the airways and lungs consist of three main types. These are slowly adapting stretch receptors, rapidly adapting stretch receptors, and C-fiber receptors. The former two are innervated by myelinated nerves. The slowly adapting receptors lie among smooth muscles of the airways and lungs and respond to lung inflation with little adaptation. The pioneering study by Breuer and Hering[15] demonstrated that changes in pulmonary volume modify respiratory outputs mainly via the slowly adapting receptors.[57] The basic action of slowly adapting receptors when they are excited is to suppress inspiratory activity. Steady lung inflation or deflation leads to a decrease or increase in respiratory frequency, respectively. Furthermore, activation of these receptors during inspiration, which occurs phasically during normal breathing, leads to earlier termination of the I phase. The second-order relay neurons are located in the NTS and are P-cells[10,19] and a type of I-AUG neuron of the DRG (Ib neuron[7]). The central pathways and mechanisms by which the slowly adapting receptor system functions are unknown.

Knowlton and Larrabee identified the rapidly adapting receptors, the firing of which adapts rapidly when stimulated by steady lung inflation.[57] The rapidly adapting receptors, also referred to as lung irritant or cough receptors, respond to both mechanical and chemical stimuli and have been implicated in a number of respiratory reflexes that result in bronchoconstriction and pronounced changes in ventilation. The real function and natural stimulants of this receptor which was originally identified as a mechanoreceptor are elusive. Rapidly adapting receptors are active in a variety of physiological and pathological conditions.[16] The second order neurons have been identified in the commissural nucleus of the NTS,[62] with their characteristic projections to the dorsolateral pons and medullary respiration-related structures.

C-fiber receptors, both bronchial and pulmonary (same as J or juxtapulmonary) receptors, respond to mechanical stimulation with higher threshold than slowly adapting receptors — they respond to hyperinflation of the lungs but very little to lung deflation.[16] They respond strongly to agents including capsaicin, histamine, bradykinin, serotonin, etc. When C-fiber receptors are stimulated, a reduction in respiration occurs followed by rapid shallow breathing, bradycardia, and a fall in blood pressure. It is generally accepted that C-fiber endings function in asthma, pulmonary congestion, embolism, and lung injury. Although the afferents from these receptors are relayed in the NTS, further details of the pathways to the respiratory system are unknown.

There are also afferents originating from proprioceptors of the diaphragm, intercostal, and abdominal muscles. Although these may work mainly at the level of the spinal cord, some project to the cerebellum and may affect the respiratory center.[100] Their functions in respiratory control remain largely speculative. In addition, vestibular inputs to the respiratory center, which possibly function to adjust respiration during changes in posture, have recently been reported[105] and are the subject of a separate chapter.

Important inputs which strongly modulate the central respiratory system arise from superior laryngeal nerve (SLN) afferents. Although SLN efferents innervate only the cricothyroid muscle, SLN afferents convey sensory information from receptors widely located in the laryngeal mucosa. Mechanical or chemical stimulation of SLN receptors evokes swallowing, coughing, or apnea. Virtually every type of respiratory neuron receives profound effects from SLN afferents.[53] SLN afferents terminate in the NTS; the second-order neurons include the majority of I-AUG neurons of the DRG.[53] A few more synapses are involved in the pathways to other respiratory neurons, which remain to be studied.

3.2 Chemoreceptors

Chemoreceptors are located both peripherally and centrally. The peripheral chemoreceptors are situated in the carotid and aortic bodies, which have a rich vascularity as well as rich innervation from the glossopharyngeal and vagus nerves. The receptors respond to decreases in arterial PO_2 and to a lesser extent to increases in arterial PCO_2. Information from these chemoreceptors is conveyed to the medial region and commissural portion of the NTS.[21] How these inputs to the NTS are then transmitted to the respiratory system is not known. However, projections from these NTS regions to the area of the rostral ventrolateral medulla (RVLM) known as the cardiovascular center suggest that part of the chemoreceptor input to the respiratory system is conveyed via the RVLM and that the respiratory and cardiovascular systems are linked at the RVLM (see Section 5).[93]

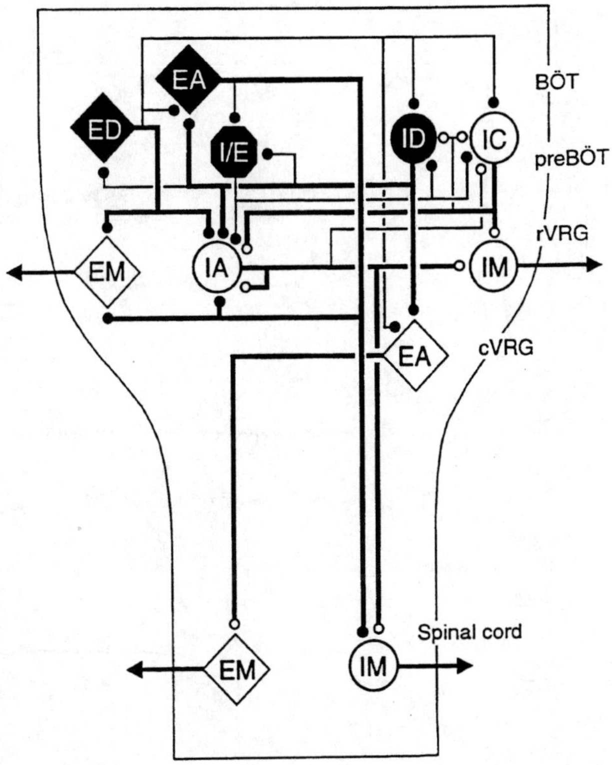

FIGURE 8
Network of medullary respiratory neurons. I neurons (circles); E neurons (diamonds); excitatory neurons (open symbols); inhibitory neurons (filled symbols). I/E, IC, ID, IA, EA, ED: I/E, I-CON, I-DEC, I-AUG, E-AUG, E-DEC neurons, respectively. IM, EM: I and E motoneurons. Connections with thick lines have been proven.

Even without afferents from the peripheral chemoreceptors, the respiratory system still responds to chemical stimulation, especially to an increase in PCO_2. After the first demonstration that perfusion of the ventriculocisternal system with high $[H^+]$ stimulated ventilation, a series of extensive studies by Loeschcke's group has localized the chemosensitive areas in the ventral surface of the medulla.[65] Three chemosensitive areas called rostral (or M), caudal (or L), and intermediate (or S) areas have been identified on or close to the ventral medullary surface. However, no specific neurons have yet been identified as chemoreceptors. It is suggested that the RVLM area, which lies in close proximity to the intermediate area of the ventral medullary surface, plays an important role in chemoreception by integrating inputs from both the peripheral and central chemoreceptors.[56,93] Definitely, the identification of central neurons that function as chemoreceptors is one of the important subjects remaining to be studied in physiology.

4 NETWORKS OF MEDULLARY RESPIRATORY NEURONS

Although there are arguments as to whether the rhythm itself is generated by the network property of respiratory neurons or by a pacemaker property of particular types of neurons, the network is certainly essential for the spatiotemporally organized activity of respiration. The following sections summarize synaptic connections between medullary respiratory neurons (Figure 8) and describe possible mechanisms of respiratory rhythm generation.

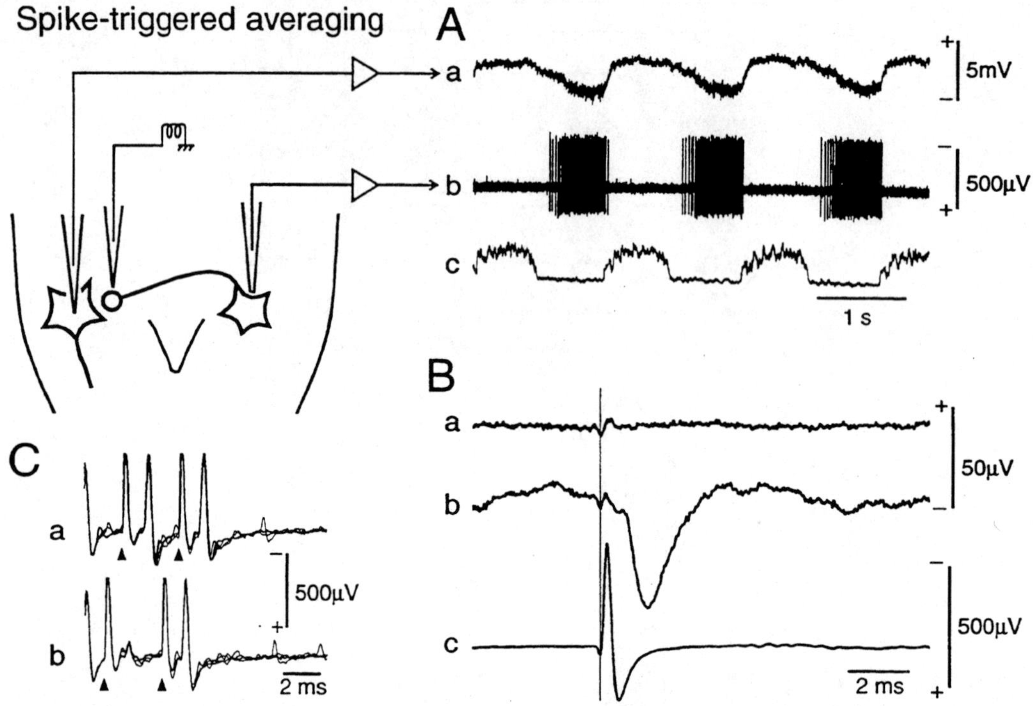

FIGURE 9
Spike-triggered averaging (STA) reveals monosynaptic IPSPs in an I-AUG neuron evoked by triggering spikes of an E-AUG neuron of the BÖT. (A) Intracellular potential from the VRG I-AUG neuron (a); extracellular spikes of the BÖT E-AUG neuron (b); integrated phrenic nerve activity (c). (B) averaged (512 times) triggering spike (c); averaged membrane potential showing hyperpolarizing potential (IPSP) locked to the trigger (b); averaged juxtacellular potential (a). (C) collision test between spontaneous spikes of the E-AUG neuron and antidromic spikes evoked by stimulation (arrowheads) of the area where the I-AUG neuron is recorded. The first shock in the lower trace induced no antidromic spikes because of collision. This test shows that the triggering E-AUG neuron really sends its axon to the vicinity of the intracellularly-recorded I-AUG neuron. Examination of antidromic and IPSP latencies indicates that the IPSP is monosynaptic.

4.1 Synaptic Connections

Synaptic connections between medullary neurons have been extensively studied by various methods. The nature of these connections could be inferred from neuronal firing patterns and intracellularly recorded synaptic potentials, the temporal relationship of neuronal activities, axonal projections, as well as effects from lung stretch receptors or from laryngeal afferents, etc. Cross-correlation studies between extracellularly recorded spikes have revealed many synaptic interactions. An important breakthrough was achieved by Merrill and co-workers[74] (see below) by using the spike-triggered averaging (STA) method, which reveals synaptic potentials induced by spikes of the triggering neuron. The STA method combined with antidromic mapping experiments provides quite convincing evidence for direct synaptic connections (Figure 9). Since the 1980s, a number of monosynaptic connections, both excitatory and inhibitory, between respiratory neurons have been found.[26] At present three types of inhibitory neurons and two types of excitatory neurons are known to make synaptic connections within the medulla. The inhibitory cells are E-AUG and E-DEC neurons of the BÖT, and I-DEC neurons of the BÖT/VRG; whereas excitatory connections are made by I-AUG neurons of the VRG and I-CON neurons of the BÖT/VRG.

4.1.1 *Inhibitory Connections*

E-AUG neurons of the BÖT were shown to inhibit I neurons of the contralateral DRG by Merrill and co-workers.[74] This is the first evidence for synaptic connections between medullary respiratory neurons. E-AUG neurons of the BÖT also inhibit phrenic motoneurons by their bulbospinal axons.[72] At present, BÖT E-AUG neurons are known to inhibit I neurons of the VRG in both the contralateral and the ipsilateral medulla. An example of a STA test which verifies monosynaptic inhibitory postsynaptic potentials (IPSPs) elicited in an I-AUG neuron of the VRG by an E-AUG neuron of the BÖT is shown in Figure 9. The E-AUG neurons inhibit not only I neurons but also E neurons: the E-AUG neuron of the BÖT inhibits vagal motoneurons with E activity on both sides and E-AUG neurons of the caudal VRG.[27,52] The same E-AUG neuron inhibits both I and E neurons. Therefore, E-AUG neurons of the BÖT exert wide inhibitory influence on various types of medullary respiratory neurons.[52]

E-DEC neurons of the BÖT are also inhibitory to various respiratory neurons in the medulla. The E-DEC neurons inhibit I neurons of the VRG and vagal motoneurons with E activity.[27] Similar to E-AUG neurons of the BÖT, the same E-DEC neuron inhibits both I and E neurons. Judging from their extensive axonal projections in the bilateral medulla including the DRG, VRG and BÖT, these E-DEC neurons also exert wide inhibitory influence on medullary respiratory neurons.[26] Since E-DEC neurons fire early in the E phase, their contribution to the transition from the I to E phase is suggested.

I-DEC neurons of the BÖT/VRG inhibit E-AUG neurons of both the caudal VRG and the BÖT.[26] These inhibitory connections explain the abrupt inhibition of E-AUG neurons at the onset of the I phase. Furthermore, I-DEC neurons have been shown to inhibit I-AUG neurons of the DRG and VRG including late-I neurons.[26,29] This connection suggests the contribution of I-DEC neurons to the formation of I-AUG patterns. Although it is very possible that the same I-DEC neuron inhibits both I and E neurons, it has not been confirmed yet.[26]

4.1.2 *Excitatory Connections*

Many I-AUG neurons of the VRG and some of DRG have medullary axon collaterals as described above. STA tests have revealed that the medullary collaterals of VRG I-AUG neurons make monosynaptic excitatory connections to bulbospinal I-AUG neurons and vagal motoneurons with I activity (Figure 10).[28] This means that the bulbospinal I-AUG neurons make excitatory connections within their populations, although this does not necessarily mean that connections are reciprocal at the level of individual neurons. These connections implement a kind of self-reexcitation, and thus might contribute to the synchronization of I neurons.[26,28] Furthermore, the augmenting firing patterns of bulbospinal I neurons might be produced partly by this reexcitatory connection (see Figure 12). Furthermore, the revealed connections indicate that some bulbospinal I neurons in the VRG drive motoneurons of upper airway muscles and phrenic or intercostal motoneurons simultaneously. At present, however, no data are available about the connectivity of medullary collaterals of DRG I neurons.

I-CON neurons of the BÖT/VRG monosynaptically excite I neurons of the VRG and the DRG.[29] These I neurons include bulbospinal I-AUG neurons, propriobulbar neurons, and vagal motoneurons. Judging from their extensive axonal projections in the medulla and from large EPSPs induced in I neurons, the I-CON neurons may be an important source of excitation for many medullary I neurons during the I phase. Since bulbospinal I-AUG neurons send collaterals also to the area of I-CON neurons (Figure 10), it is possible that I-AUG neurons make excitatory connections to I-CON neurons.[26]

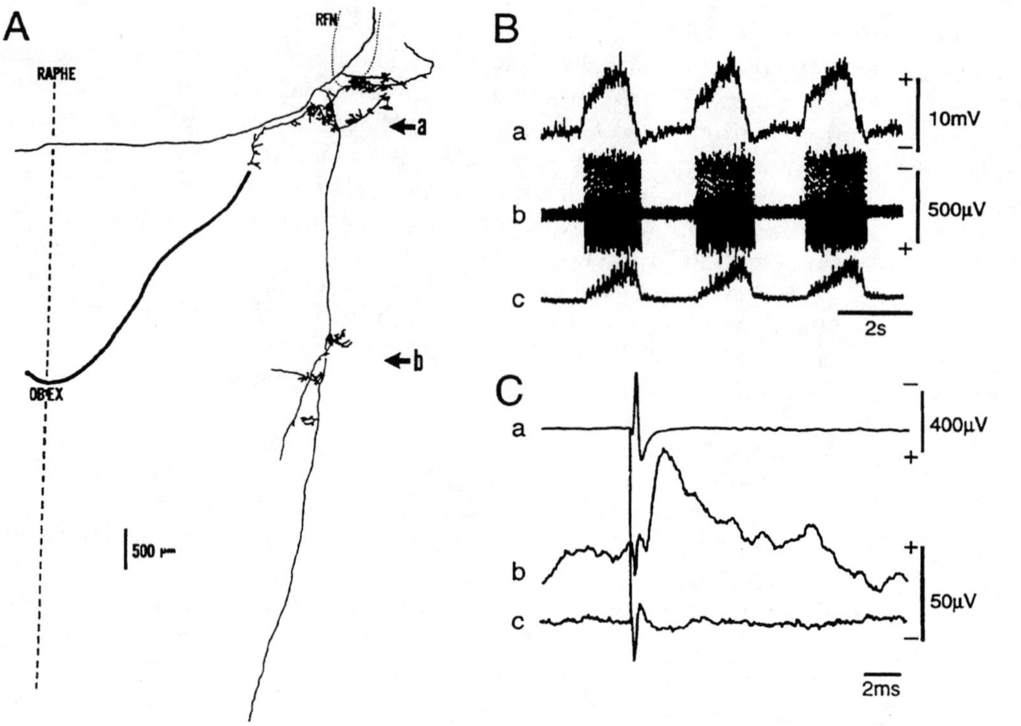

FIGURE 10
(A) HRP stained axon of I-AUG neurons of the VRG (horizontal section). Axonal collaterals are issued at the level of the pre-BÖT (arrow a) and VRG (arrow b) from the stem axon projecting to the spinal cord (modified from Sasaki et al.[95]). (B, C) STA shows that the medullary collateral of a bulbospinal I-AUG neuron makes monosynaptic excitatory connection to a vagal I motoneurons with I activity. (B) intracellular recording from a vagal motoneuron (a); extracellular recording from a bulbospinal I-AUG neuron (b); integrated phrenic nerve activity (c). (C) averaged (512 times) triggering spike (a); averaged membrane potential showing a unitary EPSP (b); averaged juxtacellular potential (C).

4.2 Respiratory Rhythmogenesis

Evidence from various transection and deafferentation experiments shows that the respiratory rhythm is produced within the medulla. In other words, the central pattern generator for respiration resides within the medulla. Since we now have quite a lot of knowledge, though far from sufficient, about the network of medullary respiratory neurons (Figure 8), we can probe the mechanisms of respiratory rhythm generation based on the hypothesis that the rhythm is produced by this network. We propose the connections extracted from Figure 8 and shown in Figure 11A as the rhythm-generating network. The network in this model is located within the rostral ventrolateral medulla in the area spanning the BÖT and the rostral part of the VRG (Figure 11B).

4.2.1 I-E Transition and E Phase

The mechanisms of the I-E transition, called I off-switch, has long been the central theme of studies on respiratory rhythm generation. Efforts have been made to search neurons that fire at the transition and shut off the on-going I activity. At present, neurons that are known to fire at the I-E transition are: E-DEC neurons, bI/E neurons (Section 1.3.3), some E-AUG neurons, late-I neurons, and early-onset E-DEC neurons (Section 4.2.4). Since axonal projections of the latter two neurons are not adequate for the transition, the former three neurons seem important for the inspiratory off-switch. The possible mechanisms are as follows.[26]

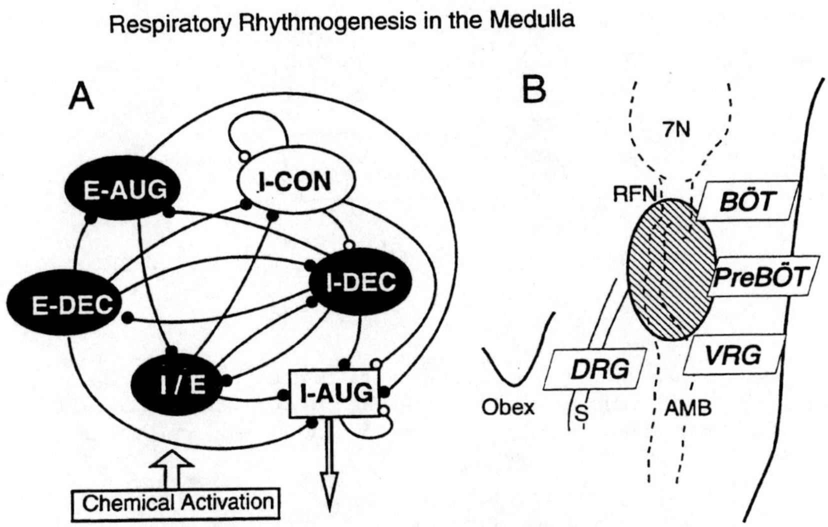

FIGURE 11
(A) a model network for respiratory rhythm generation. The network is located in the area from BÖT to the rostral part of the VRG (shaded area in [B]) including the pre-BÖT. (A) neurons (oval symbol) essential for rhythm generation have been extracted from the network shown in Figure 8. An output element (I-AUG neuron) is shown by a square.

At the later stage of the I phase, E-DEC neurons and bI/E neurons are released from the inhibitory action probably of I-DEC neurons. They then start firing, which in turn inhibits I neurons, probably including I neurons that have inhibited the E-DEC and bI/E neurons, thus accelerating the firing of E-DEC and bI/E neurons and the I-E transition. Some E-AUG neurons which start some firing during the later part of the I phase may also contribute to this stage. At this time of transition, no excitatory connections which cause synchronous bursts of E neurons are expected between these inhibitory neurons. This may be the reason why the I-E transition is not as sharp as the E-I transition in which synchronous bursting of I neurons is characteristic (see below). Once the majority of E-DEC neurons start firing, the I-E transition is almost completed. Then E-AUG neurons of the BÖT start firing and the residual I activity in the medulla is inhibited completely.

During the E phase, E-AUG neurons of the BÖT inhibit I neurons of the DRG and the VRG and also I neurons in the spinal cord to maintain the E phase. Firing of bI/E neurons may be inhibited by E-AUG neurons, but corresponding inhibition of E-DEC neurons from E-AUG neurons seems, if anything, weak. E-AUG neurons of the caudal VRG fire similarly to E-AUG neurons of the BÖT and transmit the medullary-made E activity to the spinal cord. The E phase persists as long as these E-AUG neurons continue firing.

4.2.2 E-I Transition and I Phase

The E-I transition is quite impressive when we record from respiratory neurons in the medulla. It is a sharp transition. There is substantial overlap in firing in the I and E populations at the I-E transition, but almost none at the E-I transition, as noted by Merrill.[71] The sharp transition is characterized by an abrupt start of firing of I-CON, I-DEC, and some I-AUG neurons and by an abrupt stop of firing of E-AUG neurons. The I-CON and I-DEC neurons, which start firing around several tens of milliseconds before the transition, may make essential contribution to this transition. The possible mechanisms are as follows.

At the later stage of the E phase, the I-CON and I-DEC neurons which do not receive inhibition from BÖT E-AUG neurons recover from the inhibition from E-DEC and bI/E neurons. Once some I-CON neurons start firing, the possible self-reexcitatory network

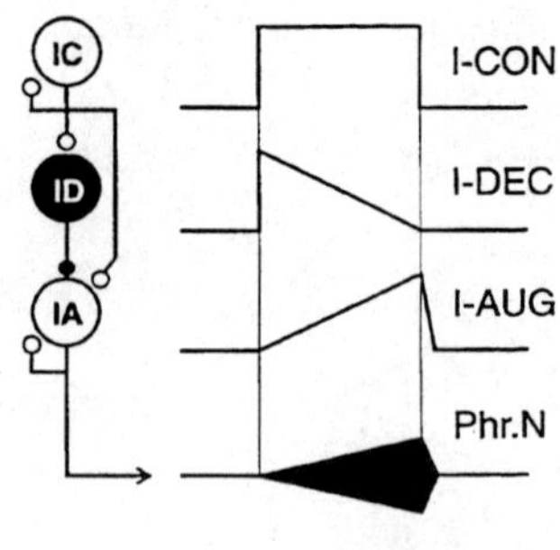

FIGURE 12
Schematic interpretation of inspiratory augmentation. The augmenting phrenic discharge reflects the augmenting discharge of bulbospinal I-AUG neurons. The latter discharge is formed by decrement of inhibition from I-DEC neurons and by self-reexcitatory connections between I-AUG neurons. (Modified from Ezure, K., *Prog. Neurobiol.*, 35, 429, 1990.)

within the population of I-CON neurons and some I-AUG neurons produces synchronous burst firing in members of the network. The firing of I-CON neurons excites I-DEC neurons almost simultaneously. The burst firing of I-DEC neurons inhibits E-AUG neurons of the BÖT and the caudal VRG to shut off their firing.[26]

During the I phase, the activity in I-AUG neurons augments, which is reflected in the augmenting discharge of the phrenic nerve and characterizes the I phase of respiration. Burns and Salmoiraghi[13] and Salmoiraghi and Baumgarten[94] postulated that self-reexcitation is the basic mechanism of augmentation and respiratory rhythmogenesis. Acting in concert with the self-reexcitatory connections between I-AUG neurons, the disinhibition of I-AUG neurons due to the decrement of inhibitory action from I-DEC neurons affords an explanation for this I augmentation (Figure 12).[26]

The firing of I-DEC neurons decreases gradually in the I phase. This is the key process by which the model presented works. However, any presently-known synaptic connections cannot explain the mechanisms responsible for this decrementing firing pattern. The situation is similar in the case of E-DEC neurons. Intrinsic properties of I-DEC and E-DEC neurons might be crucial.[70]

4.2.3 Reduced Preparations

Until recently, most studies of respiratory neurons have used *in vivo* preparations of mammalian brain and spinal cord. In the last decade, major advances in usage of mammalian *in vitro* preparations have been occurring in studies of the respiratory system. These *in vitro* preparations include brain slices at various levels, isolated and superfused brainstem-spinal cord, intra-arterially perfused brains, and even isolated cells. These methods have a great advantage in analyzing the membrane and cellular mechanisms. The progress in these reduced preparations in harmony with *in vivo* preparations, which have a great advantage in preserving a large portion of normal behavioral functions in mammals, will bring fruitful outcomes.

Feldman and co-workers have shown that the pre-BÖT area is essential for rhythm generation.[98] A isolated block of this area alone can produce a respiration-like rhythm, and there are neurons with pacemaker properties within this region in neonatal rats. Onimaru and co-workers have also isolated pacemaker-like cells from the area close to but possibly rostral to the pre-BÖT area in the isolated brainstem-spinal cord preparation of neonatal rats.[5] In these neonatal preparations, blockage of GABAergic and glycinergic inhibition do not disturb the rhythm, suggesting that pacemaker neurons but not the network of neurons are important for the rhythm.[34,81] Respiratory rhythm can also be observed in the "tilted-sagittal slice" of the brainstem in mature mammals, if the longitudinal column of the nucleus ambiguus is preserved in its entirety.[86] However, the mechanisms for respiratory rhythm generation in neonates are different from the maturing and adult animal in that inhibitory synaptic connections using glycine or GABA are essential in adult mammals.[45,86] These studies therefore indicate that the network in the BÖT/VRG area is essential for rhythm generation in the adult. The importance of network properties has also been suggested using the intra-arterially perfused preparation.[45]

4.2.4 Rostral Ventrolateral Medulla

Since neurons in the DRG and VRG are largely output neurons such as bulbospinal neurons or vagal and glossopharyngeal motoneurons, it is now widely believed that the crucial area for rhythm generation is located more rostrally in the medulla. The rostral ventrolateral medulla in the vicinity of the BÖT and the very rostral part of the rostral VRG is the focus of attention. In fact, many interesting neurons are recorded in this area in both rats[98] and cats.[17,26] The rhythm-generating circuit (Figure 11A) is modeled using neurons involved in this area.

The pre-BÖT in cats may correspond to the transitional zone between the BÖT and the rostral VRG.[17] The most conspicuous neurons in this area are I-DEC and I-CON neurons (Figure 13B).[29] Both are important key neurons in the rhythm-generating network (Figure 11) and in the formation of I augmentation (Figure 12). The center of distribution of E-AUG and E-DEC neurons is situated more rostrally in the BÖT and that of bI/E neurons more dorsomedially (Figure 13C). Therefore, the area spanning the BÖT and the pre-BÖT including the area dorsomedial to the BÖT and pre-BÖT seems to be the place of importance for rhythm generation. This view is consistent with the slice experiments in mature rats.[86]

In the same general area as bI/E neurons exist another type of respiratory neuron called early-onset E-DEC (eE-DEC) neurons (Figure 13D).[32] Their firing starts in the I phase; peak firing is around the transition from the I to E phase. Unexpectedly, however, these neurons project not to the area of respiratory neurons, but to the reticular formation dorsomedial to the nucleus ambiguus. The function of these neurons is not known, but these neurons probably play some important roles in non-respiratory but closely related activities, for instance such as coughing.[32]

Swallowing-related interneurons are distributed in a similar area (Figure 13E).[31] These neurons receive inputs from the superior laryngeal nerve, project to the hypoglossal nucleus and the nucleus ambiguus, but not to typical respiration-related areas. Many of these have some respiratory rhythm. It has been reported that neurons which are crucially involved in vomiting are located also in the area dorsomedial to the BÖT area.[35] It is therefore possible that neurons of non-respiratory but closely related activities including coughing, swallowing, or vomiting are located in this area.

Ventrally to the areas mentioned above and in the vicinity of the ventral medullary surface, the cardiovascular center is located. This area is called the rostral ventrolateral medulla (RVLM) and includes central sympathetic neurons. Many of them have respiratory rhythm as well (see next section).

5 INTERACTION WITH OTHER VITAL ACTIVITIES

5.1 Cardiorespiratory Integration

The autonomic nervous system, both sympathetic and parasympathetic, is closely linked to the respiratory system.[43] In particular, the cardiovascular system and the respiratory system, which have in common the final purpose of gas exchange between the

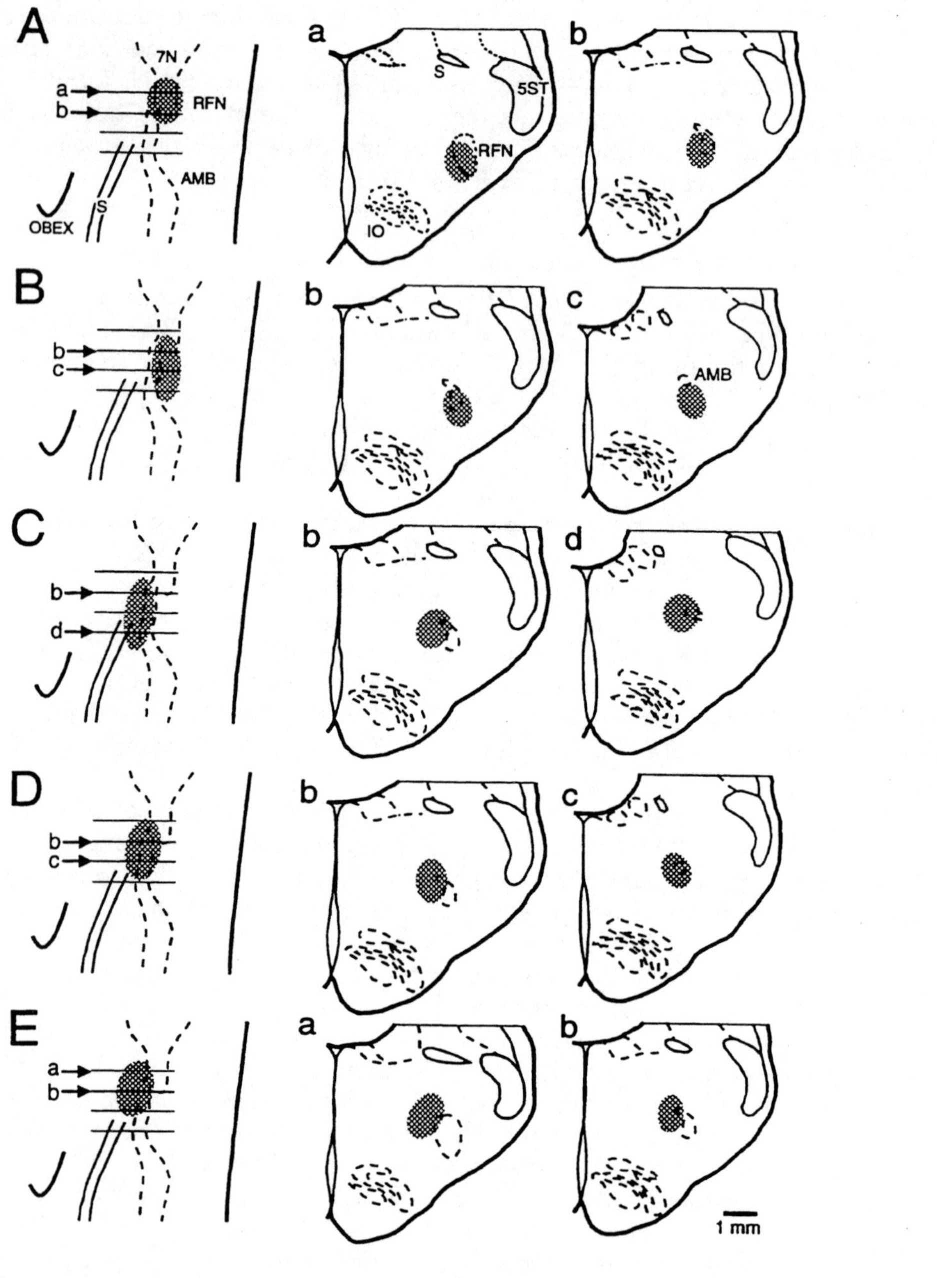

FIGURE 13
Rostral ventrolateral medulla and distributions of various respiratory and non-respiratory neurons. Transverse sections at the levels of arrows are shown on the right. (A) area of BÖT E-AUG neurons; (B) area of I-DEC and I-CON neurons; (C) area of bI/E neurons, which fire at the I-E transition and project to respiration-related areas; (D) area of early-onset E-DEC neurons, which fire at the I-E transition but do not project to respiration-related areas; (E) area of swallowing interneurons, some of which show respiratory rhythm.

external environment and the tissues, cooperate with each other. Activity in neurons in the regulation of cardiovascular function exhibits a variety of respiratory modulation. Adrian et al.[1] were the first to show that activity from sympathetic nerves was modulated with respiratory rhythm and that the rhythm was due to central respiratory drive rather than reflexes. Since then, much experimental evidence for the existence of respiratory rhythm

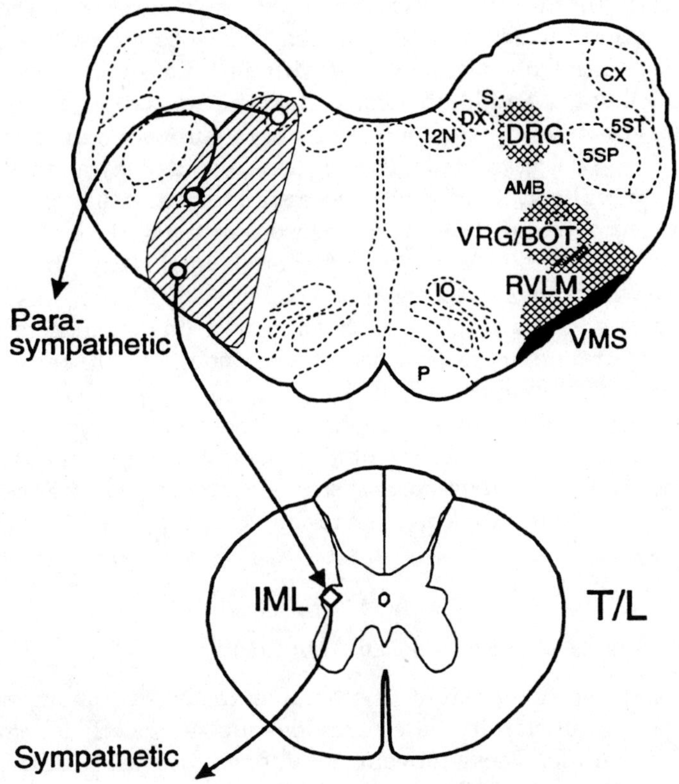

FIGURE 14
Cardiovascular-respiratory integration in the rostral ventrolateral medulla. RVLM: cardiovascular area of rostral ventrolateral medulla. VMS: chemosensitive area of the ventral medullary surface. IML: intermediolateral cell column of sympathetic preganglionic neurons. Shaded area on the left side indicates the areas of cardiovascular neurons and respiratory neurons. Respiratory rhythm is observed in both sympathetic and parasympathetic nerves.

in the autonomic system has been accumulated by recording from either peripheral nerves or central neurons.[43]

Vagal preganglionic neurons in the nucleus ambiguus often have clear respiratory modulation. Of these preganglionic cardioinhibitory neurons fire during the E phase, and neurons which project to pulmonary branches and are presumably bronchoconstrictor in function fire in the I phase.[69] Cardioinhibitory preganglionic neurons originate from the ambiguous nucleus and those from the dorsal motor nucleus of the vagus, if any, are small in number (Figure 14). The E-type rhythm, which is observed in virtually all of the cardiac inhibitory neurons examined,[51,69] has been shown to be generated by inhibitory inputs to these neurons during the I phase.[37,51] This disinhibitory mechanism explains the increase in heart rate during the I phase (the respiratory sinus arrhythmia), although the contribution of sympathetic nerves are not neglected.[43] The inhibitory input during the I phase to the cardioinhibitory neurons is possibly from some I neurons. However, inputs from I-DEC neurons, which are the only known inhibitory neurons within the I population, do not explain the firing pattern of the cardioinhibitory neurons.[43]

Thoracic sympathetic preganglionic neurons which project to the cervical sympathetic nerve show I and E rhythm in rats.[38] Similar results have been obtained in cats but predominantly I activity has been observed either from cervical nerves[89] or from neurons in the thoracic and lumbar cord.[87] It was first shown by Terui et al.[101] in rabbits and by McAllen[68] in cats that sympathetic neurons in the rostral ventrolateral medulla (RVLM) often have respiratory rhythm largely with I activity. The majority of these RVLM neurons with respiratory rhythm project to the spinal cord.[68,101] Since sympathoexcitatory neurons of the RVLM project to the intermediolateral (IML) cell column of sympathetic preganglionic neurons in the thoracic and lumbar cord, the respiratory rhythm observed in sympathetic nerves is presumably supplied by these RVLM neurons (Figure 14).[68,101] In addition, RVLM neurons which project to the locus coeruleus also display a marked central respiratory rhythmicity with I and E activity.[50] The respiratory modulation of RVLM neurons may be due to a direct synaptic input from I neurons of the BÖT and VRG.[88] Intracellular staining of I cells of the BÖT and VRG showed, though small in number, axon varicosities in the RVLM area which suggested synaptic contact with RVLM cells.[88] Moreover, the RVLM area which is situated in close association with the chemosensitive S area (see Section 3.2) of the ventral medullary surface (VMS) is suggested to be the central place of cardiovascular-respiratory integration.[93]

Cardiovascular-respiratory interaction is "relative" not "irradiation of respiratory rhythm to the cardiovascular system"; in other words, it is not a simple superposition of respiratory rhythm on the cardiovascular system.[58] Some models seeking to explain this mutually interacting system based on a common oscillator hypothesis have been proposed.[43,92]

5.2 Interaction with Respiration-Related Activities

The muscles of respiration are under both automatic and voluntary control and used during non-respiratory but closely related activities such as speech, locomotion, swallowing, coughing, or vomiting. These movements interact with each other, and almost all respiratory neurons of the medulla respond to these movements. Respiratory neurons which seem homogenous during respiration are sometimes divided into subtypes during these movements. The study of these non-respiratory but closely related activities gives new insight to the study of the respiratory system and also provides useful information on the general principle which governs mammalian central pattern generators.

The motor act of vomiting is produced primarily by changes in intra-abdominal and intra-thoracic pressures that are generated by the activation of respiratory muscles. During this activity, however, I and E muscles which contract reciprocally during normal breathing do not necessarily contract reciprocally any more. For instance, the diaphragm and the abdominal muscles contract simultaneously. The respiratory network which is built to produce reciprocal contraction of these muscles cannot work as it is. In fact, the inhibition of firing of bulbospinal I neurons is incompatible with the activity of the diaphragm. Neuronal mechanisms which integrate vomiting and respiration are presently under investigation.[42]

During coughing and swallowing, virtually all I and E neurons of the respiratory center are modulated.[80] During coughing, I and E neurons fire during the I and E phase of coughing, respectively. However, the firing of E-AUG neurons of the caudal VRG is insufficient to account for abdominal activity during coughing, suggesting that additional neurons participate in generation of the abdominal activity. During swallowing some I-DEC and I-AUG neurons are active but some are inactive.[80] This suggests that I-DEC or I-AUG neurons which behave similarly during respiration are composed of different subtypes. Two types of I-DEC neuron groups, which behave similarly during respiration but differently during vomiting, have also been reported.[75] Discharges of all E-AUG neurons and the majority of E-DEC neurons are suppressed during swallowing (Figure 15). This observation suggests that some respiratory neurons are involved in the generation of the spatiotemporally organized activities of coughing and swallowing. In other words, at least a part of the network for respiration is shared by networks for these non-respiratory but closely related activities.

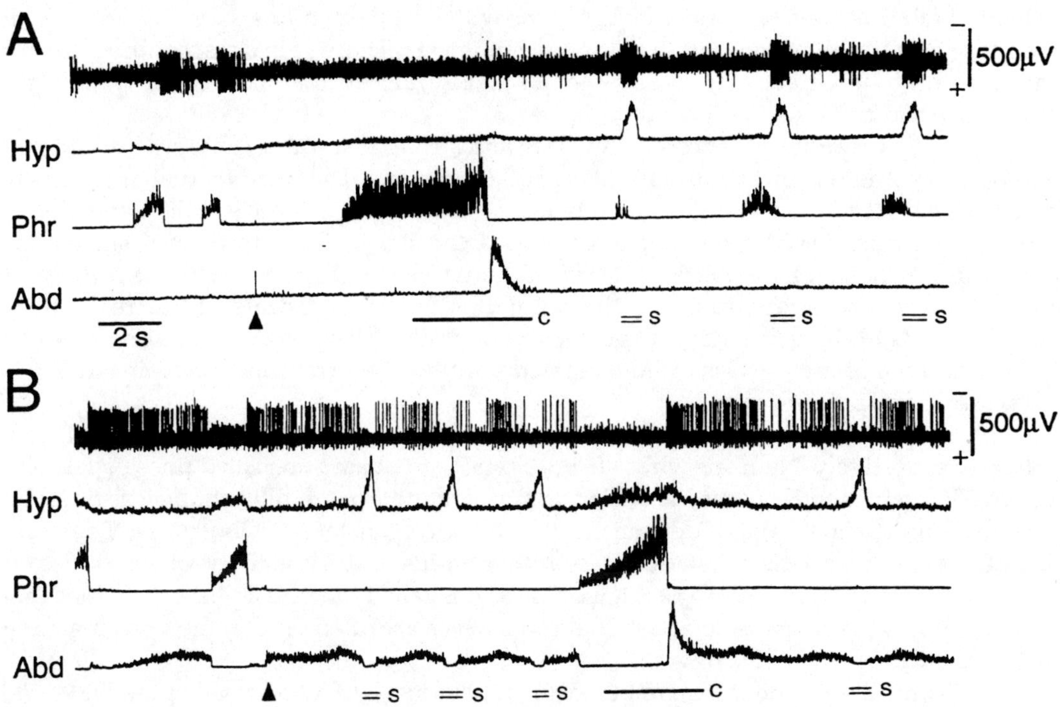

FIGURE 15
Different types of E-DEC neurons during coughing and swallowing which are induced by repetitive stimulation of the superior laryngeal nerve (SLN). Two E-DEC neurons (A and B) are shown. The neuron in (A) shows less intense firing during coughing than during control respiration; whereas firing of neuron in (B) does not change. During swallowing neuron in (A) fires but firing of neuron in (B) is suppressed. Hyp, Phr, Abd: rectified and integrated hypoglossal, phrenic, abdominal activity. (A, B) coughing (lines with letter c); swallowing (double lines with letter s); SLN stimulation (onset indicated by arrowheads) lasted beyond the end of traces. (From Oku, Y., Tanaka, I., and Ezure, K., *J. Physiol. (London)*, 480, 309, 1994.)

6 SUMMARY

The act of respiration, spatiotemporally organized contraction of respiration-related muscles which ventilate the lungs with air, is controlled by the nervous system. In the central and peripheral nervous system, there are various types of neurons that fire in synchrony with respiration. Such neurons are called respiratory neurons and include a wide variety of neurons (e.g., motoneurons, sensory neurons, interneurons, etc.). Of respiratory neurons which are distributed all over the nervous system, those in the medulla are primarily responsible for respiratory control and have been well studied. The medullary area which plays a major role in respiratory control is called the respiratory center. The respiratory center makes automatic rhythmicity of respiration, receives chemical and mechanical information from both inside and outside of the medulla, and integrates this information to generate motor outputs in such a way as to achieve the most efficient gas exchanges in the lungs.

Respiratory activity consists of two basic phases: an inspiratory (I) phase and an expiratory (E) phase. Respiratory movements are produced by respiration-related muscles which contract in phase with either the I or E phase. Close cooperation between pump muscles whose motoneurons are located in the spinal cord and upper airway muscles whose motoneurons are located in the medulla is kept during normal respiration. Medullary respiratory rhythm is transmitted to the spinal cord by bulbospinal neurons in the respiratory center. In the spinal cord, besides motoneurons of pump muscles there are segmental interneurons, propriospinal neurons with long descending axons, neurons with ascending axons, and neurons with autonomic functions. These spinal respiratory neurons may function to integrate various segmental and central information to accomplish complex behavioral patterns.

Medullary respiratory neurons with clear and strong respiratory rhythm are concentrated in dorsomedial and ventrolateral medullary regions. The dorsomedial area largely corresponds to the ventrolateral nucleus of the solitary tract and is called the dorsal respiratory group (DRG). The ventrolateral area consists of three functional subgroups extending rostrocaudally in the reticular formation in close association with the retrofacial nucleus (Bötzinger complex; BÖT), the ambiguus nucleus (rostral ventral respiratory group; VRG), and the retroambigual nucleus (caudal VRG). The transitional zone between the BÖT and the rostral VRG is recently named pre-Bötzinger complex. Each area includes characteristic respiratory neurons.

Most respiratory neurons fire during either the I or E phase and are called I or E neurons, respectively. Neurons whose firing spans both phases are called phase-spanning neurons. In both I and E phases, medullary respiratory neurons exhibit characteristic firing patterns: augmenting (AUG), decrementing (DEC), or constant (CON) types. The early part of the E phase which is associated with post-inspiratory activity of I muscles, is sometimes distinct from the rest of the E phase and is called the post-I phase. Neurons that fire in the post-I phase are called post-I neurons, which are often used in the same meaning as E-DEC neurons.

I-AUG and E-AUG neurons are bulbospinal neurons. I-AUG neurons of the DRG and VRG excite spinal I motoneurons and medullary I neurons as well via their medullary-projecting axon collaterals. E-AUG neurons of the caudal VRG have no medullary collaterals and are pre-motor neurons of spinal E motoneurons. Whereas, E-AUG neurons of the BÖT give off extensive axon collaterals in the medullary respiration-related areas and function to maintain the E phase by widely inhibiting medullary I neurons and spinal I motoneurons.

E-DEC, I-CON, and I-DEC neurons of the BÖT and the VRG are propriobulbar neurons with extensive medullary projections. E-DEC neurons are inhibitory to I neurons and contribute to the transition from the I phase to the E phase. A type of I/E phase spanning neuron may assist this transition. I-CON neurons make excitatory connections with various kinds of I neurons. Possible self-reexcitatory connections between the I-CON and I-AUG neurons might produce synchronized burst activity of I neurons at the beginning of the I phase. I-DEC neurons start burst firing at the time of the transition from the E phase to the I phase to shut off the firing of E neurons. The I-DEC neurons also inhibit I-AUG neurons, contributing to the formation of the augmenting firing pattern of the I-AUG neurons, which in turn make the augmenting activity of spinal I motoneurons.

The autonomic nervous system, both sympathetic and parasympathetic, is closely linked to the respiratory system. In particular, the cardiovascular system and the respiratory system cooperate with each other, having in common the final purpose of gas exchange between the external environment and the tissue. Activity in cardiovascular neurons, either preganglionic cardioinhibitory neurons in the nucleus ambiguus or sympathoexcitatory neurons in the rostral ventrolateral medulla (RVLM), exhibits a variety of respiratory modulation. The RVLM area which lies close to major respiratory neurons of the BÖT and VRG and also to chemosensitive areas is potentially the central place of cardiovascular-respiratory integration.

The muscles of respiration are under both automatic and voluntary control and used during non-respiratory but closely related activities such as speech, locomotion, swallowing, coughing, or vomiting. These movements interact, and almost all respiratory neurons of the medulla respond to them. Respiratory neurons which seem homogenous during respiration are sometimes divided into subtypes during these activities. It is suggested that at least a part of the network for respiration is shared by networks for these closely related activities.

REFERENCES

1. Adrian, E. D., Bronk, D. W., and Phillips, G., Discharges in mammalian sympathetic nerves, *J. Physiol. (Lond.)*, 74, 115, 1932.
2. Andrew, B. L., The respiratory displacement of the larynx: a study of innervation of accessory respiratory muscles, *J. Physiol. (Lond.)*, 130, 474, 1955.
3. Aoki, M., Mori, S., Kawahara, K., Watanabe, H., and Ebata, N., Generation of spontaneous respiratory rhythm in high spinal cats, *Brain Res.*, 202, 51, 1980.
4. Aoki, M., Watanabe, H., Ebata, N., and Mori, S., Spontaneous respiratory activity in spinal cats, *Brain Res.*, 157, 376, 1978.
5. Arata, A., Onimaru, H., and Homma, I., Respiration-related neurons in the ventral medulla of newborn rats *in vitro, Brain Res. Bull.*, 24, 599, 1990.
6. Arita, H., Kogo, N., and Koshiya, N., Morphological and physiological properties of caudal medullary expiratory neurons of the cat, *Brain Res.*, 401, 258, 1987.
7. Baumgarten, R. and von Kanzow, E., The interaction of two types of inspiratory neurons in the region of the tractus solitarius of the cat, *Arch. Ital. Biol.*, 96, 361, 1958.
8. Baumgarten, R., von Schmiedt, H., and Dodich, N., Microelectrode studies of phrenic motoneurons, *Ann. N.Y. Acad. Sci.*, 109: 536, 1963.
9. Bellingham, M. C. and Lipski, J., Respiratory interneurons in the C5 segment of the spinal cord of the cat, *Brain Res.*, 533, 141, 1990.
10. Berger, A. J., Dorsal respiratory group neurons in the medulla of the cat: spinal projections, responses to lung inflation and superior laryngeal nerve stimulation, *Brain Res.*, 135, 231, 1977.
11. Bianchi, A. L., Localization et etude des neurones respiratoires bulbaire. Mise en jeu antidromique par stimulation spinal ou vagale, *J. Physiol. (Paris)*, 63, 5, 1971.
12. Bryant, T. H., Yoshida, S., de Castro, D., and Lipski, J., Expiratory neurons of the Bötzinger complex in the rat: a morphological study following intracellular labeling with biocytin, *J. Comp. Neurol.*, 335, 267, 1993.
13. Burns, B. D. and Salmoiraghi, G. C., Repetitive firing of respiratory neurones during their burst activity, *J. Neurophysiol.*, 23, 27, 1960.
14. Coglianese, C. J., Peiss, C. N., and Wurster, R. D., Rhythmic phrenic nerve activity and respiratory activity in spinal dogs, *Resp. Physiol.*, 29, 247, 1977.
15. Cohen, M. I., Neurogenesis of respiratory rhythm in the mammal, *Physiol. Rev.*, 59, 1105, 1979.
16. Coleridge, H. M. and Coleridge, J. C. G., Reflexes evoked from tracheobronchial tree and lungs, in *Handbook of Physiology, Sec. 3: The Respiratory System. Vol. II: Control of Breathing, Part 1*, American Physiological Society, Bethesda, MD, 1986, 395.
17. Connelly, C. A., Dobbins, E. G., and Feldman, J. F., Pre-Bötzinger complex in cats: respiratory neuronal discharge patterns, *Brain Res.*, 590, 337, 1992.
18. Davies, J. G. Mc., Kirkwood, P. A., and Sears, T. A., The distribution of monosynaptic connexions from inspiratory bulbospinal neurones to inspiratory motoneurones in the cat, *J. Physiol. (Lond.)*, 368, 63, 1985.
19. Davies, R. O., Kubin, L., and Pack, A. I., Pulmonary stretch receptor relay neurones of the cat: location and contralateral medullary projections, *J. Physiol. (Lond.)*, 383, 571, 1987.
20. de Castro, D., Lipski, J., and Kanjhan, R., Electrophysiological study of dorsal respiratory neurons in the medulla oblongata of the rat, *Brain Res.*, 639, 49, 1994.
21. Donoghue, S., Felder, R. B., Jordan, D., and Spyer, K. M., The central projections of carotid baroreceptors and chemoreceptors in the cat: a neurophysiological study, *J. Physiol. (Lond.)*, 347, 397, 1984.
22. Douse, M. A. and Duffin, J., Axonal projections and synaptic connections of C5 segment expiratory interneurones in the cat, *J. Physiol. (Lond)*, 470, 431, 1993
23. Douse, M. A., Duffin, J., Brooks, D., and Fedorko, L., Role of upper cervical inspiratory neurons studied by cross-correlation in the cat, *Exp. Brain Res.*, 90, 153, 1992.
24. Duffin, J. and Lipski, J., Monosynaptic excitation of thoracic motoneurones by inspiratory neurones of the nucleus tractus solitarius in the cat, *J. Physiol. (Lond.)*, 390, 415, 1987.
25. Eyzaguirre, C. and Taylor, J. R., Respiratory discharge of some vagal motoneurons, *J. Neurophysiol.*, 26, 61, 1963.
26. Ezure, K., Synaptic connections between medullary respiratory neurons and considerations on the genesis of respiratory rhythm, *Prog. Neurobiol.*, 35, 429, 1990.
27. Ezure, K. and Manabe, M., Decrementing expiratory neurons of the Bötzinger complex. II. Direct inhibitory synaptic linkage with ventral respiratory group neurons, *Exp. Brain Res.*, 72, 159, 1988.
28. Ezure, K. and Manabe, M., Monosynaptic excitation of medullary inspiratory neurons by bulbospinal inspiratory neurons of the ventral respiratory group in the cat, *Exp. Brain Res.*, 74, 501, 1989.
29. Ezure, K., Manabe, M., and Otake, K., Excitation and inhibition of medullary inspiratory neurons by two types of burst inspiratory neurons in the cat, *Neurosci. Lett.*, 104, 303, 1989.

30. Ezure, K., Manabe, M., and Yamada, H., Distribution of medullary respiratory neurons in the rat, *Brain Res.*, 455, 262, 1988.
31. Ezure, K., Oku, Y., and Tanaka, I., Location and axonal projection of one type of swallowing interneurons in cat medulla, *Brain Res.*, 632, 216, 1993.
32. Ezure, K., Tanaka, I., and Oku, Y., Location and axonal projection of early-onset decrementing expiratory neurons in the cat, *Neurosci. Lett.*, 163, 105, 1993.
33. Fedorko, L., Merrill, E. G., and Lipski, J., Two descending medullary inspiratory pathways to phrenic motoneurones, *Neurosci. Lett.*, 43, 285, 1983.
34. Feldman, J. L. and Smith, J. L., Cellular mechanisms underlying modulation of breathing pattern in mammals, *Ann. N.Y. Acad. Sci.*, 563, 114, 1989.
35. Fukuda, H. and Koga, T., Non-respiratory neurons in the Bötzinger complex exhibiting appropriate firing patterns to generate the emetic act in dogs, *Neurosci. Res.*, 14, 180, 1992.
36. Gesell, R., Bricker, J., and Magee, C., Structural and functional organization of the central mechanism controlling breathing, *Am. J. Physiol.*, 117, 423, 1936.
37. Gilbey, M. P., Jordan, D., Richter, D. W., and Spyer, K. M., Synaptic mechanisms involved in the inspiratory modulation of vagal cardio-inhibitory neurones in the cat, *J. Physiol. (Lond.)*, 356, 65, 1984.
38. Gilbey, M. P., Numao, Y., and Spyer, K. M., Discharge patterns of cervical sympathetic preganglionic neurones related to central respiratory drive in the rat, *J. Physiol. (Lond.)*, 378, 253, 1986.
39. Green, J. H. and Neil, E., The respiratory function of the laryngeal muscles, *J. Physiol. (Lond.)*, 129, 134, 1955.
40. Grélot, L., Barillot, J. C., and Bianchi, A. L., Pharyngeal motoneurons: respiratory-related activity and responses to laryngeal afferents in the decerebrate cat, *Exp. Brain Res.*, 78, 336, 1989
41. Grélot, L., Milano, S., Portillo, F., and Miller, A. D., Respiratory interneurons of the lower cervical (C4-C5) cord: membrane potential changes during fictive coughing, vomiting, and swallowing in the decerebrate cat, *Pflügers Arch.*, 425, 313, 1993.
42. Grélot, L. and Miller A. D., Vomiting — its ins and outs, *News Physiol. Sci.*, 9, 142, 1994.
43. Häbler, H.-J., Jänig, W., and Michaelis, M., Respiratory modulation in the activity of sympathetic neurones, *Prog. Neurobiol.*, 43, 567, 1994.
44. Harber, E., Kohn, K. W., Ngai, S. H., Holaday, D. A., and Wang, S. C., Localization of spontaneous respiratory neuronal activities in the medulla oblongata of the cat: a new location of the expiratory center, *Am. J. Physiol.*, 190, 350, 1957.
45. Hayashi, F. and Lipski, J., The role of inhibitory aminoacids in control of respiratory motor output in an arterially perfused rat, *Respir. Physiol.*, 89, 47, 1992.
46. Hoskin, R. and Duffin, J., Excitation of upper cervical inspiratory neurons by inspiratory neurons of the nucleus tractus solitarius in the cat, *Exp. Neurol.*, 95, 126, 1987.
47. Hoskin, R. and Duffin, J., Excitation of upper cervical inspiratory neurons by inspiratory neurons of the nucleus retroambigualis in the cat, *Exp. Neurol.*, 98, 404, 1987.
48. Hoskin, R. W., Fedorko, L. M., and Duffin, J., Projections from upper cervical inspiratory neurons to thoracic and lumbar expiratory motor nuclei in the cat, *Exp. Neurol.*, 99, 544, 1988.
49. Hosogai, M., Matsuo, S., and Nakao, S., Firing pattern and location of respiratory neurons in cat medullary raphe nuclei, *Neurosci. Lett.*, 161, 149, 1993.
50. Huangfu, D., Verberne, A. J. M., and Guyenet, P. G., Rostral ventrolateral medullary neurons projecting to locus coeruleus have cardiorespiratory inputs, *Brain Res.*, 598, 67, 1992.
51. Iriuchijima, J. and Kumada, M., Activity of single vagal fibers efferent to the heart, *Jpn. J. Physiol.*, 14, 479, 1964
52. Jiang, C. and Lipski, J., Extensive monosynaptic inhibition of ventral respiratory group neurons by augmenting neurons in the Bötzinger complex in the cat, *Exp. Brain Res.*, 81, 631, 1990
53. Jiang, C. and Lipski, J., Synaptic inputs to medullary respiratory neurons from superior laryngeal afferents in the cat, *Brain Res.*, 584, 197, 1992.
54. Kirkwood, P. A., Munson, J. B., Sears, T. A., and Westgaard, R. H., Respiratory interneurones in the thoracic spinal cord of the cat, *J. Physiol. (Lond.)*, 395, 161, 1988.
55. Kirkwood, P.A. and Sears, T.A., Monosynaptic excitation of thoracic expiratory motoneurones from lateral respiratory neurones in the medulla of the cat, *J. Physiol. (Lond.)*, 234, 87, 1973.
56. Kita, I., Sakamoto, M., and Arita, H., Adrenergic cell group in rostral ventrolateral medulla of cat: its correlation with central chemoreceptors, *Neurosci. Res.*, 20, 265, 1994.
57. Knowlton, G. C. and Larrabee, M. G., A unitary analysis of pulmonary volume receptors, *Am. J. Physiol.*, 147, 100, 1946.
58. Koepchen, H. P., Respiratory and cardiovascular "centers": functional entirety or separate structures? in *Central Environment and the Control Systems of Breathing and Circulation*, Schlafke, E. M., Koepchen, H. P., and See, W. R., Eds., Springer-Verlag, Berlin, 1983, 221.

59. Lindsey, B. G., Segers, L. S., Morris, K. F., Hernandez, Y. M., Saporta, S., and Shannon, R., Distributed actions and dynamic associations in respiratory-related neuronal assemblies of the ventrolateral medulla and brain stem midline: evidence from spike train analysis, *J. Neurophysiol.*, 72, 1830, 1994.
60. Lipski, J. and Duffin, J., An electrophysiological investigation of propriospinal inspiratory neurons in the upper cervical cord of the cat, *Exp. Brain Res.*, 61, 625, 1986.
61. Lipski, J., Duffin, J., Kruszewska, B., and Zhang, X., Upper cervical inspiratory neurons in the rat: an electrophysiological and morphological study, *Exp. Brain Res.*, 95, 477, 1993.
62. Lipski, J., Ezure, K., and Wong She, R. B., Identification of neurons receiving input from pulmonary rapidly adapting receptors in the cat, *J. Physiol. (Lond)*, 443, 55, 1991.
63. Lipski, J., Kubin, L., and Jodkowski, J., Synaptic action of Rb neurons on phrenic motoneurons studied with spike-triggered averaging, *Brain Res.*, 288, 105, 1983.
64. Lipski, J., Zhang, X., Kruszewska, B., and Kanjhan, R., Morphological study of long axonal projections of ventral medullary inspiratory neurons in the rat, *Brain Res.*, 640, 171, 1994.
65. Loeschcke, H. H., Central chemosensitivity and the reaction theory, *J. Physiol. (Lond.)*, 332, 1, 1982.
66. Lowe, A. A., The neural regulation of tongue movements, *Prog. Neurobiol.*, 15, 295, 1981.
67. Lumsden, T., Observations on the respiratory centres in the cat, *J. Physiol. (Lond.)*, 57, 153, 1923.
68. McAllen, R. M., Central respiratory modulation of subretrofacial bulbospinal neurones in the cat, *J. Physiol. (Lond.)*, 388, 533, 1987.
69. McAllen, R. M. and Spyer, K. M., Two types of vagal preganglionic motoneurones projecting to the heart and lungs, *J. Physiol. (Lond.)*, 282, 353, 1978.
70. Merrill, E. G., Finding a respiratory function for the medullary respiratory neurons, in *Essays on the Nervous System*, Bellairs, R. and Gray, E. G., Eds., Clarendon, Oxford, 1974, 451.
71. Merrill, E. G., Where are the real respiratory neurons? *Fed. Proc.*, 40, 2389, 1981.
72. Merrill, E. G. and Frdorko, L., Monosynaptic inhibition of phrenic motoneurons: a long descending projection from Bötzinger neurons, *J. Neuroscience*, 4, 2350, 1984.
73. Merrill, E. G. and Lipski, J., Inputs to intercostal motoneurons from ventrolateral medullary respiratory neurons in the cat, *J. Neurophysiol.*, 57, 1837, 1987.
74. Merrill, E. G., Lipski, J., Kubin, L., and Fedorko, L., Origin of the expiratory inhibition of the nucleus tractus solitarius inspiratory neurones, *Brain Res.*, 263, 43, 1983.
75. Miller, A. D. and Ezure, K., Behavior of inhibitory and excitatory propriobulbar respiratory neurons during fictive vomiting, *Brain Res.*, 578, 168, 1992.
76. Miller, A. D., Ezure, K., and Suzuki, I., Control of abdominal muscles by brain stem respiratory neurons in the cat, *J. Neurophysiol.*, 54, 155, 1985.
77. Monteau, R., Khatib, M., and Hilaire, G., Central determination of recruitment order: intracellular study of phrenic motoneurons, *Neurosci. Lett.*, 56, 341, 1985.
78. Nesland, R. and Plum, F., Subtypes of medullary respiratory neurons, *Exp. Neurol.*, 12, 337, 1963.
79. Oku, Y., Tanaka, I., and Ezure, K., Possible inspiratory off-switch neurones in the ventrolateral medulla of the cat, *Neuro. Report*, 3, 933, 1992.
80. Oku, Y., Tanaka, I., and Ezure, K., Activity of bulbar respiratory neurons during fictive coughing and swallowing in the decerebrate cat, *J. Physiol. (Lond.)*, 480, 309, 1994.
81. Onimaru, H., Arata, A., and Homma, I., Inhibitory synaptic inputs to the respiratory rhythm generator in the medulla isolated from newborn rats, *Pflugers Arch.*, 417, 425, 1990.
82. Otake, K., Sasaki, H., Ezure, K., and Manabe, M., Axonal trajectory and terminal distribution of inspiratory neurons of the dorsal respiratory group in the cat's medulla, *J. Comp. Neurol.*, 286, 218, 1989.
83. Otake, K., Sasaki, H., Ezure, K., and Manabe, M., Medullary projection of non-augmenting inspiratory neurons of the ventrolateral medulla in the cat, *J. Comp. Neurol.*, 302, 485, 1990.
84. Otake, K., Sasaki, H., Mannen, H., and Ezure, K., Morphology of expiratory neurons of the Bötzinger complex: an HRP study in the cat, *J. Comp. Neurol.*, 258, 565, 1987.
85. Palisses, R., Persegol, L., and Viala, D., Evidence for respiratory interneurones in the C3-C5 cervical spinal cord in the decorticate rabbit, *Exp. Brain Res.*, 78, 624, 1989.
86. Paton, J. F. R., Ramirez, J.-M., and Richter, D. W., Functionally intact *in vitro* preparation generating respiratory activity in neonatal and mature mammals, *Pflügers Arch.*, 428, 250, 1994.
87. Pilowsky, P., Llewellyn-Smith, I. J., Arnolda, L., Minson, J., and Chalmers, J., Intracellular recording from sympathetic preganglionic neurons in cat lumbar spinal cord, *Brain Res.*, 656, 319, 1994.
88. Pilowsky, P., Llewellyn-Smith, I. J., Lipski, J., Minson, J., Arnolda, L., and Chalmers, J., Projections from inspiratory neurons of the ventral respiratory group to the subretrofacial nucleus of the cat, *Brain Res.*, 633, 63, 1994.
89. Preiss, G. and Polosa, C., The relation between end-tidal CO2 and discharge patterns of sympathetic preganglionic neurons, *Brain Res.*, 122, 255, 1977.

90. Remmers, J. E., deGroot, W. J., Sauerland, E. K., and Anch, A. M., Pathogenesis of upper airway occlusion during sleep, *J. Appl. Physiol.*, 44, 931, 1978.
91. Richter, D. W., Generation and maintenance of the respiratory rhythm, *J. Exp. Biol.*, 100, 93, 1982.
92. Richter, D. W. and Spyer, K. M., Cardiorespiratory control, in *Central Regulation of Autonomic Functions*, Loewy, A.D. and Spyer, K.M., Eds., Oxford University Press, Oxford, 1990, 189.
93. Ruggiero, D. A., Gomez, R. E., Cravo, S. L., Mtui, E., Anwar, M., and Reis, D. J., The rostral ventrolateral medulla: anatomical substrates of cardiopulmonary integration, in *Cardiorespiratory and Motor Coordination*, Koepchen, H. -P. and Huopaniemi, T., Eds., Springer-Verlag, Berlin, 1991, 89.
94. Salmoiraghi, G. C. and Baumgarten, R., von, Intracellular potentials from respiratory neurones in brain-stem of cat and mechanism of rhythmic respiration, *J. Neurophysiol.*, 24, 203, 1961.
95. Sasaki, H., Otake, K., Mannen, H., Ezure, K., and Manabe, M., Morphology of augmenting inspiratory neurons of the ventral respiratory group in the cat, *J. Comp. Neurol.*, 282, 157, 1989.
96. Sasaki, S.-I., Uchino, H., Imagawa, M., Miyake, T., and Uchino, Y., Lower lumbar branching of caudal medullary expiratory neurons of the cat, *Brain Res.*, 553, 159, 1991.
97. Sasaki, S.-I., Uchino, H., and Uchino, Y., Axon branching of medullary expiratory neurons in the lumbar and the sacral spinal cord of the cat, *Brain Res.*, 648, 229, 1994.
98. Smith, J. C., Ellenberger, H. H., Ballanyi, K., Richter, D. W., and Feldman, J. L., Pre-Bötzinger complex: a brainstem region that may generate respiratory rhythm in mammals, *Science*, 254, 726, 1991
99. Sumi, T., Spinal respiratory neurons and their reaction to stimulation of intercostal nerves, *Pflügers Arch.*, 278, 127, 1963.
100. Tanaka, Y. and Hirai, N., Physiological studies of thoracic spinocerebellar tract neurons in relation to respiratory movement, *Neurosci. Res.*, 19, 317, 1994.
101. Terui, N., Saeki, Y., and Kumada, M., Barosensory neurons in the ventrolateral medulla in rabbits and their responses to various afferent inputs from peripheral and central sources, *Jpn. J. Physiol.*, 36, 1141, 1986.
102. Van Lunteren, E., Haxhiu, M. A., Prabhakar, N. R., Mitra, J., and Cherniack, N. S., Increased duration of postinspiratory inspiratory activity during augmented breaths in cats, *Neurosci. Lett.*, 71, 83, 1986.
103. Viala, D. and Freton, E., Evidence for respiratory and locomotor pattern generators in the rabbit cervico-thoracic cord and for their interaction, *Exp. Brain Res.*, 49, 247, 1983.
104. Withington-Wray, D. J., Mifflin, S. W., and Spyer, K. M., Intracellular analysis of respiratory-modulated hypoglossal motoneurons in the cat, *Neuroscience*, 25, 1041, 1988.
105. Yates, B. J., Jakus, J., and Miller, A. D., Vestibular effects on respiratory outflow in the decerebrate cat, *Brain Res.*, 629, 209, 1993.

5

The Neural Basis of Nausea and Vomiting

Alan D. Miller and Laurent Grélot

CONTENTS

1 INTRODUCTION

Nausea is an unpleasant sensation of discomfort, generally not life threatening, which markedly reduces the quality of life. It often culminates in vomiting, i.e., the violent expulsion of gastrointestinal contents through the mouth. Although vomiting (emesis) can lead to serious physiological disorders (e.g., dehydration and alkalosis) when attacks are recurrent, it is primarily a protective reflex preventing the blood circulatory system from invasion by an ingested poison. Nausea and vomiting also occur in various circumstances for which the benefit is unclear. For example, nausea and vomiting occurring during motion and space sickness induce a debilitating state that limits the inclination for work.

The responses of nausea and vomiting can be thought of as consisting of three components — inputs, central coordinating circuitry, and pre-motor and motor outputs.[3,22,33] The last two components are believed to be common to all acts of vomiting, irrespective of the

0-8493-7668-1/96/$0.00+$.50

precipitating condition. In contrast, nausea and vomiting can be triggered under different conditions by a variety of inputs that are mediated by different brain regions. For example, the vestibular system is essential for motion sickness. Abdominal visceral afferents and the chemoreceptor trigger zone in the area postrema are important for eliciting nausea and vomiting associated with anticancer chemotherapy. Higher brain centers presumably trigger psychogenic vomiting. The inputs responsible for some conditions, such as pregnancy sickness and post-operative nausea and vomiting, remain unknown and may be due to multiple factors.

2 NEURAL BASIS OF MOTION SICKNESS

The vestibular system is essential for motion sickness, as shown by studies involving people with vestibular defects and animals with lesions of the vestibular periphery.[13,52,57] Surgical procedures selectively affecting either the semicircular canals[53] or otolith organs[7,24] have shown that each of these vestibular endorgans can be indispensable for eliciting motion sickness. Labyrinthine-defective subjects are also immune to sickness induced by moving visual fields.[10] Blind people, however, have susceptibility to motion sickness similar to that of subjects with normal vision.[19]

Motion sickness is generally thought to result from a "sensory conflict" arising between vestibular, visual, and other sensory input and the input that is expected on the basis of past experience.[12,57] As discussed in detail by Reason and Brand,[57] different types of conflict situations are possible. For example, a conflict can occur when different sensory inputs simultaneously signal conflicting information or when one type of sensory input is received in the absence of an expected input from another source. The latter is thought to be important for eliciting space motion sickness. Head tilts in microgravity would result in an assumed normal input from the vertical semicircular canals in the absence of the usual expected otolith input.

The underlying neuroanatomical or physiological basis for the sensory conflict process has not been identified.[56] However, the vestibular nuclei may play a role since both visual[14,26,61] and somatic,[65,66] as well as vestibular, inputs converge onto vestibular nuclear neurons. As discussed in Chapter 9, once the appropriate stimulus conditions occur, the nausea and vomiting of motion sickness may result from activation of the body's defense mechanisms against accidental poisoning. Thus, motion sickness may be mediated at least in part by vestibular inputs to neurons in the medullary nucleus of the solitary tract (NTS) that are also excited by other emetic inputs and may represent the beginning of a final common pathway by which different inputs elicit nausea and vomiting.[68]

In the 1950s it was proposed that motion-induced vomiting is produced by vestibular signals that traverse the nodulus and uvula of the vestibulocerebellum and then somehow activate the area postrema.[63] However, more recent studies have indicated that these structures are not essential. Vomiting and related prodromal activity can be readily obtained both by electrical stimulation of the vestibular labyrinths of decerebrate cats following large lesions of the posterior cerebellar vermis including the nodulus and uvula[50] and by natural vestibular stimulation after chronic ablation of the area postrema in cats and monkeys.[5,17,64]

Several other brain regions have also been shown to be non-essential for at least some aspects of motion sickness. Removal of the pituitary gland has no effect on the latency to motion-induced vomiting in dogs.[54] Motion sickness has been produced in a functionally decorticate man.[15] Animals decerebrated at the brainstem collicular level vomit, and exhibit other typical prodromal signs, in response to natural vestibular stimulation[2] and to sinusoidal electrical stimulation of the vestibular periphery.[50]

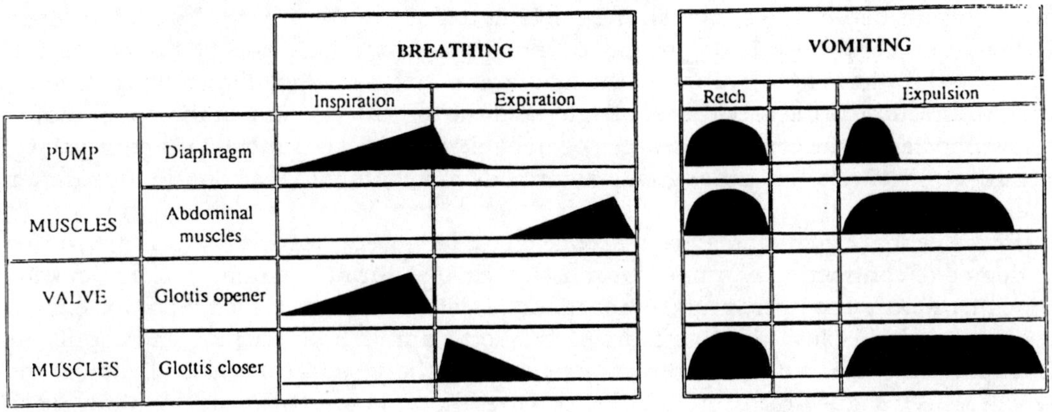

FIGURE 1
Schematic representation of muscle activity during breathing and vomiting. The timing of muscle activation is indicated in black. For simplicity, only the major pump muscles and those of the larynx are illustrated; see Grélot and Miller[22] for a more complete description. Innervation: diaphragm — phrenic nerve; abdominal muscles — abdominal muscle nerves; glottis opener — recurrent and superior laryngeal nerves; glottis closer — recurrent laryngeal nerve.

Although the cerebral cortex is not essential for vomiting, it is presumably necessary for the sensation of nausea. Magnetic source imaging studies indicate that the inferior frontal gyrus of the human cerebral cortex may be important for nausea.[58] There are also motion sickness-related changes in the electroencephalogram recorded from the temporofrontal region.[9]

3 RESPIRATORY AND GASTROINTESTINAL MUSCLE ACTIVATION DURING VOMITING

Emesis is usually preceded by a variety of prodromal signs, such as pallor, salivation, and hyperventilation, together with the sensation of nausea. Vomiting is produced by the sequential and highly coordinated contraction of numerous smooth and somatic muscles. Precisely, it results from drastic changes in the activity of the gastrointestinal (GI) smooth muscles, upper airway muscles, and the main "respiratory pump" muscles (i.e., the diaphragm, abdominal, and intercostal muscles).

The motor act of vomiting can be divided into three stages, although the first two are not essential for expelling the bolus: (1) the GI events, which precede the somatic motor changes, (2) retching, and (3) expulsion.

It is well established since the early work of Magendie[41] and Eggleston and Hatcher[16] that the GI events (stomach relaxation and retrograde giant contraction in the proximal small intestine), although important, are not essential for vomiting. The occurrence of the retrograde giant contraction during conditions when there are no toxins in the gastrointestinal tract (e.g., during motion sickness) is thought to serve to buffer and dilute the acidic contents of the stomach before their passage through the esophagus and mouth.[35] The GI, as well as other autonomic, correlates of vomiting are more fully discussed in Chapter 9.

Retching and expulsion persist after complete extrinsic denervation of the upper GI tract[35] but are compromised if respiratory muscle activation is prevented.[42] Thus, the motor act of vomiting is clearly dependent on changes in intraabdominal and intrathoracic pressures that are generated by the main respiratory pump (i.e., thoraco-abdominal) muscles, in concert with activation of the upper airway (i.e., valve) muscles (Figure 1).

Several events serve to distinguish the retching and expulsion phases from each other. During expulsion, there is an increased and prolonged contraction of the abdominal muscles, relaxation of the portion of the diaphragm that surrounds the esophagus, retrograde contraction of the cervical esophagus, and the forceful opening of the upper esophageal sphincter by the contraction of the geniohyoid muscle. The functional significance of retching remains unclear, although it may provide momentum to the bolus in preparation for expulsion.[34]

In the cephalic region, the oropharyngeal and laryngeal patencies, which depend on the degree of contraction of numerous muscles, change during vomiting.[20] The relaxation of the inspiratory laryngeal muscles concomitant with the potent activation of the expiratory laryngeal muscles leads to closure of the glottis during both retching and expulsion. Simultaneously, the rhythmic contraction of the stylopharyngeus (pharyngeal dilator) and genioglossus (tongue protrusor) muscles increases the oropharyngeal patency, allowing an easier expulsion of gastrointestinal contents. The tonic contraction of the digastricus muscle depresses the jaw and thus opens the mouth during vomiting. Following the transit of GI contents through the mouth, several episodes of deglutition occur to clean the oropharyngeal cavity.

In the thoraco-abdominal region, the main respiratory (i.e., diaphragm and abdominal) muscles, which contract alternately to generate airflow in the airways during breathing, co-contract to expel the GI contents during vomiting. The sternal and costal parts of the diaphragm co-contract rhythmically with the abdominal wall (i.e., internal and external oblique muscles, transverse abdominis and rectus abdominis). In contrast, the crural portion of the diaphragm adjacent to the esophagus exhibits brief activation at the onset of each retch but relaxes completely during expulsion to facilitate the rostral movement of GI contents.[44,55]

The behavior of the intercostal muscles is still controversial despite numerous investigations using electrophysiological[23,25,27,42,51] or more recently sonomicrometric[1] approaches. All investigations have revealed that intercostal muscles contract rhythmically during vomiting; however, the observed phase of their activation with respect to the diaphragm and abdominal muscles differed among various studies. It is possible that the discrepancies are due to interspecies (cat vs. dog) variations and drastic differences in experimental procedures (i.e., the supine or prone position of the animal, tracheotomy, pneumothorax, or paralysis). However, such differences do not affect the phase relationship between the diaphragm and abdominal muscles. Given the apparent variability of intercostal discharge during vomiting, it would seem that these muscles play a role secondary to those of the diaphragm and abdominal muscles in generating the thoraco-abdominal pressure changes essential for vomiting.

In addition to changes in GI and respiratory-related activity, vomiting is accompanied by more widespread changes in muscle discharge. Recent investigations revealed that vomiting develops with postural adjustments in shoulder (i.e., trapezius) and limb (i.e., triceps brachii) muscles.[21] Such an activation during both the retching and expulsion phases of vomiting is also observed in the pudendal nerve, which innervates the external anal and urethral sphincters, thus serving to maintain continence during the large increase in abdominal pressure that occurs during vomiting.[27,47]

4 CENTRAL COORDINATING CIRCUITRY FOR NAUSEA AND VOMITING

In the 1950s, vomiting was thought to be produced by a "vomiting center" located in the dorsolateral reticular formation of the brainstem.[6,62] More recent studies have been unable to localize a unique, neuroanatomically distinct area that might function as a center and have proposed that nausea and vomiting may be coordinated by more distributed brain areas.[46,48,49] Studies involving the expression of the immediate-early gene c-fos, a marker of cellular activation, indicate that neurons involved in coordinating vomiting are distributed in an arc that extends from the area postrema and nucleus of the solitary tract (NTS) through the lateral tegmental field to the ventrolateral medulla (Figure 2).[48] The NTS may represent the beginning of a final common pathway by which different inputs trigger nausea and vomiting. The NTS receives input from the area postrema, which is a chemoreceptor trigger zone for vomiting.[4,45] Other inputs that trigger vomiting also activate neurons in the NTS. For example, some NTS neurons receive convergent inputs from the vestibular system and abdominal vagal afferents (Figure 3).[68] The respiratory-related components of vomiting may be mediated via projections from the NTS to the region of the retrofacial nucleus in the ventrolateral medulla.[28] The ventrolateral medulla contains respiratory-related pre-motor circuitry that is activated during vomiting.[22] Lesions in this region abolish the respiratory components of vomiting;[18,46] however, possible effects on other (e.g., gastrointestinal) components of vomiting and on nausea are unknown. The gastrointestinal changes during vomiting, which are abolished by vagotomy,[35] are presumably mediated by projections from the NTS to the dorsal motor nucleus of the vagus and the nucleus ambiguus.

Almost nothing is known about the neural substrate that produces the sensation of nausea. Recently, however, preliminary studies using the non-invasive technique of magnetic source imaging have indicated that a region of the human inferior frontal gyrus is activated during nausea induced by vestibular stimulation or by ingestion of syrup of ipecac.[58]

5 PHARMACOLOGY OF MOTION SICKNESS

Effective drugs are available to reduce the development of motion sickness, but used individually most of them are associated with at least some adverse effects. These medications have been tested in both humans and animals, and a fairly definitive consensus has been established concerning their relative effectiveness. In humans, an association of scopolamine (i.e., a competitive antagonist at the muscarinic receptors of acetylcholine) and amphetamine (i.e., an enhancer of central noradrenergic and dopaminergic transmission) is one of the most effective preparations; however, the potential abuse of amphetamine clearly limits its usefulness.[67] The mechanism underlying the therapeutic synergism of the two compounds may occur within the basal ganglia[32] or the limbic system[29] rather than in brainstem structures. Recent results suggest that the antimotion sickness effect of acetylcholine antagonists may be mediated through M3 and/or M5 receptors.[40,60] The association of promethazine (i.e., a phenothiazine acting as an antagonist at dopaminergic D2, histaminergic H1, and muscarinic receptors) plus ephedrine (i.e., a sympathomimetic drug) appears as potent as the former treatment.[67] The most effective over-the-counter medication is the antihistamine dimenhydrinate (Dramamine®). Antihistaminergic drugs effective against motion sickness act centrally at a site that remains to be elucidated. Peripherally acting antihistamines have either no[31] or marginal[30] efficacy against motion sickness.

Other pharmacological agents such as a glucocorticoid (i.e., dexamethasone[29]), the anticonvulsant phenytoin,[8] vasopressin V1 receptor antagonists,[11] and calcium channel blockers (e.g., flunarizine[36]) exhibit anti-motion-sickness properties. Unlike other families of drugs, flunarizine is thought to depress the peripheral vestibular system. Finally, buspirone and 8-OH-DPAT, two agonists at serotonin 1A receptors, also exert potent anti-motion-sickness effects in animal models.[38,39] Since serotonin 1A receptor agonists block vomiting triggered under various circumstances (e.g., anticancer chemotherapy and xylazine administration), their wide antiemetic spectrum might be due to their ability to block the

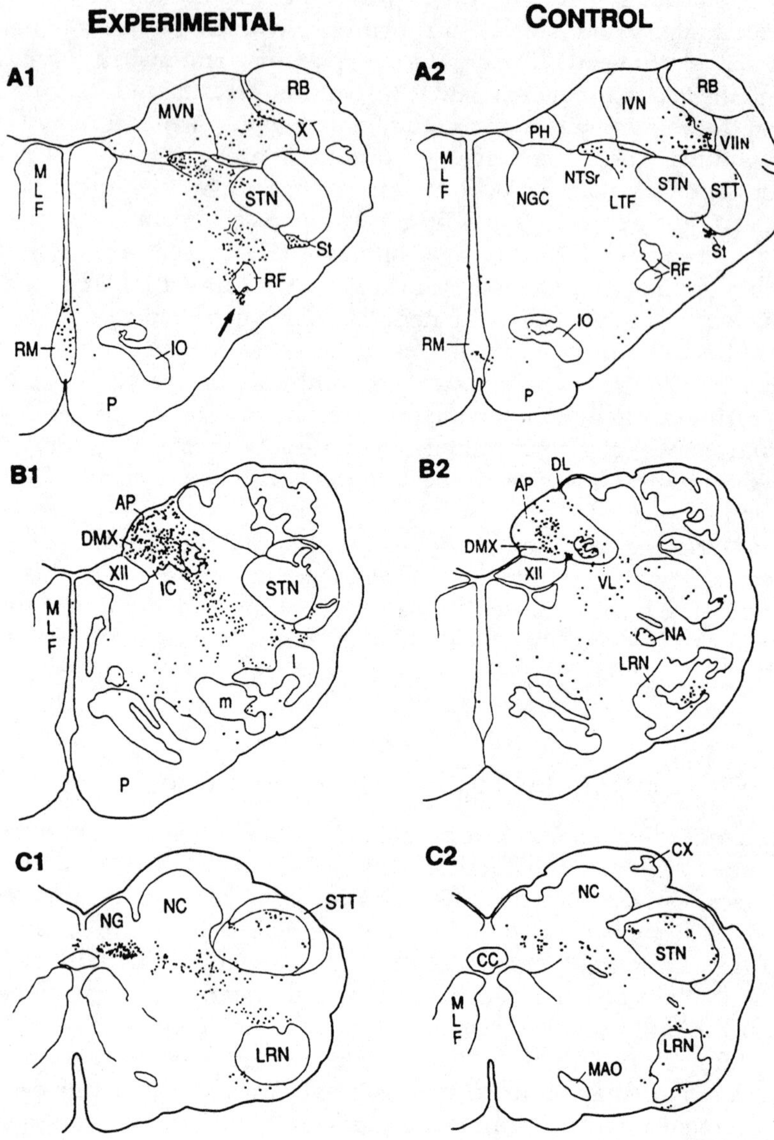

FIGURE 2
Distribution of Fos-like immunoreactivity following vomiting induced by a combination of emetic drugs, illustrated at three levels of the cat medulla (A1, rostral, to C1, caudal). Data from a control animal are shown in A2-C2. Labeled cells were densely distributed in experimental animals in the area postrema (AP), nucleus of the solitary tract (NTS), and intermediate reticular zone of the lateral tegmental field (LTF). Abbreviations: CC, central canal; CX, external cuneate nucleus; DL, dorsolateral subnucleus of NTS; DMX, dorsal motor nucleus of the vagus; IC, nucleus intercalatus; IO, inferior olive; IVN, inferior vestibular nucleus; LRN, lateral reticular nucleus (l, lateral division; m, medial division); MAO, medial accessory olive; MLF, medial longitudinal fasciculus; MVN, medial vestibular nucleus; NA, nucleus ambiguus; NC, cuneate nucleus; NG, gracilis nucleus; NGC, nucleus gigantocellularis; NTSr, rostral NTS; P, pyramidal tract; RB, restiform body; RF, retrofacial nucleus; RM, raphe magnus; St, subtrigeminal nucleus; STN, spinal trigeminal nucleus; STT, spinal trigeminal tract; VL, ventrolateral subnucleus of NTS; X, subnucleus X of IVN; XII, hypoglossal nucleus. (From Miller, A. O. and Ruggiero, D. A., *J. Neurosci.*, 14, 871, 1994. With permission.)

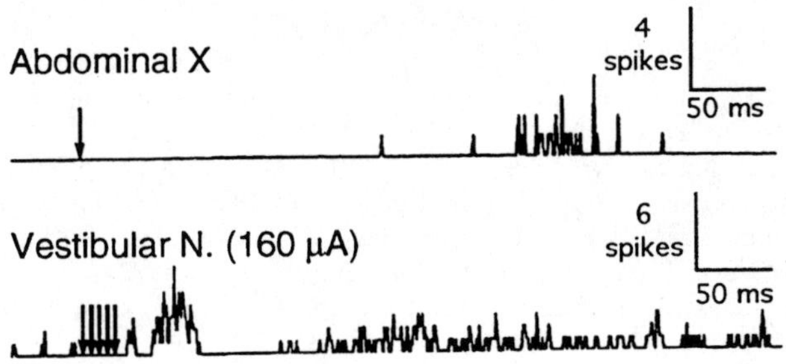

FIGURE 3
Convergence of abdominal vagal and vestibular inputs to a neuron in the nucleus of the solitary tract. Such neurons may be important for mediating motion sickness. Top: Response of the neuron to single shocks applied to the abdominal vagus (X) nerve; histogram consists of 38 sweeps. Bottom: Combined excitation and inhibition produced in the same cell by five stimuli applied to the ipsilateral vestibular nerve (N); response consists of 59 sweeps. Arrows indicate times when stimuli were delivered. (From Yates, B. J. et al., *Am. J. Physiol.*, 267, R974, 1994. With permission.)

neural network responsible for the elicitation of vomiting. In contrast, several antagonists of the serotonin type 3 receptor (i.e., tropisetron, MDL 72222, zacopride, ondansetron, and granisetron), which are potent antiemetics against chemotherapy-induced vomiting, are ineffective for preventing the development of motion sickness either in human[59] and animal models.[37,43]

6 SUMMARY

Nausea and vomiting (emesis) can be triggered by a variety of different inputs under various conditions. The vestibular system is essential for motion sickness. Neurons in the medullary nucleus of the solitary tract, which are excited by both vestibular and other inputs capable of eliciting vomiting, may represent the beginning of a final common pathway by which different inputs trigger nausea and vomiting. Vomiting is produced by coordinated changes in gastrointestinal and respiratory-related activity. However, the motor act of vomiting results primarily from the powerful activation of the major respiratory muscles. Most of the respiratory-related pre-motor circuitry is located in the ventrolateral medulla. Neurons involved in coordinating vomiting may be distributed in an arc that extends from the nucleus of the solitary tract through the lateral tegmental field of the reticular formation to the ventrolateral medulla. Recent studies have identified a region in the inferior frontal gyrus of the human cerebral cortex that may be important for nausea. The most widely used treatment against motion sickness includes drugs having an antagonist action either at the muscarinic receptors of acetylcholine or at the histaminergic H1 receptor. Other drugs such as sympathomimetic agents (e.g., amphetamine) enhance the action of the former compounds. Serotonergic drugs that are agonists at type 1A receptors are potent anti-motion-sickness agents, as tested in animal models.

ACKNOWLEDGMENTS

The authors' work is supported by grants from the National Institute of Neurological Disorders and Stroke, U.S.A. (NS20585) (A.D.M.), and the CNRS (URA 1832), France, and the Direction des Recherches, Etudes et Techniques (DRET, 87/110 and 90/108), France (L.G.).

REFERENCES

1. Abe, T., Kieser, T. M., Tomita, T., and Easton, P. A., Respiratory muscle function during emesis in awake canines, *J. Appl. Physiol.*, 76, 2552, 1994.
2. Bard, P., Woolsey, C. N., Snider, R. S., Mountcastle, V. B., and Bromiley, R. B., Delimitation of central nervous mechanisms involved in motion sickness, *Fed. Proc.*, 6, 72, 1947.
3. Bianchi, A. L., Grélot, L., Miller, A. D., and King, G. L., Eds., *Mechanisms and Control of Emesis*, Colloques INSERM/John Libbey Eurotext, Montrouge, France, 1992.
4. Borison, H. L., Area postrema: chemoreceptor circumventricular organ of the medulla oblongata, *Prog. Neurobiol.*, 32, 351, 1989.
5. Borison, H. L. and Borison, R., Motion sickness reflex arc bypasses the area postrema in cats, *Exp. Neurol.*, 92, 723, 1986.
6. Borison, H. L. and Wang, S. C., Functional localization of central coordinating mechanism for emesis in cat, *J. Neurophysiol.*, 12, 305, 1949.
7. Brizzee, K. R. and Igarashi, M., Effect of macular ablation on frequency and latency of motion-induced emesis in the monkey, *Aviat. Space Environ. Med.*, 57, 1066, 1986.
8. Chelen, W., Kabrisky, M., Hatsell, C., Morales, R., Fix, E., and Scott, M., Use of phenytoin in the prevention of motion sickness, *Aviat. Space Environ. Med.*, 61, 1022, 1990.
9. Chelen, W. E., Kabrisky, M., and Rogers, S. K., Spectral anaylsis of the electroencephalographic response to motion sickness, *Aviat. Space Environ. Med.*, 64, 24, 1993.
10. Cheung, B. S. K., Howard, I. P., and Money, K. E., Visually-induced sickness in normal and bilaterally labyrinthine-defective subjects, *Aviat. Space Environ. Med.*, 62, 527, 1991.
11. Cheung, B. S. K., Kohl, R. L., Money, K. E., and Kinter, L. B., Etiologic significance of arginine vasopressin in motion sickness, *J. Clin. Pharmacol.*, 34, 664, 1994.
12. Claremont, C. A., The psychology of sea-sickness, *Psyche*, 11, 86, 1930-31.
13. Crampton, G. H., Ed., *Motion and Space Sickness*, CRC Press, Boca Raton, FL, 1990.
14. Daunton, N. and Thomsen, D., Visual modulation of otolith units in cat vestibular nuclei, *Exp. Brain Res.*, 37, 173, 1979.
15. Doig, R. K., Wolf, S., and Wolff, H. G., Study of gastric function in a "decorticate" man with gastric fistula, *Gastroenterology*, 23, 40, 1953.
16. Eggleston, C. and Hatcher, R. A., The site of the emetic action of various drugs, *J. Pharmacol. Exp. Ther.*, 7, 225, 1912.
17. Fox, R. A., Corcoran, M., and Brizzee, K. R., Conditioned taste aversion and motion sickness in cats and squirrel monkeys, *Can. J. Physiol. Pharmacol.*, 68, 269, 1990.
18. Fukuda, H. and Koga, T., The Bötzinger complex as the pattern generator for retching and vomiting in the dog, *Neurosci. Res.*, 12, 471, 1991.
19. Graybiel, A., Susceptibility to acute motion sickness in blind persons, *Aerospace Med.*, 41, 650, 1970.
20. Grélot, L., Barillot, J. C., and Bianchi, A. L., Activity of respiratory-related oropharyngeal and laryngeal motoneurones during fictive vomiting in the decerebrate cat, *Brain Res.*, 513, 101, 1990.
21. Grélot, L. and Bianchi, A. L., Activities of cranial and spinal respiratory-related motoneurons during vomiting, in *New Vistas on Mechanisms and Control of Emesis*, Vol. 223, Bianchi, A. L., Grélot, L., Miller, A. D., and King, G. L., Eds., Colloque INSERM/John Libbey Eurotext, Montrouge, France, 1992, 29.
22. Grélot, L. and Miller, A. D., Vomiting — its ins and outs, *News Physiol. Sci.*, 9, 142, 1994.
23. Hukuhara, T., Okada, H., and Yamagami, M., On the behavior of the respiratory muscles during vomiting, *Acta Med. Okayama*, 11, 117, 1957.
24. Igarashi, M., Himi, T., Kulecz, W. B., and Kobayashi, K., Role of otolith endorgans in the genesis of vestibular-visual conflict sickness (pitch) in the squirrel monkey (First Report), *Aviat. Space Environ. Med.*, 58 (Suppl.), A207, 1987.
25. Iscoe, S. and Grélot, L., Regional intercostal activity during coughing and vomiting in decerebrate cats, *Can. J. Physiol. Pharmacol.*, 70, 1995, 1992.
26. Keller, E. L. and Precht, W., Visual-vestibular responses in vestibular nuclear neurons in the intact and cerebellectomized, alert cat, *Neuroscience*, 4, 1599, 1979.
27. Koga, T. and Fukuda, H., Characteristic behavior of the respiratory muscles, esophagus, and external anal and urethral sphincters during straining, retching and vomiting in the decerebrate dog, *Jpn. J. Physiol.*, 40, 789, 1990.
28. Koga, T. and Fukuda, H., Neurons in the nucleus of the solitary tract mediating inputs from emetic vagal afferents and the area postrema to the pattern generator for the emetic act in dogs, *Neurosci. Res.*, 14, 166, 1992.
29. Kohl, R. L., The antimotion sickness efficacy of dexamethasone, amphetamine and scopolamine implicates the limbic system in the etiology of motion sickness, *Spec. Sci. Technol.*, 8, 305, 1986.

30. Kohl, R. L., Calkins, D. S., and Robinson, R. E., Control of nausea and autonomic dysfunction with terfenadine, a peripherally acting antihistamine, *Aviat. Space Environ. Med.*, 62, 392, 1991.
31. Kohl, R. L., Homick, J. L., Cintron, N., and Calkins, D. S., Lack of effects of astemizole on vestibular ocular reflex, motion sickness, and cognitive performance in man, *Aviat. Space Environ. Med.*, 58, 1171, 1987.
32. Kohl, R. L. and Lewis M. R., Mechanisms underlying the antimotion sickness effects of psychostimulants, *Aviat. Space Environ. Med.*, 58, 1215, 1987.
33. Kucharczyk, J., Stewart, D. J., and Miller, A. D., Eds., *Nausea and Vomiting: Recent Research and Clinical Advances*, CRC Press, Boca Raton FL, 1991.
34. Lang, I. M., New perspective on the mechanisms controlling vomitus expulsion, in *New Vistas on Mechanisms and Control of Emesis*, Vol. 223, Bianchi, A. L., Grélot, L., Miller, A. D., and King, G. L., Eds., Colloque INSERM/John Libbey Eurotext, Montrouge, France, 1992, 71.
35. Lang, I. M. and Sarna, S. K., Motor and myoelectric activity associated with vomiting, regurgitation, and nausea, in *Handbook of Physiology, the Gastrointestinal System. Vol. I: Motility and Circulation*, Wood, J. D., Ed., American Physiological Society, Bethesda, MD, 1989, 1179.
36. Lee, J. A., Watson, L. A., and Boothy, G., Calcium antagonists in the prevention of motion sickness, *Aviat. Space Environ. Med.*, 57, 45, 1986.
37. Lucot, J. B., Blockade of 5-hydroxytryptamine 3 receptors prevents cisplatin-induced but not motion- or xyzline-induced emesis in the cat, *Pharmacol. Biochem. Behav.*, 32, 207, 1989.
38. Lucot, J. B. and Crampton, G. H., Buspirone blocks motion sickness and xylazine-induced emesis in the cat, *Aviat. Space Environ. Med.*, 58, 989, 1987.
39. Lucot, J. B. and Crampton, G. H., 8-OH-DPAT suppresses vomiting in the cat elicited by motion, cisplatin or xylazine, *Pharmacol. Biochem. Behav.*, 33, 627, 1989.
40. Lucot, J. B., van Charldorp, K. J., and Tulp, M. Th. M., Idaverine, an M_2- vs. M_3-selective muscarinic antagonist, does not prevent motion sickness in cats, *Pharmacol. Biochem. Behav.*, 40, 345, 1991.
41. Magendie, F., Précis élémentaire de physiologie, 2nd ed., Méquignon-Marvis, Coale, E. J., Ed., Paris, 1824.
42. McCarthy, L. E. and Borison, H. L., Respiratory mechanics of vomiting in decerebrate cats, *Am. J. Physiol.*, 226, 738, 1974.
43. Milano, S. and Grélot, L., personal communication, 1995.
44. Miller, A. D., Lakos, S. F., and Tan, L. K., Central motor program for relaxation of periesophageal diaphragm during the expulsive phase of vomiting, *Brain Res.*, 456, 367, 1988.
45. Miller, A. D. and Leslie, R. A., The area postrema and vomiting, *Frontiers Neuroendocrin.*, 15, 301, 1994.
46. Miller, A. D., Nonaka, S., and Jakus, J., Brain areas essential or non-essential for emesis, *Brain Res.*, 647, 255, 1994.
47. Miller, A. D., Nonaka, S., Siniaia, M. S., and Jakus, J., Multifunctional ventral respiratory group: Bulbospinal expiratory neurons play a role in pudendal discharge during vomiting, *J. Auton. Nerv. Syst.*, 54, 253, 1995.
48. Miller, A. D. and Ruggiero, D. A., Emetic reflex arc revealed by expression of the immediate-early gene c-fos in the cat, *J. Neuroscience*, 14, 871, 1994.
49. Miller, A. D. and Wilson, V. J., "Vomiting center" reanalyzed: an electrical stimulation study, *Brain Res.*, 270, 154, 1983.
50. Miller, A. D. and Wilson, V. J., Vestibular-induced vomiting after vestibulocerebellar lesions, *Brain Behav. Evol.*, 23, 26, 1983.
51. Miller, A. D. and Yates, B. J., Evaluation of role of the upper cervical inspiratory neurones in respiration, emesis and cough. *Brain Res.*, 606, 143, 1993.
52. Money, K. E., Motion sickness, *Physiol. Rev.*, 50, 1, 1970.
53. Money, K. E. and Friedberg, J., The role of the semicircular canals in causation of motion sickness and nystagmus in the dog, *Can. J. Physiol. Pharm.*, 42, 793, 1964.
54. Money, K. E. and Wood, J. D., Neural mechanisms underlying the symptomatology of motion sickness, in *4th Symp. Role of the Vestibular Organs in Space Exploration*, NASA Spec. Publ. No. 187, 1970, 35.
55. Monges, H., Salducci, J., and Naudy, B., Dissociation between the electrical activity of the diaphragmatic dome and crura muscular fibers during esophageal distension, vomiting and eructation: an electromyographic study in the dog, *J. Physiol. (Paris)*, 74, 541, 1978.
56. Oman, C. M., Motion sickness: a synthesis and evaluation of the sensory conflict theory, *Can. J. Physiol. Pharmacol.*, 64, 294, 1990.
57. Reason, J. T. and Brand, J. J., *Motion Sickness*, Academic Press, New York, 1975.
58. Roberts, T., Miller, A., Rowley, H., and Kucharczyk, J., Functional neural imaging of nausea in humans, *Soc. Neurosci. Abstr.*, 20, 1417, 1994.

59. Stott, J. R. R., Barnes, G. R., Wright, R. J., and Ruddock, C. J. S., The effect on motion sickness and oculomotor function of GR 38032F, a 5-HT3-receptor antagonist with antiemetic properties, *Br. J. Clin. Pharmacol.*, 27, 147, 1989.
60. Stott, J. R. R. and Golding, J. F., The effect of a selective muscarinic receptor antagonist and scopolamine on motion sickness, skin conductance and heart rate in humans, *Proc. Physiol. Soc. Abstr.*, 47P, 1994.
61. Waespe, W. and Henn, V., The velocity response of vestibular nucleus neurons during vestibular, visual, and combined angular acceleration, *Exp. Brain Res.*, 37, 337, 1979.
62. Wang, S. C. and Borison, H. L., The vomiting center: its destruction by radon implantation in dog medulla oblongata, *Am. J. Physiol.*, 166, 712, 1951.
63. Wang, S. C. and Chinn, H. I., Experimental motion sickness in dogs. Importance of labyrinth and vestibular cerebellum, *Am. J. Physiol.*, 185, 617, 1956.
64. Wilpizeski, C. R., Lowry, L. D., and Goldman, W. S., Motion-induced sickness following bilateral ablation of area postrema in squirrel monkeys, *Laryngoscope*, 96, 1221, 1986.
65. Wilson V. J., Vestibulospinal and neck reflexes: interaction in the vestibular nuclei, *Arch. Ital. Biol.*, 129, 43, 1991.
66. Wilson, V. J. and Melvill Jones, G., *Mammalian Vestibular Physiology*, Plenum Press, New York, 1979.
67. Wood, C. D., Antimotion sickness and antiemetic drugs, *Drugs*, 17, 471, 1979.
68. Yates, B. J., Grélot, L., Kerman, I. A., Balaban, C. D., Jakus, J., and Miller, A. D., Organization of vestibular inputs to nucleus tractus solitarius and adjacent structures in cat brain stem, *Am. J. Physiol.*, 267, R974, 1994.

Section II

Vestibular Influences on Autonomic Regulation

6

Vestibular Influences on Cardiovascular Control

Bill J. Yates

CONTENTS

1 INTRODUCTION

The cardiovascular system is profoundly susceptible to gravitational accelerations; simple changes in posture initiate decreases in venous return to the heart that are a dramatic challenge to the circulation. For example, standing from a supine position in a

0-8493-7668-1/96/$0.00+$.50

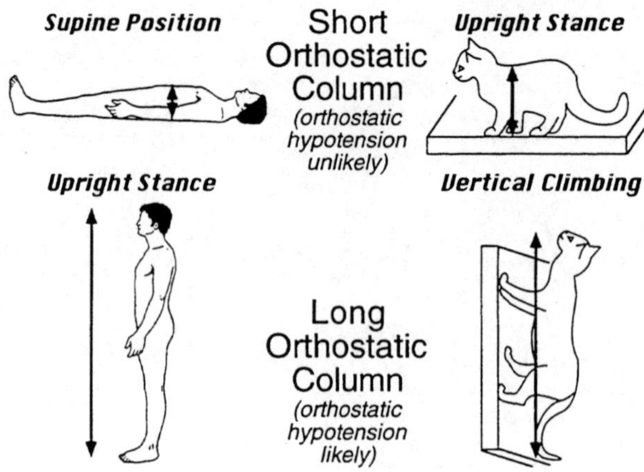

FIGURE 1
The height of the orthostatic column that blood must overcome in flowing from the distal extremities to the heart. The height of the column is dramatically increased in a human who stands from a supine position, and in the cat that climbs vertically. Increases in the length of the orthostatic column can compromise the return of blood to the heart, resulting in decreased cardiac filling and decreased stroke volume.

man of average height results in over a sixfold increase in the height of the orthostatic column that must be overcome for blood to flow from the distal extremities back to the heart (Figure 1). Quadrupeds that engage in such movements as vertical climbing can also experience posturally related cardiovascular effects. Orthostatic hypotension is particularly problematic in animals with a long longitudinal axis, such as cats. The consequences of postural hypotension are decreased cardiac filling, decreased cardiac output, and decreased perfusion of the brain, which can result in syncope (fainting) if the condition is not corrected quickly.[40]

A number of mechanisms act to resist posturally related hypotension. Contraction of leg muscles reduces the capacitance of large intramuscular veins, and helps to push blood toward the heart.[28] Contraction of the diaphragm acts to lower intra-thoracic pressure, which produces a favorable pressure gradient for return of blood to the heart.[25,32,45] Enhanced intra-abdominal pressure produced by contraction of abdominal muscles also tends to force blood into the thorax, and toward the heart.[32,33] However, the most important short-latency mechanism in compensating for orthostatic hypotension is to produce increased sympathetic outflow to selected arterial beds, which serves to shut down blood flow to unnecessary regions and reserve blood for the systemic circulation.[16] For example, the abdominal organs typically have a large blood flow which subserves the process of digestion and circulation of nutrients, but this blood flow can be temporarily diminished greatly if needed to maintain systemic blood pressure.[17] Shunting blood flow away from the limbs in which blood pooling is occurring would also be advantageous in abating orthostatic hypotension.

Several sensory mechanisms can detect the presence of orthostasis and trigger one of the compensatory mechanisms described above. Baroreceptors, located in the carotid sinus and aortic arch, signal changes in blood pressure[37] and provide powerful inputs to the brainstem regions that regulate the sympathetic nervous system and thus control vasoconstriction in peripheral arterial beds,[43] as described in detail in Chapter 3. However, the baroreceptor mechanisms do not begin to counteract orthostasis until a sizable drop in blood pressure has already occurred. Thus, although this feedback mechanism helps to prevent a long-term decrease in blood pressure following a change in posture, it has limited effects on the early phases of orthostatic hypotension. The walls of large veins in the limbs also contain stretch receptors that provide inputs to the nervous system which indicate the presence of peripheral blood pooling.[10,47,56] Activation of limb venous afferents results in the contraction of muscles surrounding the vein that has been distended by pooling of blood, which aids in pumping the blood toward the heart.[46,57] Thus, the orthostatic compensatory mechanisms elicited by venous afferent activation act to abate long-term hypotension, but become effective only after a substantial accumulation of blood has occurred in the limbs.

A number of older studies, as well as considerable recent data, suggest that reflex adjustments in the circulation elicited by the vestibular system are also involved in preventing orthostatic hypotension. The existence of vestibulo-cardiovascular reflexes is not surprising, as the vestibular system provides inputs which signal when changes in body position, including those that can affect circulation, are occurring. In fact, it seems practical for the vestibular system to participate in adjustments for orthostatic hypotension, because it can act to stabilize the circulation during movement even before blood pressure has changed significantly. This chapter will (1) review the evidence suggesting that the vestibular system has effects on cardiovascular control and (2) describe the current knowledge of the neural circuitry that links the vestibular system with brainstem cardiovascular regulatory mechanisms.

2 EVIDENCE THAT THE VESTIBULAR SYSTEM INFLUENCES CARDIOVASCULAR CONTROL

2.1 Animal Studies

Some of the first work which indicated that the vestibular system has influences on cardiovascular control was published in the early 1900s by Spiegel and Démétriades (reviewed in Reference 42). These investigators showed that activation of vestibular afferents by electrical stimulation or perfusing the ear canal with hot or cold water (caloric stimulation) produced changes in blood pressure in cats, dogs, and guinea pigs. This observation has been confirmed and extended in many subsequent studies. A number of experiments in cats have demonstrated that electrical stimulation of the vestibular nerve evokes changes in activity in sympathetic nerves, including those innervating the heart,[6] kidney,[20,21,48] and abdominal digestive organs.[29,55,57] Examples of potentials recorded from the major sympathetic nerve innervating the abdominal viscera, the splanchnic nerve, following stimulation of the vestibular nerve at several intensities, are shown in the left panel of Figure 2. Increased sympathetic outflow to visceral organs results in vasoconstriction of arterioles providing their blood supply, and thus perfusion of these organs is diminished (allowing more blood to remain in the systemic circulation). Effects on sympathetic outflow produced by stimulation of the vestibular nerve can be eliminated by placing lesions in the vestibular nuclei (see Section 3.1), and it is thus clear that these effects are due to activation of the vestibular system.[55]

Natural vestibular stimulation can also affect sympathetic outflow. This was demonstrated in a recent study in which sinusoidal rotations of the head that combined both roll (ear-down movements) and pitch (fore-aft movements) were shown to produce sinusoidal modulation of splanchnic nerve activity.[57] These experiments were performed on cats whose upper cervical dorsal roots were transected to remove influences from neck afferents, so that the only sensory inputs elicited by the head rotations came from the vestibular system. The right panel of Figure 2 shows sympathetic outflow produced by clockwise and counterclockwise vestibular stimulation. Increased activity in the splanchnic nerve was elicited by nose-up rotations, whereas decreased sympathetic outflow was produced by

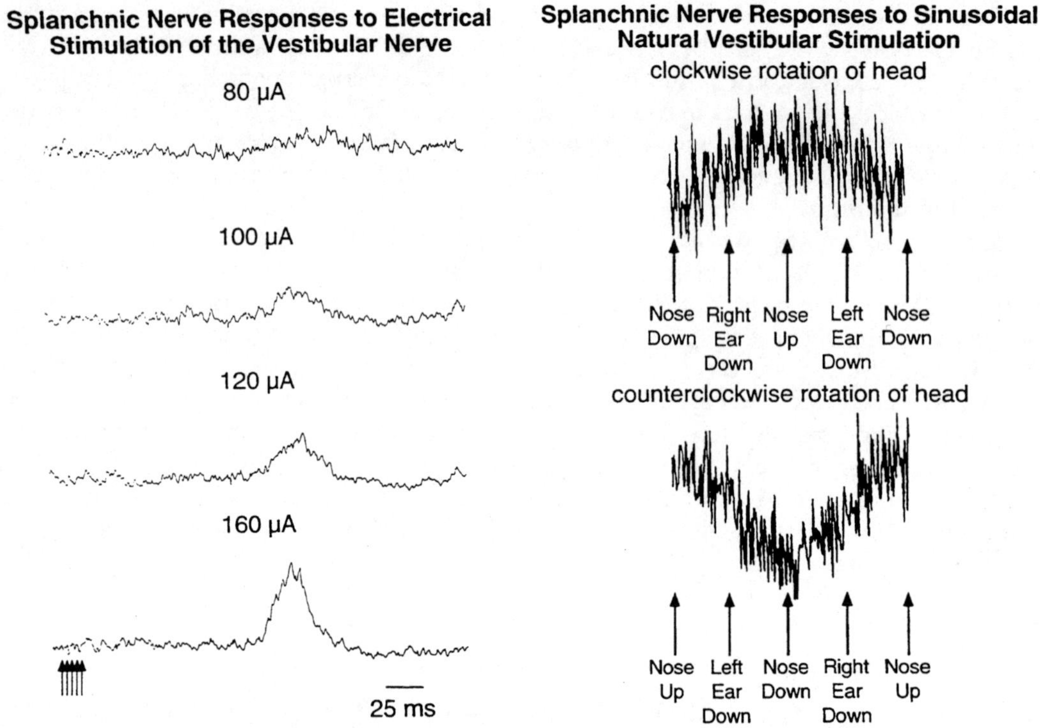

FIGURE 2

Effects of vestibular nerve stimulation on activity recorded from the splanchnic nerve, the largest sympathetic nerve innervating the abdominal viscera. Left panel: averaged responses elicited by electrical stimulation of the vestibular nerve at a number of intensities; each trace was evoked by a train of five shocks, as indicated by arrows beneath the bottom waveform. The threshold of the response was approximately 80 μA, which was less than two times the minimal intensity for producing field potentials in the vestibular nuclei. Furthermore, a higher stimulus strength (200 μA) was required to activate fibers in the closest non-target nerve, the facial nerve. Thus, it is certain that these responses were due to activation of the vestibular system. Right panel: averaged splanchnic nerve responses to 15° sinusoidal head rotations that incorporated roll and pitch. The activity was recorded from cats the upper cervical dorsal roots of which were transected to remove inputs from the neck, so that the head rotations selectively activated vestibular afferents. Activity in the splanchnic nerve was dependent on head position, and was maximal during nose-up head rotations and minimal during nose-down rotations. (Data in the right panel from Yates, B. J. and Miller, A. D., *J. Neurophysiol.*, 71, 2087, 1994. With permission.)

nose-down tilt of the head. This response pattern is concordant with the hypothesis that vestibular-elicited sympathetic nerve activity serves to offset orthostasis, since the long longitudinal axis of the cat makes it susceptible to hypotension during movements involving nose-up body rotations.

More direct evidence that the vestibular system acts to counteract orthostasis comes from the work of Doba and Reis.[11] These investigators showed that bilateral transection of the vestibular nerves in paralyzed, chloralose-anesthetized cats significantly impaired compensation for the hypotension produced by 30 or 60° head-up body tilt, suggesting that the vestibular system is involved in correcting posturally related changes in blood pressure.

The observation that orthostatic tolerance is decreased for many seconds or minutes by vestibular nerve lesions[11] suggests that the otolith organs play an important role in vestibulo-cardiovascular responses. As discussed in Chapter 2, the otolith organs comprise one of the two types of vestibular endorgans, the other being semicircular canals. The otolith organs are activated by changes in head position with respect to gravity, and would be responsible for any long-duration effects of vestibular stimulation. Our recent studies[57] confirm the presumption that the otolith organs have influences on the sympathetic nervous system. The gain of vestibulo-sympathetic reflexes elicited by head rotations in the pitch plane is constant across stimulus frequencies, and the responses are in phase with head position. These are characteristics typically associated with otolith afferents,[14] and not with semicircular canal afferents.[15] Thus, a restricted group of vestibular receptors — otolith receptors that respond to pitch — appear to be selectively involved in producing vestibulo-sympathetic responses.

2.2 Human Studies

A definitive analysis of vestibular effects on cardiovascular control is harder in human subjects than in animals, as it is difficult to do the appropriate controls to prove that effects on the sympathetic nervous system are due to activation of labyrinthine receptors, and not visceral receptors or baroreceptors.[31] For example, a number of studies have shown that caloric stimulation can result in changes in blood pressure,[39] but this stimulus can also result in nausea and discomfort that could lead to psychologically induced cardiovascular effects. Hemmingway[19] showed that 150° swinging of human subjects has effects on both heart rate and blood pressure, but did not account for the possibility that activation of baroreceptors or other visceral receptors could have produced the responses. A better-controlled study was performed by Essandoh et al.,[13] who showed that head-down neck flexion in prone subjects produces a decrease in blood flow in the forearm and calf, as would occur to offset orthostatic hypotension. Controls were done to assure that the effects were not due to activation of baroreceptors or receptors in neck muscles, the most likely non-vestibular sources of inputs that could have produced these responses.

Some anecdotal evidence also suggests the existence of vestibular effects on sympathetic outflow in humans. In one study, patients with peripheral vestibular disease were tested for hypotension several minutes after standing from a supine position, and it was shown that a significant fraction was susceptible to long-lasting orthostasis.[36] However, short-term orthostatic hypotension is more likely to be affected by loss of vestibular inputs than the hypotension that persists over minutes, as non-labyrinthine mechanisms (such as baroreceptors) should provide for long-term compensation for the condition. If the initial drops in blood pressure produced by head-up tilt were compared in labyrinthine-defective and labyrinth-intact subjects, it is possible that a considerable difference would exist, although this remains to be determined.

3 THE NEURAL CIRCUITRY THAT MEDIATES VESTIBULAR INFLUENCES ON CARDIOVASCULAR CONTROL

3.1 Vestibular Nuclei

As discussed in Chapter 2, vestibular afferents predominantly terminate in the vestibular nuclei, which are located dorsally in the pons and medulla. The main vestibular nucleus complex can be differentiated into four nuclei on the basis of cytoarchitecture, as well as on functional and physiological differences: the superior, lateral (or Deiters'), medial, and descending (or inferior) nuclei.[49] Since the vestibular nuclei receive the primary vestibular inputs, one or more of these structures must be involved in mediating vestibulo-sympathetic reflexes.

Several studies have shown that electrolytic lesions placed in regions of the medial vestibular nucleus just caudal to Deiters' nucleus abolish vestibular-elicited activity in the splanchnic or renal nerves.[38,42,48] However, it can be argued that these studies are not definitive, because the lesions could have destroyed fibers of passage (such as vestibular afferents coursing toward other vestibular nucleus regions) in addition to cell bodies at the lesion site. Recently, we have repeated these experiments, with the exception that the lesions were created by injecting the neurotoxin kainic acid into the brainstem. Kainic acid must bind to receptors to produce toxic effects, and it inactivates cell bodies without affecting fibers of passage.[7] The kainic acid injections that eliminated splanchnic nerve responses elicited by electrical or natural vestibular stimulation are shown in Figure 3. Injections into other regions of the vestibular nuclei had little effect on these responses. The results attained using chemical lesioning techniques agree with those obtained using electrolytic lesions, and show that vestibulo-sympathetic responses are mediated by regions of the vestibular nucleus complex just caudal to Deiters' nucleus, including both the medial nucleus and the medial parts of the inferior nucleus.[55,57]

Another approach that has been used to establish which regions of the vestibular nuclei mediate vestibulo-sympathetic reflexes is to determine the locations of neurons that have response characteristics matching those of the reflex. Extensive studies have been made of the response properties of neurons in the medal, inferior, and lateral vestibular nuclei. Figure 4 shows the best direction of tilt (roll or pitch) that activated vestibular nucleus neurons with predominant otolith inputs. Neurons in the *caudal* half of the inferior vestibular nucleus and the adjacent medial nucleus mainly received otolith inputs from afferents activated by roll tilts.[12] The otolith inputs to Deiters' nucleus are similar, in that few cells respond to pitch rotations[23]. In contrast, a majority of cells in the *rostral* parts of the inferior vestibular nucleus and the adjacent medial nucleus receive otolith inputs from afferents that respond better to pitch than to roll rotations;[23,41] thus, many neurons in this region have response properties similar to those of the vestibulo-sympathetic reflex.[57] Since lesions in the same area abolish vestibulo-sympathetic reflexes, there is strong evidence to suggest that these responses are mediated by this specific part of the vestibular nuclei located just caudal to Deiters' nucleus.

3.2 Descending Pathways from the Brainstem to Sympathetic Preganglionic Neurons

As described in Chapter 3, sympathetic outflow to arterioles and the heart arises from preganglionic neurons located predominantly in the intermediolateral gray matter of the thoracic and upper lumbar spinal cord.[26] A number of studies conducted in the cat have sought to determine which descending pathways from the brainstem relay vestibular signals to sympathetic preganglionic neurons. Sympathetic outflow in this species is mainly controlled by spinally projecting neurons from a restricted area of the reticular formation of the rostral ventrolateral medulla located ventral to the retrofacial nucleus, which is referred to as the subretrofacial nucleus.[8] Additional modulatory influences on sympathetic outflow are produced by the caudal medullary raphe nuclei, predominantly raphe obscuruus and pallidus.[27,34,35] More than two-thirds of the spinally projecting neurons from both the subretrofacial nucleus and the caudal medullary raphe nuclei that presumably make connections with sympathetic preganglionic neurons respond to stimulation of the vestibular nerve;[51,60] an example of such a response is shown in the left panel of Figure 5. Raphespinal and subretrofacial neurons respond powerfully to natural vestibular stimulation, as illustrated in the right panel of Figure 5; the labyrinthine inputs to these cells come almost exclusively from otolith organs, and predominantly from receptors activated by pitch rotations.[52,53] Thus, the response properties of neurons in regions of the brainstem that regulate sympathetic outflow are similar to those of the vestibulo-sympathetic reflex, suggesting that these neurons are involved in mediating this reflex.

Injections of the neurotoxin kainic acid into the subretrofacial rostral ventrolateral medulla abolishes splanchnic nerve responses to electrical stimulation of the vestibular nerve, indicating that this area is critically involved in relaying vestibular signals to sympathetic preganglionic neurons.[58] On the other hand, lesions of the raphe nuclei had little effect on vestibulo-sympathetic reflexes,[58] despite the fact that a majority of raphespinal neurons receive vestibular input.[51,53] The best explanation for this apparent discrepancy is that raphespinal neurons with vestibular inputs serve only to regulate the background excitability of sympathetic preganglionic neurons. If this is the case, then it is not unexpected that responses elicited by strong, synchronous electrical stimulation of the vestibular nerve are not dependent on the presence of the raphe nuclei.

3.3 Interneurons That Relay Vestibular Signals to Descending Cardiovascular Regulatory Pathways

Spinally projecting neurons in the subretrofacial rostral ventrolateral medulla and the caudal medullary raphe nuclei respond at latencies >4 ms from electrical stimuli applied to the vestibular nerve.[51,60] Since neurons in the vestibular nuclei are activated at latencies <1.5 ms following electrical stimulation of the vestibular nerve,[49] it seems likely that one or more interneurons are interposed between the vestibular nuclei and cells that make connections with sympathetic preganglionic neurons. Furthermore, injections of retrograde neuroanatomical tracers into the rostral ventrolateral medulla produce only scanty labeling in the vestibular nuclei, confirming that there is little direct vestibular nucleus input to this region.[9] A number of studies have sought to determine which interneurons mediate vestibulo-sympathetic responses.

One potential relay site for vestibular inputs to both the raphe nuclei and the subretrofacial rostral ventrolateral medulla is nucleus tractus solitarius (solitary nucleus). The solitary nucleus receives primary signals from cardiovascular receptors, as well as other visceral receptors,[2,5,22] and thus provides inputs to most other cardiovascular regulatory neurons in the brainstem. Anatomical studies in both the cat and the rabbit have shown that regions of the vestibular nuclei that mediate vestibulo-sympathetic responses have projections to the solitary nucleus.[4,54] Electrophysiological studies confirmed that solitary neurons receive vestibular inputs,[54] but only 5 of 28 cells with baroreceptor inputs responded to vestibular nerve stimulation.[54] The responses of one of these cells is illustrated in Figure 6. The latencies of vestibular nerve effects on solitary neurons were no shorter than for neurons in the subretrofacial rostral ventrolateral medulla or raphe nuclei; thus, neurons in the solitary nucleus are unlikely to mediate the early components of vestibulo-sympathetic reflexes, although they could be involved in producing the longer-latency aspects of these responses.

Several other regions of the brainstem that contain interneurons involved in cardiovascular regulation have also been examined for the presence of vestibular inputs. These regions include both the lateral parvicellular reticular formation of the caudal medulla (lateral tegmental field) and the cell groups adjacent to the superior cerebellar peduncle, which include both the parabrachial and Kölliker-Fuse nuclei. Both the lateral tegmental field and parabrachial-Kölliker-Fuse nuclei receive projections from the regions of the vestibular nuclei that mediate vestibulo-sympathetic responses,[3,50] and both areas contain neurons that respond at short latency to electrical stimulation of the vestibular nerve.[50,59] Lesion studies have also suggested that both the lateral tegmental field and parabrachial nuclei, and particularly the former, play an important role in mediating vestibulo-sympathetic reflexes.[44]

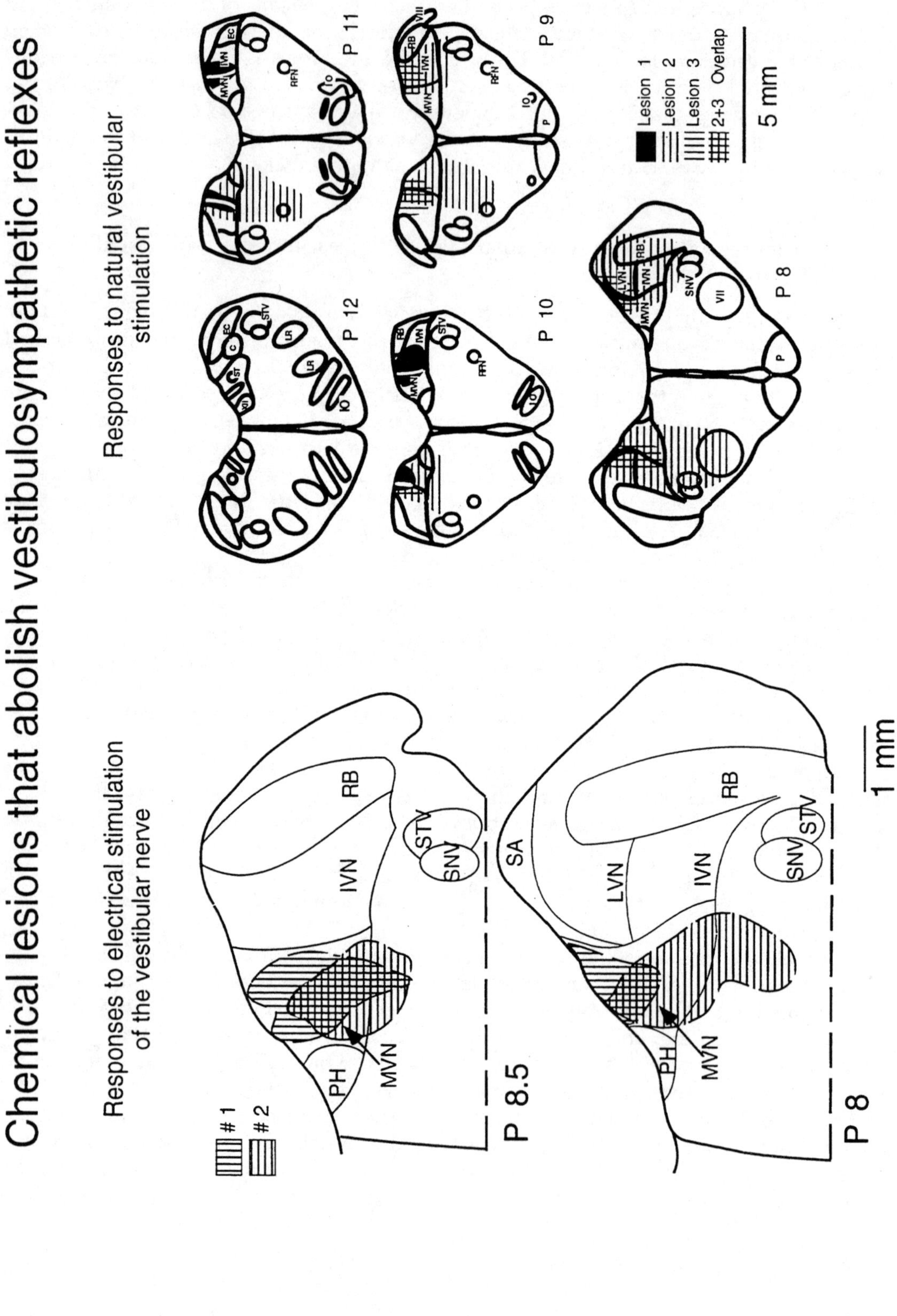
Chemical lesions that abolish vestibulosympathetic reflexes
Responses to electrical stimulation of the vestibular nerve
#1
#2
PH
MVN
IVN
RB
STV
SNV
P 8.5
SA
LVN
P 8
1 mm
Responses to natural vestibular stimulation
P 12
P 11
P 10
P 9
P 8
Lesion 1
Lesion 2
Lesion 3
2+3 Overlap
5 mm

FIGURE 3 (Facing page)
Kainic acid injections into the vestibular nuclei that abolished vestibular-elicited responses recorded from the splanchnic nerve. Fast green dye was dissolved in the kainic acid solution, and the size of the injections is based upon the location of dye in histological sections. The number beside each section indicates its location posterior to stereotaxic zero. Left panel: Injections that abolished responses to electrical stimulation of the vestibular nerve. These lesions were on the same side as the nerve that was stimulated. Data from Reference 55. Right panel: Injections that abolished responses to nose-up tilt of the head (in animals with upper cervical dorsal root rhizotomies). (From Yates, B. J. and Miller, A. D., *J. Neurophysiol.*, 71, 2087, 1994. With permission.)

Abbreviations: C, cuneate nucleus; EC, external cuneate nucleus; IO, inferior olivary nucleus; LR, lateral reticular nucleus; MVN, medial vestibular nucleus; IVN, inferior vestibular nucleus; LVN, lateral vestibular nucleus; P, pyramid; PH, prepositus hypoglossi; RB, restiform body; RFN, retrofacial nucleus; SA, stria acoustica; SNV, spinal trigeminal nucleus; ST, solitary tract; STV, spinal trigeminal tract; VII, facial nucleus; VIII, vestibulo-cochlear nerve; XII, hypoglossal nucleus.

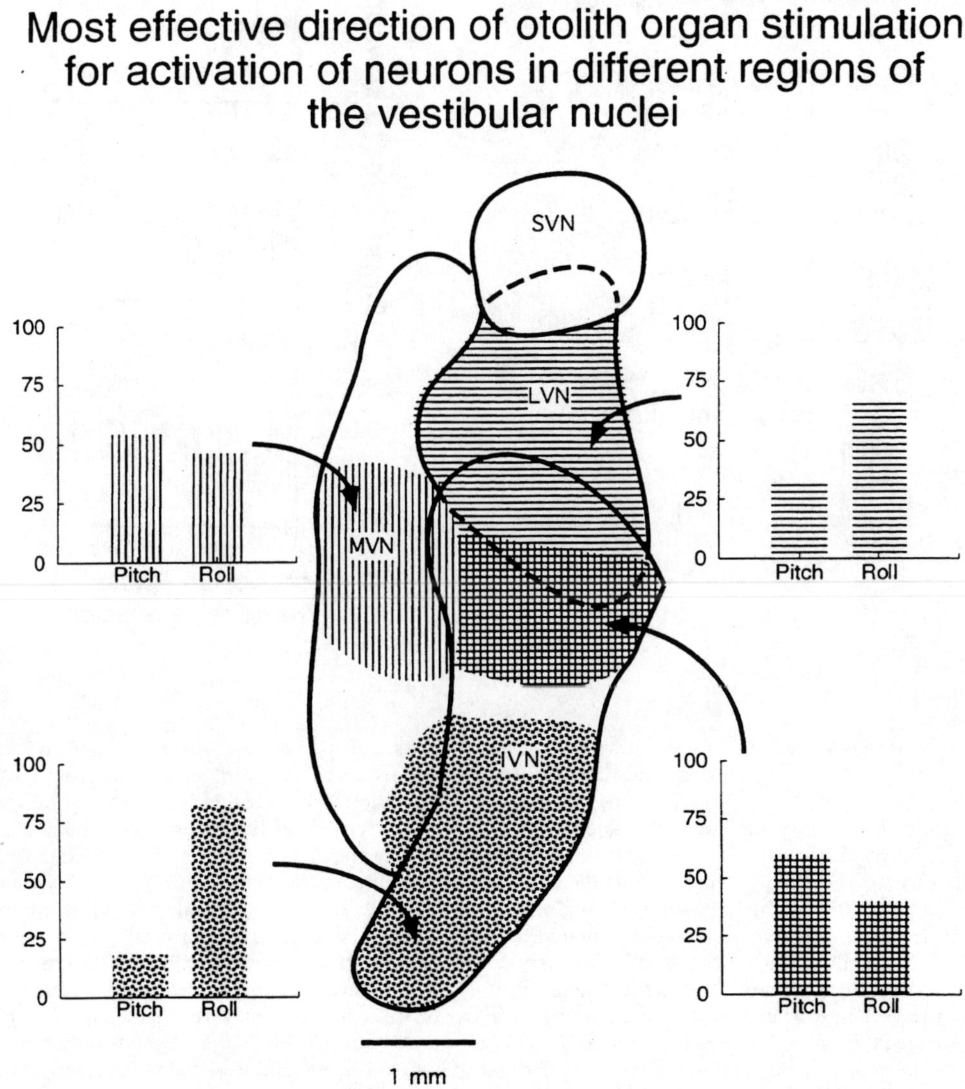

FIGURE 4
Most effective direction of vertical tilt for activating neurons having a predominant otolith input, shown for different regions of the vestibular nuclei. The histograms indicate the percentage of neurons in each region that responded better to tilt in the roll or pitch planes. The regions of the vestibular nuclei that were studied are shown on a horizontal section through the nuclei. Abbreviations: IVN, inferior vestibular nucleus; LVN, lateral vestibular nucleus; MVN, medial vestibular nucleus; SVN, superior vestibular nucleus.

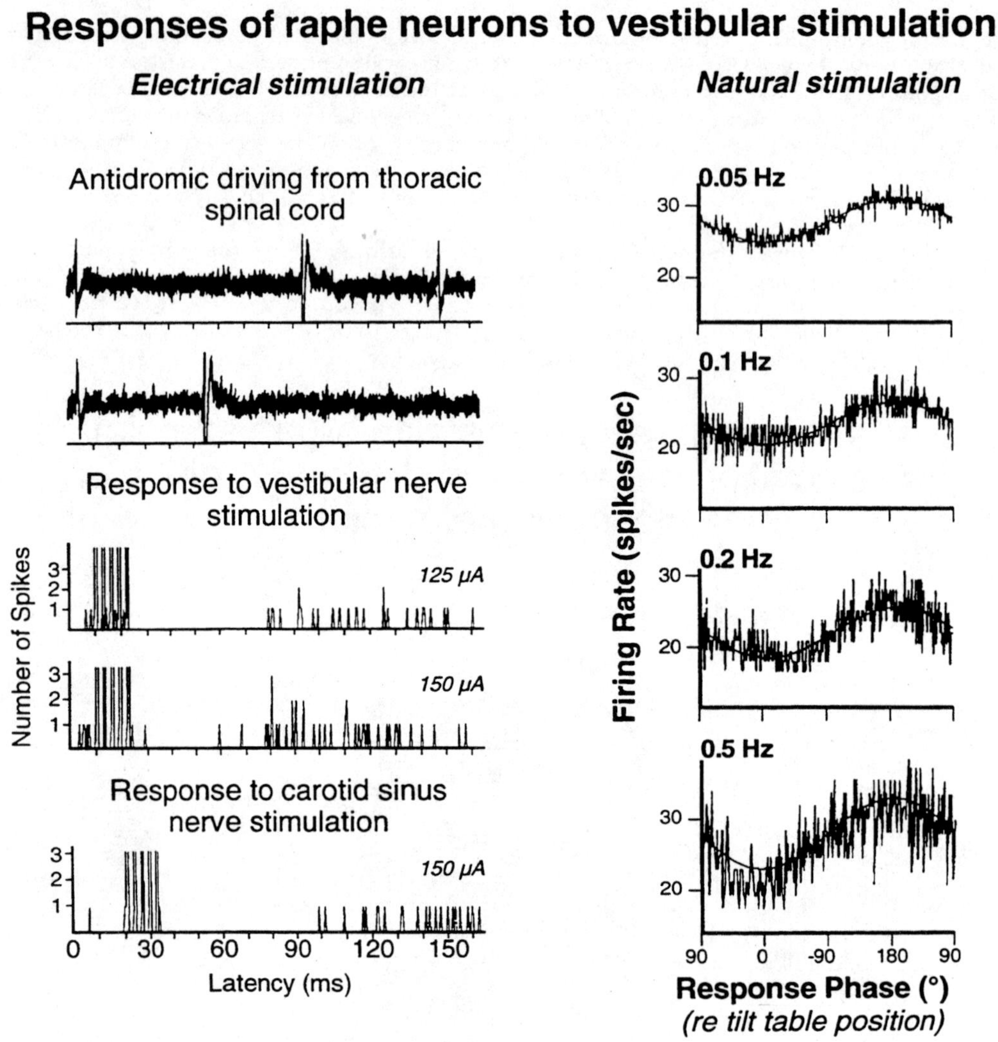

FIGURE 5
Responses of two raphespinal neurons with vestibular inputs. Left panel: Responses of a neuron located in raphe pallidus to electrical stimulation of the vestibular nerve, carotid sinus nerve, and upper thoracic spinal cord. The top traces show collision of spikes elicited by stimulation of the white matter of the upper thoracic spinal cord with spontaneous spikes, proving that the neuron has a projection to the spinal cord. Each trace is composed of 10 overlapping sweeps. The neuron had a slow conduction velocity (2.1 m/s), which is typical of raphe neurons that influence sympathetic outflow.[34,35] The middle traces are poststimulus histograms showing effects of two intensities of vestibular stimulation; shock artifacts indicate the five pulses that elicited the responses. The unit's firing rate was powerfully inhibited by stimulation of the vestibular nerve. The bottom trace shows inhibition of the neuron produced by stimulation of the carotid sinus nerve. The presence of inputs from the carotid sinus suggests that the raphespinal neuron participates in cardiovascular control. (From Yates, B. J., Goto, T., and Bolton, P. S., *Exp. Brain Res.*, 89, 323, 1992. With permission.) Right panel: Responses of another raphespinal neuron (conduction velocity of 7.5 m/s) to sinusoidal natural vestibular stimulation incorporating both roll and pitch. The traces are averages of binned spike data indicating the response to several frequencies of tilt; the stimulus frequency tested is indicated above each waveform. Superimposed on each trace is the sinusoid fit to the fundamental of the response. The stimulus amplitude was 7.5° in all runs. The response phase was within 30° of table position at all frequencies, as indicated by a response phase near 180° for all averages. In addition, the response gain increased by a factor of only 1.6 over this decade range of frequencies. This behavior is similar to that of otolith afferents.[14] (From Yates, B. J. et al., *J. Neurophysiol.*, 70, 938, 1993. With permission.)

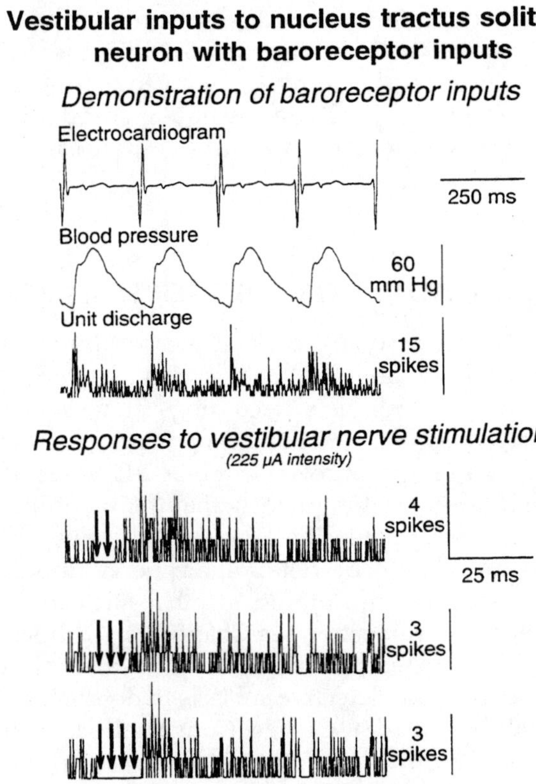

FIGURE 6
Responses of a baroreceptor-sensitive neuron in nucleus tractus solitarius to stimulation of the vestibular nerve. The top three traces show the sensitivity of the unit to baroreceptor inputs. By triggering from the electrocardiogram (top trace), blood pressure fluctuations during the cardiac cycle were entrained (second trace). The unit's activity, shown in the third trace, was related to these blood pressure fluctuations. The bottom three traces show the responses to three different train lengths of vestibular nerve stimulation. Arrows indicate when the shocks were delivered. (From Yates, B. J. et al., *Am. J. Physiol.*, 267, R974, 1994. With permission.)

3.4 Regions Rostral to the Brainstem

Powerful vestibulo-sympathetic reflexes can be elicited in decerebrate animals,[55,57] so it is clear that neural structures higher than the brainstem are not required to produce these responses. Furthermore, decerebration has little effect on vestibular-elicited activity recorded from sympathetic nerves of chloralose-anesthetized cats.[20,42] This is not to say, however, that higher centers play no role in vestibulo-sympathetic responses, particularly in awake, anesthetized animals. There is some evidence to suggest, for example, that regions of the hypothalamus that regulate cardiovascular functioning (the lateral, paraventricular, and supraoptic nuclei) receive vestibular inputs.[1,24] Some cells in the paraventricular and supraoptic hypothalamus are responsible for the release of vasopressin into the blood stream, which produces long-term increases in blood pressure.[18] Thus, it is conceivable that vestibular stimulation could result in hormonal as well as neural effects that act to stabilize blood pressure during movement and changes in posture. This possibility is yet to be explored in detail.

3.5 The Cerebellum

The role of the cerebellum in regulation of autonomic functions will be considered thoroughly in Chapter 8. There is considerable evidence to suggest that the cerebellum is involved in some aspects of cardiovascular control, and perhaps in vestibulo-sympathetic reflexes. However, the cerebellum is not required for these responses to occur in decerebrate animals,[55,57] and cerebellectomy does not alter the spatial and temporal properties of sympathetic responses to natural vestibular stimulation.[57] Nonetheless, the role of the cerebellum in vestibulo-cardiovascular responses could be important in awake animals.

4 DIRECTIONS FOR FURTHER RESEARCH

There is considerable evidence to suggest that the vestibular system participates in cardiovascular control, but further data must be obtained to determine the significance of these responses in counteracting orthostasis. Recordings from single sympathetic neurons identified as mediating cardiovascular function (as opposed to whole sympathetic nerves containing fibers subserving a variety of roles) and blood flow recordings during natural vestibular stimulation will be essential in proving that the vestibular system participates in compensating for orthostatic hypotension. The role of inputs from neck afferents in cardiovascular control is also yet to be determined; neck inputs are integrated with vestibular signals in the vestibular nuclei and at other sites to help distinguish head movements from whole-body movements[49] (see Chapter 2). Additional experiments must also be conducted in humans to determine the clinical relevance of this research. It would be interesting to look at labyrinthine-defective subjects to determine their susceptibility to orthostasis following a sudden postural change. Conversely, it would be informative to study patients with idiopathic orthostatic hypotension to determine if they have subtle vestibular deficits that might be responsible for this condition.

5 SUMMARY

There is considerable evidence from animal models, as well as some data from humans, to suggest that the vestibular system has influences on the sympathetic nervous system and cardiovascular control. The inputs that appear to be critical in producing these responses come from otolith receptors that are activated by pitch. The circuitry that mediates vestibulo-sympathetic reflexes has been studied extensively, and is summarized in Figure 7. These responses are mediated by (1) a restricted part of the inferior and medial vestibular nuclei located just caudal to Deiters' nucleus; (2) interneurons in the solitary nucleus, lateral tegmental field, and lateral parabrachial nuclei; and (3) neurons in the subretrofacial rostral ventrolateral medulla that make connections with sympathetic preganglionic neurons in the spinal cord. However, further experiments must be conducted to more fully establish the physiological and clinical significance of vestibular actions on the neural circuitry that regulates circulation.

ACKNOWLEDGMENTS

The author's research is supported by NIH grant DC 00693.

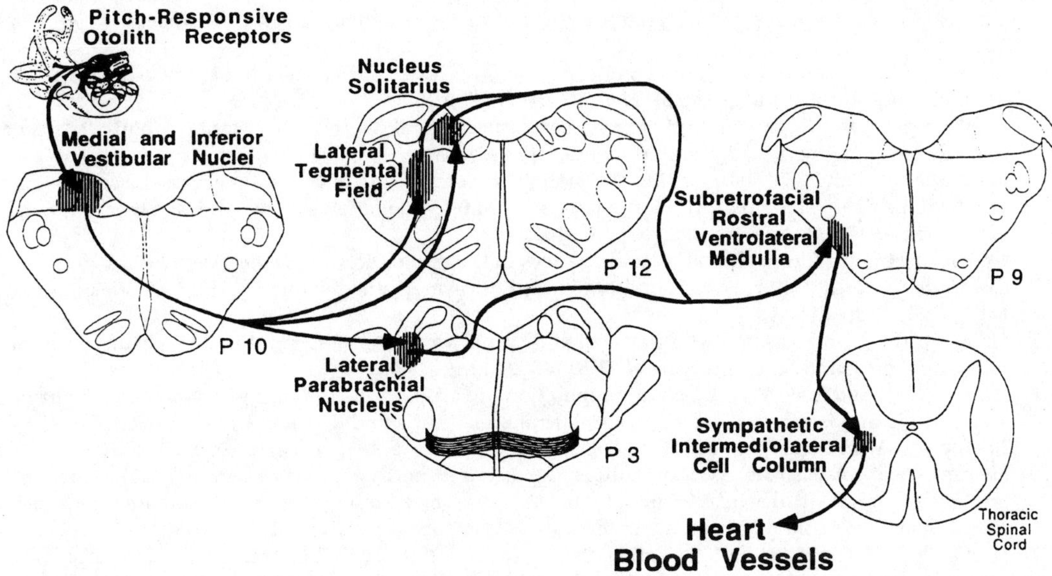

FIGURE 7
The neural pathways that mediate vestibulo-sympathetic reflexes. These responses are elicited by pitch rotations that activate otolith afferents, and are mediated by interneurons in the lateral tegmental field, nucleus solitarius, and the lateral parabrachial nuclei with effects on spinally projecting neurons in the subretrofacial rostral ventrolateral medulla.

REFERENCES

1. Azzena, G. B., Melis, F., Caria, M. A., Teatini, G. P., and Bozzo, G., Vestibular projections to hypothalamic supraoptic and paraventricular nuclei, *Arch. Ital. Biol.*, 131, 127, 1993.
2. Altschuler, S. M., Bao, X., Bieger, D., Hopkins, D. A., and Miselis, R. R., Viscerotopic representation of the upper alimentary tract in the rat: sensory ganglia and nuclei of the solitary and spinal trigeminal tracts, *J. Comp. Neurol.*, 283, 248, 1989.
3. Balaban, C. D., Vestibular nucleus projections to the parabrachial nucleus in rabbits: implications for vestibulo-autonomic interactions, *Assoc. Res. Otolaryngol. Abstr.*, 101, 1995.
4. Balaban, C. D. and Beryozkin, G., Vestibular nucleus projections to nucleus tractus solitarius and the dorsal motor nucleus of the vagus nerve: potential substrates for vestibulo autonomic interactions, *Exp. Brain Res.*, 98, 200, 1994.
5. Berger, A. J., Distribution of carotid sinus nerve afferent fibers to solitary tract nuclei of the cat using transganglionic transport of horseradish peroxidase, *Neurosci. Lett.*, 14, 153, 1979.
6. Cobbold, A. F., Megirian, D., and Sherrey, J. H., Vestibular evoked activity in autonomic motor outflows, *Arch. Ital. Biol.*, 106, 113, 1968.
7. Coyle, J. T. and Schwartz, R., The use of excitatory amino acids as selective neurotoxins, in *Handbook of Chemical Neuroanatomy*, Björklund, A. and Hökfelt, T., Eds., Elsevier, Amsterdam, 1983, chap. 13.
8. Dampney, R. A. L., Functional organization of central pathways regulating the cardiovascular system, *Physiol. Rev.*, 74, 323, 1994.
9. Dampney, R. A. L., Czachurski, J., Dembowsky, K., Goodchild, A. K., and Seller, H., Afferent connections and spinal projections of the pressor region in the rostral ventrolateral medulla of the cat. *J. Autonom. Nerv. Syst.*, 20, 73, 1987.
10. Davenport, P. W. and Thompson, F. J., Mechanosensitive afferents of femoral-saphenous vein, *Am. J. Physiol.*, 252, R367, 1987.
11. Doba, N. and Reis, D. J., Role of the cerebellum and vestibular apparatus in regulation of orthostatic reflexes in the cat, *Circ. Res.*, 34, 9, 1974.

12. Endo, K., Thomson, D. B., Wilson, V. J., Yamaguchi, T., and Yates, B. J. Vertical vestibular input to and projections from the caudal vestibular nuclei of the decerebrate cat, *J. Neurophysiol.*, 74, 428, 1995.
13. Essandoh, L. K., Duprez, D. A., and Shepherd, J. T., Reflex constriction of human resistance vessels to head-down neck flexion, *J. Appl. Physiol.*, 64, 767, 1988.
14. Fernández, C. and Goldberg, J. M., Physiology of peripheral neurons innervating otolith organs of the squirrel monkey. III. Response dynamics, *J. Neurophysiol.*, 39, 996, 1976.
15. Fernández, C. and Goldberg, J. M., Physiology of peripheral neurons innervating semicircular canals of the squirrel monkey. II. Response to sinusoidal stimulation and dynamics of the peripheral vestibular system, *J. Neurophysiol.*, 34, 661, 1971.
16. Gauer, O. H. and Thron, H. L., Postural changes in the circulation, in *Handbook of Physiology, Sec. 2: Circulation. Vol. III*, Dow, P. and Hamilton, W. F., Eds., American Physiological Society, Washington, D.C., 1965, chap. 19.
17. Hainsworth, R. and Karim, F., Response of abdominal vascular capacitance in the anesthetized dog to change in carotid sinus pressure, *J. Physiol.*, 262, 659, 1976.
18. Harris, M. C. and Loewy, A. D., Neural regulation of vasopressin-containing hypothalamic neurons and the role of vasopressin in cardiovascular function, in *Central Regulation of Autonomic Functions*, Loewy, A. D. and Spyer, K. M., Eds., Oxford University Press, New York, 1990, chap. 13.
19. Hemmingway, A., Cardiovascular changes in motion sickness, *J. Aviation Med.*, 16, 417, 1945.
20. Ishikawa, T. and Miyazawa, T., Sympathetic responses evoked by vestibular stimulation and their interactions with somato-sympathetic reflexes, *J. Autonom. Nerv. Syst.*, 1, 243, 1980.
21. Ishikawa, T., Miyazawa, T., Shimizu, I., and Tomita, H., Similarity between vestibulo-sympathetic response and supraspinal sympathetic reflex, *Nihon Univ. J. Med.*, 21, 201, 1979.
22. Kalia, M. and Mesulam, M.-M., Brain stem projections of sensory and motor components of the vagus complex in the cat. II. Laryngeal, tracheobronchial, pulmonary, cardiac and gastrointestinal branches, *J. Comp. Neurol.*, 193, 467, 1980.
23. Kasper, J., Schor, R. H., and Wilson, V. J., Response of vestibular neurons to head rotations in vertical planes. I. Response to vestibular stimulation, *J. Neurophysiol.*, 60, 1753, 1988.
24. Katafuchi, T., Puthuraya, K. P., Yoshimatsu, H., and Oomura, Y., Responses of rat lateral hypothalamic neuron activity to vestibular nuclei stimulation, *Brain Res.*, 400, 62, 1987.
25. Lloyd, T. C., Effect of inspiration on inferior vena caval blood flow in dogs, *J. Appl. Physiol.*, 55, 1701, 1983.
26. Loewy, A. D., Central autonomic pathways, in *Central Regulation of Autonomic Functions*, Loewy, A. D. and Spyer, K. M., Eds., Oxford University Press, New York, 1990, chap. 6.
27. Loewy, A. D., Raphe pallidus and obscurus projections to the intermediolateral cell column in the rat, *Brain Res.*, 222, 129, 1981.
28. Ludbrook, J., The musculovenous pumps of the human lower limb, *Am. Heart J.*, 71, 635, 1966.
29. Megirian, D. and Manning, J. W., Input-output relations in the vestibular system, *Arch. Ital. Biol.*, 105, 15, 1967.
30. Miller, A. D., Yamaguchi, T., Siniaia, M. S., and Yates, B. J., Ventral respiratory group bulbospinal inspiratory neurons participate in vestibular-respiratory reflexes, *J. Neurophysiol.*, 73, 1303, 1995.
31. Mittelstaedt, H., Somatic versus vestibular gravity reception in man, *Ann. N.Y. Acad. Sci.*, 656, 124, 1992.
32. Mixter, G., Respiratory augmentation of inferior venal caval flow demonstrated by a low-resistance phasic flowmeter, *Am. J. Physiol.*, 172, 446, 1953.
33. Moreno, A. H., Burchell, A. R., and Burke, J. H., Respiratory regulation of splanchnic and systemic venous return, *Am. J. Physiol.*, 213, 455, 1967.
34. Morrison, S. F. and Gebber, G. L., Classification of raphe neurons with cardiac-related activity, *Am. J. Physiol.*, 243, R49, 1982.
35. Morrison, S. F. and Gebber, G. L., Raphe neurons with sympathetic-related activity: baroreceptor responses and spinal connections, *Am. J. Physiol.*, 246, R338, 1984.
36. Ohashi, N., Imamura, J., Nakagawa, H., and Mizukoshi, K., Blood pressure abnormalities as background roles for vertigo, dizziness and disequilibrium, *Otorhinolaryngology*, 52, 355, 1990.
37. Paintal, A. S., Cardiovascular receptors, in *Handbook of Sensory Physiology. Vol. III/1: Enteroceptors*, Neil, E., Ed., Springer-Verlag, Berlin, 1972, chap. 1.
38. Pan, P. S., Zhang, Y. S., and Chen, Y. Z., Role of nucleus vestibularis medialis in vestibulo-sympathetic response in rats, *Acta Physiol. Sin.*, 43, 184, 1991.
39. Preber, L., Vegetative reactions in caloric and rotatory tests, *Acta Otolaryngol.*, Suppl. 144, 1, 1958.
40. Rushmer, R. F., *Cardiovascular Dynamics*, 4th ed., W. B. Saunders, Philadelphia, 1976, chap. 6.
41. Schor, R. H., Steinbacher, B. C., and Yates, B. J., Response of medial vestibular nucleus neurons to horizontal linear and angular stimulation of the decerebrate cat, *Soc. Neurosi. Abstr.*, 21, 1913, 1995.

42. Spiegel, E. A., Effect of labyrinthine reflexes on the vegetative nervous system, *Arch. Otolaryngol.*, 44, 61, 1946.
43. Spyer, K. M., The central nervous organization of reflex cardiovascular control, in *Central Regulation of Autonomic Functions*, Loewy, A. D. and Spyer, K. M., Eds., Oxford University Press, New York, 1990, chap. 10.
44. Steinbacher, B. C. and Yates, B. J., Brainstem interneurons necessary for vestibular influences on sympathetic outflow, *Soc. Neurosci. Abstr.*, 21, 872, 1995.
45. Takata, M. and Robotham, J. L., Effects of diaphragmatic descent on inferior vena caval venous return, *J. Appl. Physiol.*, 72, 597, 1992.
46. Thompson, F. J., Barnes, C. D., and Wald, J. R., Interactions between femoral venous afferents and lumbar spinal reflex pathways, *J. Autonom. Nerv. Syst.*, 6, 113, 1982.
47. Thompson, F. J., Yates, B. J., Franzen, O., and Wald, J. R., Lumbar spinal cord responses to limb vein distention, *J. Autonom. Nerv. Syst.*, 9, 531, 1983.
48. Uchino, Y., Kudo, N., Tsuda, K., and Iwamura, Y., Vestibular inhibition of sympathetic nerve activities, *Brain Res.*, 22, 195, 1970 .
49. Wilson, V. J. and Melvill Jones, G., *Mammalian Vestibular Physiology*, Plenum Press, New York, 1979, chap. 5.
50. Yates, B. J., Balaban, C. D., Miller, A. D., Endo, K., and Yamaguchi, Y., Vestibular inputs to the lateral tegmental field of the cat: potential role in autonomic control, *Brain Res.*, 689, 197, 1995.
51. Yates, B. J., Goto, T., and Bolton, P. S., Responses of neurons in the caudal medullary raphe nuclei of the cat to stimulation of the vestibular nerve, *Exp. Brain Res.*, 89, 323, 1992.
52. Yates, B. J., Goto, T., and Bolton, P. S., Responses of neurons in the rostral ventrolateral medulla of the cat to natural vestibular stimulation, *Brain Res.*, 601, 255, 1993.
53. Yates, B. J., Goto, T., Kerman, I., and Bolton, P. S., Responses of caudal medullary raphe neurons to natural vestibular stimulation, *J. Neurophysiol.*, 70, 938, 1993.
54. Yates, B. J., Grélot, L., Kerman, I. A., Balaban, C. D., Jakuš, J., and Miller, A. D., Organization of vestibular inputs to nucleus tractus solitarius and adjacent structures in cat brain stem, *Am. J. Physiol.*, 267, R974, 1994.
55. Yates, B. J., Jakuš, J., and Miller, A. D., Vestibular effects on respiratory outflow in the decerebrate cat, *Brain Res.*, 629, 209, 1993.
56. Yates, B. J., Mickle, J. P., Hedden, W. J., and Thompson, F. J., Tracing of afferent pathways from the femoral-saphenous vein to the dorsal root ganglia using transport of horseradish peroxidase, *J. Autonom. Nerv. Syst.*, 20, 1, 1987.
57. Yates, B. J. and Miller, A. D., Properties of sympathetic reflexes elicited by natural vestibular stimulation: implications for cardiovascular control, *J. Neurophysiol.*, 71, 2087, 1994.
58. Yates, B. J., Siniaia, M. S., and Miller, A. D., Descending pathways necessary for vestibular influences on sympathetic and inspiratory outflow, *Am. J. Physiol.*, 268, R1381, 1995.
59. Yates, B. J. and Steinbacher, B.C., unpublished data.
60. Yates, B. J., Yamagata, Y., and Bolton, P. S., The ventrolateral medulla of the cat mediates vestibulosympathetic reflexes., *Brain Res.*, 552, 265, 1991.

7

VESTIBULAR EFFECTS ON RESPIRATORY ACTIVITY

Alan D. Miller and Bill J. Yates

CONTENTS

0-8493-7668-1/96/$0.00+$.50

1 INTRODUCTION

The vestibular system produces a number of compensatory responses to accelerations imposed on the head, as reviewed in Chapter 2. Among these are vestibular influences on the respiratory system. As discussed below, the vestibular system has actions on all the major "pump" muscles of respiration: the diaphragm (the principal muscle of inspiration), the abdominal muscles (the major muscles of expiration), and the intercostal muscles (which have both inspiratory and expiratory roles). In addition, the vestibular system affects the "accessory" respiratory muscles of the upper airway, which maintain airway patency and act as "valves" that help regulate air flow. The functions of these muscle groups in the control of breathing were described in Chapter 4. The roles of vestibular effects on respiration are not well understood, mainly because the respiratory muscles are multifunctional and subserve several tasks in addition to breathing, including maintaining posture and assisting in venous return to the heart.

This chapter will review the evidence suggesting that the vestibular system has powerful influences on respiratory muscles, discuss the possible functions of vestibulo-respiratory connections, and describe the current knowledge of the neural circuitry that relays vestibular signals to respiratory motoneurons.

2 EVIDENCE FOR VESTIBULAR EFFECTS ON THE RESPIRATORY SYSTEM

There are three main types of evidence indicating that the vestibular system has influences on the respiratory system: (1) observations following changes in body position, (2) responses to stimulation and lesion of the vestibular nerve or nuclei, and (3) neuroanatomical studies.

2.1 Changes in Respiratory Muscle Activity During Movement and Changes in Posture

It is well established that alterations in thoracic and abdominal respiratory muscle activity accompany changes in body position in both humans and animals.[13,16,22,32,48,49,59] However, caution must be used in assessing the extent to which these responses are due to vestibular input. For example, the increase in abdominal muscle activity induced by static head-up tilt can be greatly reduced or abolished by cervical vagotomy or by blocking slowly adapting pulmonary mechanoreceptors, and thus it could be mainly of non-vestibular origin.[12,31,32,48] Abdominal muscle stretch receptors also probably contribute to the increased activity of these muscles observed during head-up body tilt.[12,13,48]

However, not all of the changes in respiratory muscle activity during movement and changes in posture can be attributed to non-vestibular receptors. For example, tilt-induced changes in the activity of the diaphragm, parasternal intercostal, and triangularis sterni muscles are largely independent of vagal input.[14,32] Increases in the depth and frequency of respiration produced by vertical linear acceleration (up and down movements), which are observed in humans[10] and animals,[56] are markedly reduced by transection of the vestibular nerve, suggesting that they are mainly due to activation of labyrinthine receptors.[56]

Studies using natural vestibular stimulation have also reported that the vestibular system can influence accessory respiratory muscles of the upper airway. The activity of hypoglossal nucleus neurons is modulated both by static ear-down tilts (roll), which activate the utricle (one of the otolith organs; see Chapter 2), and by thermal (caloric) stimulation of the labyrinth, which affects mainly the semicircular canals.[42] In addition, the activity of laryngeal and tongue muscles is influenced by static head tilt, a stimulus that has little effect on the activity of the diaphragm.[6] However, the responses elicited by head tilts could also arise from activation of neck receptors or mechanoreceptors in the upper airway in addition to labyrinthine receptors.

This paucity of data regarding the responses of respiratory muscles to natural vestibular stimulation does not necessarily indicate that vestibular effects on the respiratory muscles are absent or weak. Only a few studies have attempted to examine vestibulo-respiratory reflexes, mainly because these responses are difficult to investigate. In order to appropriately study vestibular effects on breathing, natural vestibular stimulation in a variety of directions must be utilized in preparations in which non-labyrinthine receptors that could be activated by these stimuli are denervated. Preliminary data collected by the authors suggest that vertical tilts combining ear-down (roll) and nose-up and -down (pitch) rotations in cats with vagotomies and upper airway denervations produce modulation of the activity of phrenic and abdominal nerves. Further studies are required, however, to determine which particular vestibular receptors (semicircular canals or otolith organs) are responsible for these effects and which direction of tilt is most effective in eliciting the responses.

2.2 Changes in Respiratory Neuron Activity Produced by Electrical Activation and Lesion of Vestibular Inputs

Electrical stimulation of the vestibular nerve can be used to selectively excite vestibular afferents, without inadvertently eliciting inputs from other sources. The major drawback in using this stimulus paradigm is that a non-physiological signal is delivered to the brainstem; in effect, the "message" that is sent indicates that the head is simultaneously moving in all directions. Thus, whereas electrical-stimulation studies are useful in selectively mapping vestibular connections in the nervous system, they are not advantageous in determining the role of the responses.

Electrical stimulation of the vestibular nerve affects the discharge of a number of respiratory-related motor neurons, as has been shown by recording from the phrenic (diaphragmatic),[28,64] intercostal,[64] abdominal,[64] and recurrent laryngeal[1,43] nerves and from hypoglossal nucleus neurons[42] and tongue protrusive muscles.[21] Examples of potentials recorded from phrenic and abdominal nerves following vestibular stimulation are shown in Figure 1. Vestibular-evoked responses were recorded bilaterally from respiratory nerves at relatively short latencies of 10–25 ms.[1,21,28,43,64] Thus, pathways involving only a few synapses interconnect vestibular afferents with respiratory motoneurons. Control studies have demonstrated that respiratory responses to vestibular nerve stimulation are due to activation of vestibular afferents, and not to stimulus spread to non-target nerves.[21,28,64]

The amplitude of respiratory nerve responses to vestibular nerve stimulation is reasonably large: 15% of the amplitude of spontaneous respiratory discharge in the case of the phrenic nerve and 100% in the case of the abdominal nerve.[64] The vestibular-evoked responses recorded from the phrenic and abdominal nerves were also 20–30% of the amplitude of responses elicited by stimulation of the sciatic nerve at intensities sufficient to activate unmyelinated C-fibers.[64] Thus, the vestibular system appears to be capable of producing substantial changes in the excitability of respiratory motoneurons.

In addition to evoking activity recordable from respiratory nerves, electrical stimulation of the vestibular system can lead to changes in spontaneous respiratory discharge. For example, stimulation of the vestibular nuclei can lead to an early onset of phrenic inspiratory discharge and to an entrainment of the respiratory rhythm.[5] Furthermore, the discharge of both expiratory and inspiratory spinal nerves can be modified by stimulation or lesion of the vestibular nuclei.[36]

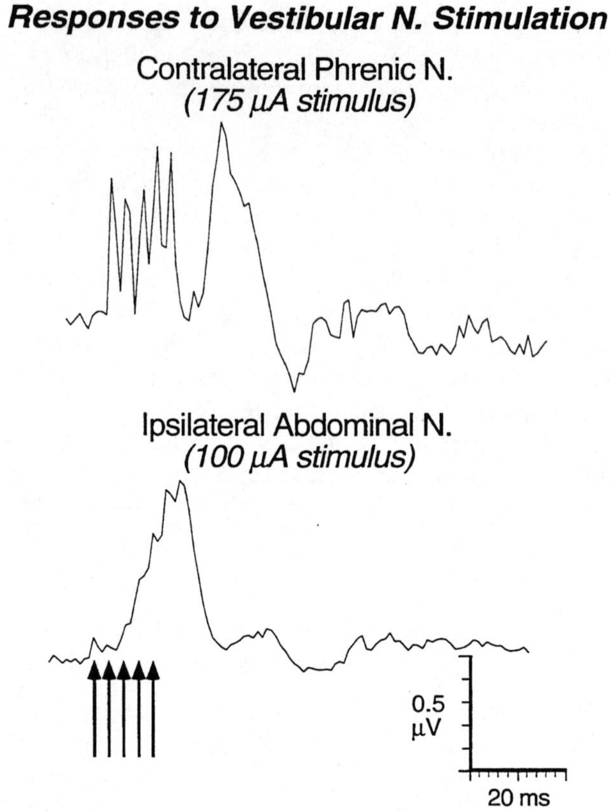

FIGURE 1
Vestibular nerve-elicited responses recorded from the phrenic and abdominal nerves of two cats. The responses were produced by a train of five shocks delivered to the vestibular nerve in the periphery; the timing of the stimuli is indicated by arrows beneath the bottom trace, as well as by shock artifacts in the top trace. The stimulus intensities delivered to the vestibular nerve were approximately half those required to produce current spread to the closest non-target nerve, the facial nerve. The phrenic and abdominal nerve responses are the average of 330 and 378 sweeps, respectively. This amount of averaging was sufficient to cancel out spontaneous rhythmic nerve discharges, so that the responses could be clearly resolved.

2.3 Neuroanatomical Studies Indicating the Presence of Vestibulo-Respiratory Connections

Neuroanatomical studies involving injections of the retrograde tracer horseradish peroxidase (HRP) into motor pools supplying the diaphragm,[19,50,51,53] intercostal,[54] and abdominal[44,54] muscles label projections from the vestibular nuclei. These studies must be interpreted with caution, however, since respiratory motoneurons are located near motoneurons with other functions. For example, the phrenic motor nucleus is adjacent to motoneurons that innervate shoulder muscles,[35] and labeling in the vestibular nuclei following injection of anatomical tracers into the vicinity of respiratory motoneurons could be due to pick up of these tracers by projections to other cells.

The potential methodological limitations presented by classical neuroanatomical methods can be circumvented using a new technique: the trans-synaptic transport of pseudorabies virus. In a recent study, this technique was used to localize neurons that make connections with phrenic motoneurons in the rat.[15] No retrograde labeling was observed in the vestibular nuclei even though numerous other bulbospinal neurons were labeled, as were propriobulbar neurons. However, it is possible that different results would have been obtained if another species had been used in this study. Vestibular influences on the respiratory muscles may not be as critical in rats as in some other species, because these animals dwell on the ground and may not place the same demands on respiratory muscles as animals with a more dynamic behavioral repertoire. Furthermore, the rat lacks the ability to vomit[7] and is unlikely to have specialized connections that might contribute to control of respiratory motoneurons during motion-induced vomiting.

3 POSSIBLE PHYSIOLOGICAL ROLES

Since the vestibular system signals when movement and changes in posture are occurring, it seems reasonable to assume that vestibular influences on respiration serve to counteract movement-related changes in homeostasis. Movement can compromise the maintenance of stable blood oxygenation and blood pressure, and changes in the activity of respiratory muscles can assist in correction of these disturbances. Furthermore, enhanced activity of some respiratory muscles contributes directly to the stabilization of posture. These conditions are described in more detail below.

3.1 Mechanical Constraints on Respiration Imposed by Postural Changes

Movement can produce mechanical constraints on the process of respiration; however, alveolar ventilation remains relatively constant as the activity of respiratory muscles is rapidly altered to produce compensation.[2,6,8,13,14,16,22,32,34,49,59] The most dramatic challenges to respiration in both quadrupeds and humans result from changing from a supine to a head-up posture. Standing in humans or nose-up pitch in quadrupeds results in the diaphragm descending into the abdominal cavity, due to the force of gravity and the fact that this muscle is no longer as well supported by the abdominal viscera. As a result, the resting length of the diaphragm is altered considerably.[39,49,59]

A complex series of changes in the activity of respiratory muscles occurs during changes from supine to head-up posture. As would be expected, activity of muscle fibers increases in the diaphragm to compensate for the muscle's reduced mechanical advantage in the upright posture,[16,49,59] with the greatest effects in the crural region.[59] Another important mechanism to counteract gravity effects is to produce increased intra-abdominal pressure through the activity of the abdominal muscles.[16,23-26] In particular, the drive on one of the abdominal muscles, *transversis abdominis,* seems to be particularly important in offsetting the effects of head-up posture.[23] Thus, selected muscles play a predominant role in correcting for the postural effects on respiration, indicating that very specific changes occur in the excitability of the brainstem circuitry that regulates breathing.

The adjustments in respiratory muscle activity during postural changes typically occur very rapidly, within the first breath, suggesting that these responses are independent of blood oxygenation.[4,32,33,52] Furthermore, this reflex is maintained in patients with complete transection of the lower cervical spinal cord[4] and those with heart-lung transplants,[37] indicating that proprioceptors located in the rib cage and abdominal wall (the afferents of which enter the thoracic and lumbar spinal cord) and vagal pulmonary receptors are not required. Cervical vagotomy together with transection of cervical dorsal roots to cut phrenic nerve afferents, as well as vagotomy plus lower cervical cordotomy, do not abolish the rapid changes in respiratory muscle activity during postural change in dogs.[9] Thus, inputs from sources in addition to pulmonary receptors and receptors in the respiratory muscles are likely to play a role in correcting the effects of gravity on respiration. The most likely candidate is labyrinthine receptors, although this is yet to be demonstrated experimentally.

3.2 Role of the Respiratory Muscles in Maintaining Venous Return During Postural Changes

Changing from a supine to a head-up posture results in reduced venous return to the heart, which can produce orthostatic hypotension if rapid corrections in blood pressure are not executed. Compensation for postural hypotension is accomplished mainly by increasing sympathetic outflow to the heart and arterioles (reviewed in Chapter 6). However, the respiratory muscles also play an important role in cardiac return. Contraction of the diaphragm produces negative intra-thoracic pressure, which pulls blood into the chest from the abdomen, whereas abdominal muscle contraction pushes blood into the thorax.[40,46,47,57,66] Thus, it is feasible that vestibulo-respiratory reflexes work in concert with vestibulo-sympathetic responses in regulating blood pressure during postural changes.

3.3 Role of the Respiratory Muscles in Producing Movement

Respiratory muscles are skeletal muscles, and their contraction can lead to movement and postural stabilization. Among the respiratory muscles, the abdominal muscles are the most involved in the execution of movements, as they are utilized in the flexion and rotation of the trunk.[11,13,41,58] Furthermore, the abdominal muscles have a postural role, particularly during standing in humans, as they serve to counteract trans-abdominal pressure.[13,41] The intercostal muscles also have a secondary "movement" function, as they are involved in rotation of the thorax.[60] Even the diaphragm has a role in postural control; the only muscle that spans the body cavity, it can assist in stabilizing the trunk.[34] Thus, in some ways vestibulo-respiratory reflexes may function similar to other vestibulo-spinal reflexes, in that they act to stabilize posture and produce compensatory movements to perturbations.

3.4 Importance of the Vestibular System in Providing Facilitatory Inputs to Respiratory Neurons

The vestibular system may be involved in providing facilitatory inputs to brainstem respiratory circuits under some conditions during which the background drive on respiratory neurons must be altered. One circumstance in which inputs from the labyrinth could serve to enhance respiratory drive is during exercise. Rhythmic exercise starting from rest elicits a rapid change in ventilation that is independent of arterial levels of oxygen or carbon dioxide.[17] It is likely that proprioceptors in limbs are involved in eliciting these changes in respiration,[17] but it is also feasible that the vestibular system plays a role in increasing ventilation during rhythmic exercise. Rhythmic movements will elicit a stereotyped pattern of vestibular inputs in addition to a fixed pattern of proprioceptor inputs from the body. It is conceivable that a neural integrator in the brainstem could detect a rhythmic pattern of vestibular inputs and provide facilitatory inputs to brainstem respiratory neurons. However, this hypothesis remains to be tested.

Background influences of the vestibular system on respiratory neurons may be particularly important in infants. Rocking is one of the most efficient mechanisms for stabilizing respiration in young children who are subject to apnea and sudden infant death syndrome.[27,38] Respiration becomes entrained to rocking frequency, at least in infants.[20,55] Although rocking produces modulation in the activity of vestibular afferents, other receptors can be activated by this movement (including cutaneous receptors, muscle proprioceptors, etc.). It will be necessary to perform control studies to demonstrate that vestibular inputs are responsible for the effects on respiration produced by rocking.

4 BRAINSTEM NEURONAL CIRCUITRY MEDIATING VESTIBULAR EFFECTS ON THE RESPIRATORY SYSTEM

4.1 Vestibular Nuclei

Extensive studies have been undertaken to determine which brainstem structures mediate the reflex responses recorded from respiratory muscle nerves in response to electrical stimulation of the vestibular nerve. The first synapse in the vestibulo-respiratory reflex arc is undoubtedly in the vestibular nuclei, since virtually all vestibular afferents terminate in this complex of four main nuclei located in the dorsal medulla and pons (see Chapter 2). Chemical lesions of a restricted portion of the vestibular nuclei, the medial and inferior vestibular nuclei just caudal to the lateral vestibular (Deiters') nucleus, abolish the effects of vestibular stimulation on both the phrenic and abdominal nerves.[64] Similar lesions also abolish vestibulo-sympathetic reflexes (see Chapter 6), suggesting that a restricted part of the vestibular nucleus complex mediates a number of vestibulo-autonomic responses.

4.2 Brainstem Bulbospinal Neurons

There are two clusters of neurons in the medulla that are responsible for imparting rhythmic discharge characteristics on spinal respiratory motoneurons (reviewed in Chapter 4): the dorsal (DRG) and ventral (VRG) respiratory groups. The DRG is located in the ventrolateral nucleus of the solitary tract (NTS), whereas the VRG lies in the ventrolateral reticular formation between the first cervical segment and the retrofacial nucleus. Since the DRG and VRG relay central respiratory drive to spinal motoneurons, they are potentially the most promising candidates for mediating vestibulo-respiratory reflexes.

In one series of studies, neuroanatomical methods were used to determine if regions of the vestibular nuclei that mediate vestibulo-respiratory responses have projections to the DRG and VRG. Microinjections of the anterograde neuroanatomical tracer *Phaseolus vulgaris* leucoagglutinin (PHA-L) into the medial and inferior vestibular nuclei labeled projections to the nucleus of the solitary tract, but only sparse labeling was observed in the ventrolateral NTS where the DRG is located.[3,63] In contrast, labeling of terminals occurred in the VRG near nucleus ambiguus.[62] Thus, neuroanatomical studies suggest that neurons in the VRG may receive vestibular inputs, whereas those in the DRG do not.

Cell recording studies have shown that almost one half of inspiratory bulbospinal neurons in the VRG, which presumably make mono- or oligosynaptic connections with respiratory motoneurons in the spinal cord, receive vestibular input.[45] An example of the response of a VRG neuron to vestibular stimulation is shown in Figure 2. In contrast, as would be expected from the PHA-L studies, DRG inspiratory neurons received only sparse and weak vestibular input in experiments where robust vestibular effects were simultaneously recorded from the phrenic nerve.[63] Thus, there appears to be a difference between inspiratory neurons in the DRG and VRG with regard to their role in vestibulo-respiratory reflexes.

Medullary neurons with inspiratory or expiratory activity have also been tested for responses to static roll tilts that activate a subset of receptors in the utriculus (one of the otolith organs; see Chapter 2). The firing of these neurons, which were mainly located in the nucleus ambiguus and in the reticular formation between the nucleus ambiguus and NTS, was unaffected by maintained body tilt.[29,30] Presumably, the sample of neurons in this study overlapped the sample of VRG neurons shown to receive vestibular inputs.[45] If this is the case, then vestibular receptors other than those activated by static ear-down tilt are

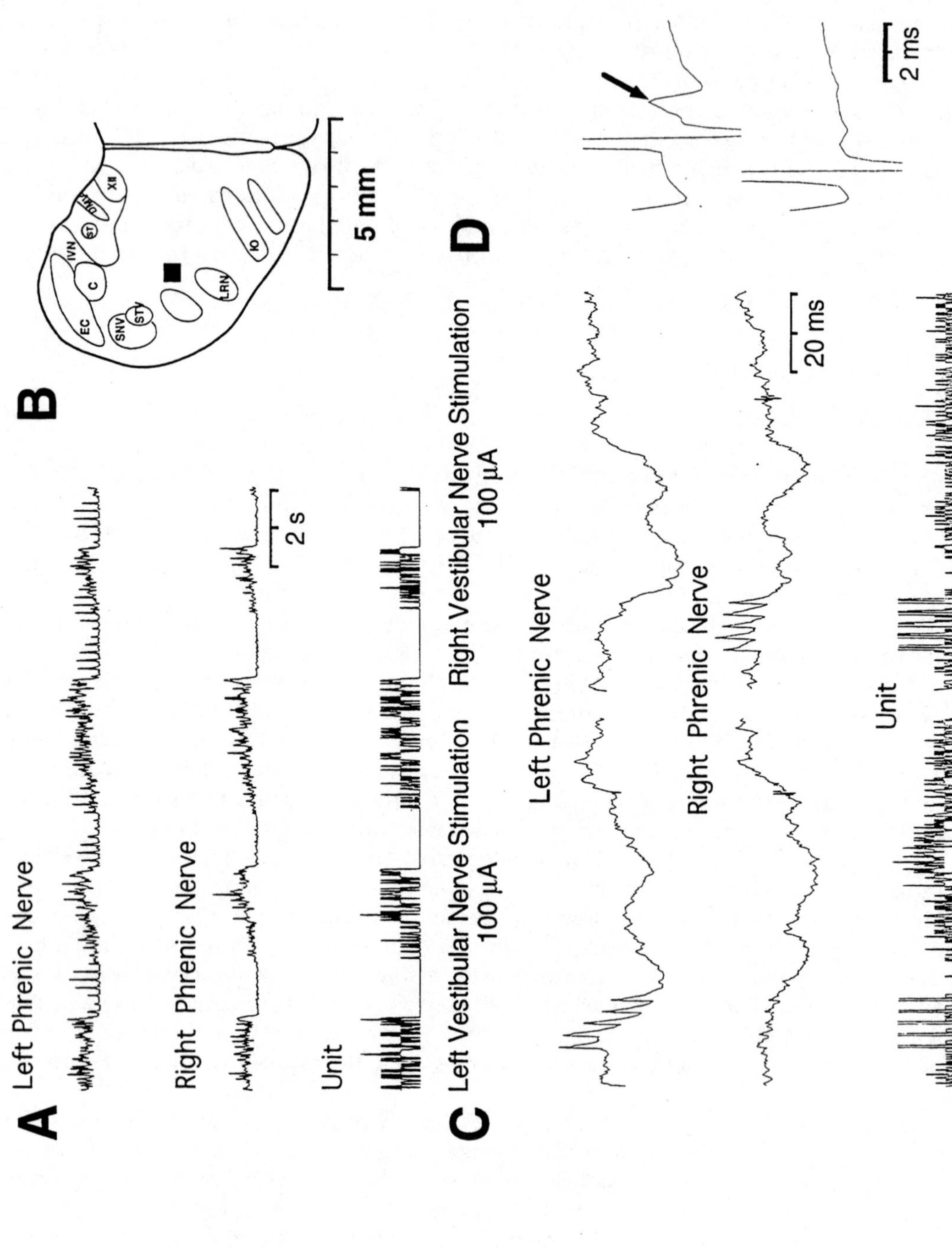
A
Left Phrenic Nerve
Right Phrenic Nerve
Unit
2 s
B
EC
SNV
STV
C
IVN
ST
DMV
XII
LRN
IO
5 mm
C
Left Vestibular Nerve Stimulation
100 µA
Right Vestibular Nerve Stimulation
100 µA
Left Phrenic Nerve
Right Phrenic Nerve
Unit
20 ms
D
2 ms

FIGURE 2 (Facing page)
Responses recorded from a feline bulbospinal VRG neuron with inspiratory-related activity. (A) Comparison of binned spike data recorded from the neuron with spontaneous discharges recorded from the phrenic nerve. The unit clearly had inspiratory-related activity. (B) Location of the neuron (indicated by ■) in the ventrolateral reticular formation, approximately 2 mm rostral to the obex. Abbreviations: C, cuneate n.; DMV, dorsal motor nucleus of vagus; EC, external cuneate nucleus; IVN, inferior vestibular nucleus; LRN, lateral reticular nucleus; IO, inferior olivary nucleus; SNV, spinal trigeminal nucleus; ST, solitary tract; STV, spinal trigeminal tract; XII, hypoglossal nucleus. (C) Responses recorded from the unit and from the phrenic nerves following electrical stimulation of the vestibular nerve. The responses were elicited by a train of five shocks, as indicated by arrows beneath the bottom traces. Note that vestibular nerve stimulation produced a short-latency inhibition of the unit's firing, which was correlated with inhibition recorded from the phrenic nerve. A total of 195 sweeps were averaged in the left panel (left vestibular nerve stimulation), and 204 in the right. (D) Antidromic response of the unit (indicated by arrow in top panel) to electrical stimulation of the ipsilateral lateral funiculus at C_3. The threshold was 400 μA. When the latency between a spontaneous spike and the stimulus was decreased (D, bottom panel), there was a collision block of the response, proving that it had been antidromic. (Panel A and and the left portion of panel C are reproduced from Miller, A. D. et al., *J. Neurophysiol.*, 73, 1303, 1995. With permission.)

apparently involved in providing inputs to the VRG. These alternate vestibular receptors could include the semicircular canals and otolith receptors activated by pitch (nose-up/down tilt) or vertical linear acceleration (up-and-down movements).

The vestibulo-phrenic reflex is not abolished by kainic acid injections into either the DRG or the VRG (illustrated in Figure 3), indicating that other descending pathways are important for producing the response.[65] These pathways are yet to be identified, but could include projections from the vestibular nuclei and reticular formation that participate in vestibulospinal reflexes. The vestibulospinal and reticulospinal tracts could potentially influence phrenic motoneurons directly or indirectly via spinal interneurons (see Chapter 4 for a description of spinal respiratory interneurons).

4.3 Other Pathways

Much of the circuitry that mediates vestibulo-respiratory responses remains to be elucidated. Because the vestibular nuclei have direct projections to the vicinity of the VRG, which contains inspiratory bulbospinal neurons, propriobulbar neurons, and laryngeal and pharyngeal motoneurons, it seems likely that labyrinthine signals reach some bulbospinal respiratory neurons, as well as some upper airway motoneurons, fairly directly. However, it is also conceivable that brainstem interneurons that are yet to be identified also participate in relaying vestibular signals to respiratory spinal and upper airway motoneurons. Considerable further research will be necessary to elucidate the neural circuitry that produces vestibulo-respiratory responses.

5 SUMMARY

The vestibular system produces short-latency effects on the respiratory system. Vestibulo-respiratory reflexes are mediated by the medial and inferior vestibular nuclei and in part by neurons located in the caudal ventrolateral medulla that participate in imparting patterned rhythmic discharges on respiratory motoneurons (i.e., spinal projecting inspiratory neurons located in the para-ambigual VRG). However, large lesions of the caudal medulla that abolish spontaneous respiratory discharge have little effect on vestibulo-respiratory reflexes, suggesting that pathways other than those that typically control respiration are also involved in producing these responses. Vestibular effects on the respiratory system could subserve a number of functions including compensation for

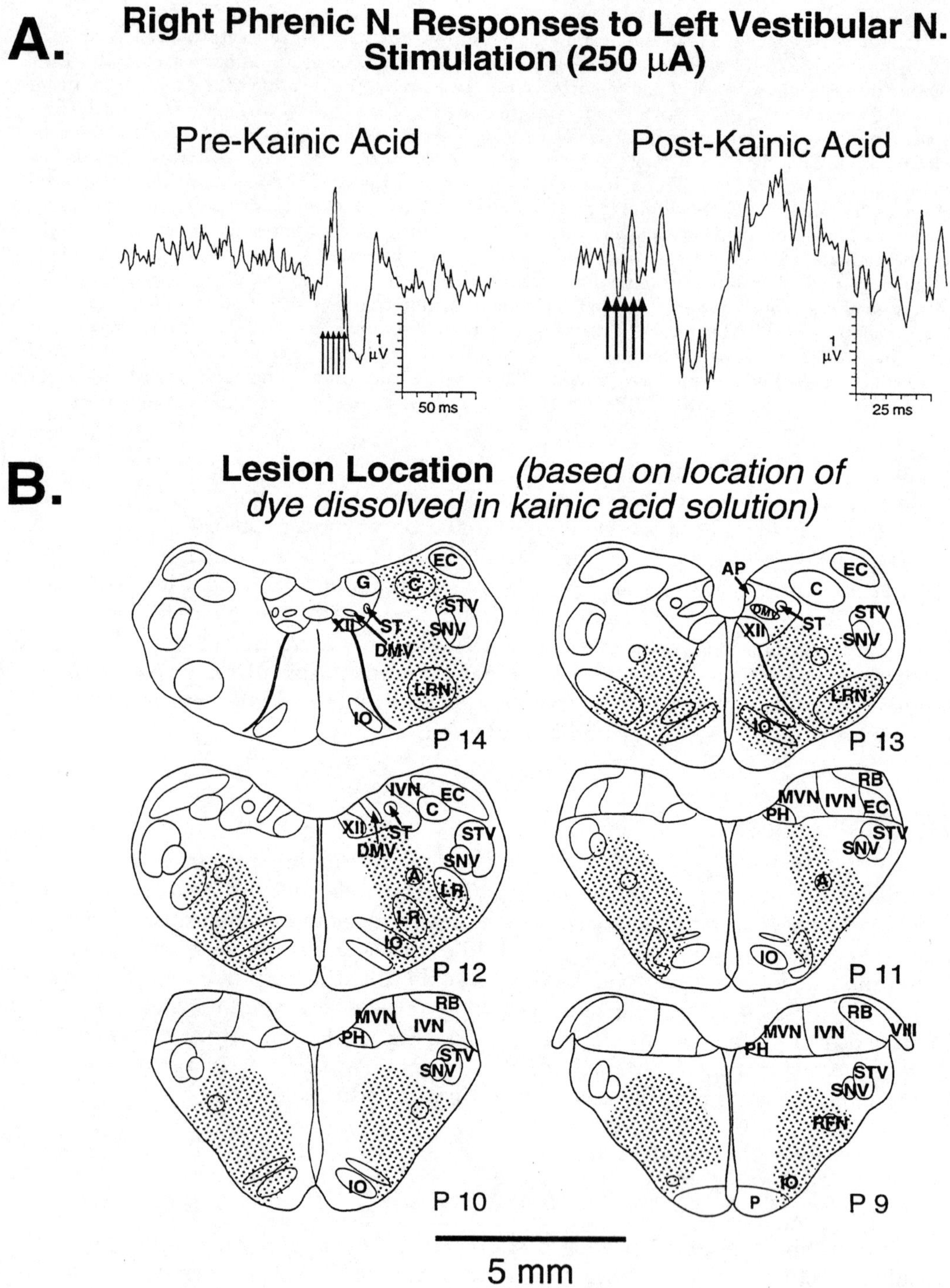

mechanical constraints on respiration or orthostatic hypotension induced by changes in body position, maintenance of posture, rotation of the thorax, and reduction of apnea in infants. However, the fact that the vestibular system has widespread effects on a large number of respiratory muscles with a variety of roles (including the diaphragm, intercostal muscles, and upper airways muscles) suggests that at least one purpose of these connections is to maintain airway patency and make adjustments in ventilation during movement and postural changes.

FIGURE 3 (Facing page)
Effects of a chemical lesion of the VRG on vestibulo-phrenic reflexes. (A) Phrenic nerve responses to a 250-μA stimulus delivered to the contralateral vestibular nerve. The responses were recorded before and 30 min after a large injection of kainic acid into the ventrolateral medullary reticular formation (lesion location shown in B). Arrows indicate the occurrence of the five shocks that produced each response. Note that the kainic acid injection produced no decrement in the response amplitude. (B) Location of kainic acid injection into the vicinity of the VRG. The location is based on the location of dye (which had been dissolved in the injection solution) in histological sections. The dye filled the para-ambiguual area of the caudal medulla. The coordinate beside each section indicates distance (in mm) posterior to stereotaxic zero. *Abbreviations:* A, nucleus ambiguus; AP, area postrema; C, cuneate nucleus; DMV, dorsal motor nucleus of the vagus nerve; EC, external cuneate nucleus; IO, inferior olivary nucleus; IVN, inferior vestibular nucleus; LR, lateral reticular nucleus; LVN, lateral vestibular nucleus; MVN, medial vestibular nucleus; P, pyramid; PH, prepositus hypoglossi; RB, restiform body; RFN, retrofacial nucleus; SA, stria acoustica; SNV, spinal trigeminal nucleus; ST, solitary tract; STV, spinal trigeminal tract; VII, facial nucleus; VIII, vestibulocochlear (VIIIth) nerve; XII, hypoglossal nucleus. (Adapted from Yates, B. J. et al., *Am. J. Physiol.* [*Reg. Integr. Comp. Physiol.*, 37], 268, R1381, 1995. With permission.)

ACKNOWLEDGMENTS

The authors' work is supported by grants from the National Institute on Deafness and other Communication Disorders, U.S.A. (DC02644, DC00693), and the National Institute of Neurological Disorders and Stroke, U.S.A. (NS20585).

REFERENCES

1. Akert, K. and Gernandt, B. E., Neurophysiological study of vestibular and limbic influences upon vagal outflow, *Electroenceph. Clin. Neurophysiol.*, 14, 904, 1962.
2. Amis, T. C., Jones, H. A., and Hughes, J. M. B., Effect of posture on inter-regional distribution of pulmonary ventilation in man, *Resp. Physiol.*, 56, 145, 1984.
3. Balaban, C. D. and Beryozkin, G., Vestibular nucleus projections to nucleus tractus solitarius and the dorsal motor nucleus of the vagus nerve: potential substrates for vestibulo-autonomic interactions, *Exp. Brain Res.*, 98, 200, 1994.
4. Banzett, R. B., Inbar, G. F., Brown, R., Goldman, M., Rossiter, A., and Mead, J., Diaphragm electrical activity during negative lower torso pressure in quadriplegic men, *J. Appl. Physiol.*, 51, 654, 1981.
5. Bassal, M. and Bianchi, A. L., Inspiratory onset or termination induced by electrical stimulation of the brain, *Resp. Physiol.*, 50, 23, 1982.
6. Bonora, M., Bartlett, D., Jr., and Knuth, S. L., Changes in upper airway muscle activity related to head position in awake cats, *Resp. Physiol.*, 60, 181, 1985.
7. Borison, H. L., Area postrema: chemoreceptor cirumventricular organ of the medulla oblongata, *Prog. Neurobiol.*, 32, 351, 1989.
8. Boynton, B. R., Barnas, G. M., Dadmun, J. T., and Fredberg, J. J., Mechanical coupling of the rib cage, abdomen, and diaphragm through their area of apposition, *J. Appl. Physiol.*, 70, 1235, 1991.
9. Brancatisano, A., Kelly, S. M., Tully, A., Loring, S. H., and Engel, L. A., Postural changes in spontaneous and evoked regional diaphragmatic activity in dogs, *J. Appl. Physiol.*, 66, 1699, 1989.
10. Crampton, G. H., Studies of motion sickness. XVII. Physiological changes accompanying sickness in man, *J. Appl. Physiol.*, 7, 501, 1955.
11. Davis, P. R. and Troup, J. D. G., Pressure in the trunk cavities when pulling, pushing and lifting, *Ergonomics*, 7, 465, 1964.
12. Davies, A., Sant'Ambrogio, F. B., and Sant'Ambrogio, G., Control of postural changes of end expiratory volume (FRC) by airways slowly adapting mechanoreceptors, *Resp. Physiol.*, 41, 211, 1980.
13. De Troyer, A., Mechanical role of the abdominal muscles in relation to posture, *Resp. Physiol.*, 53, 341, 1983.
14. De Troyer, A. and Ninane, W., Effect of posture on expiratory muscle use during breathing in the dog, *Resp. Physiol.*, 67, 311, 1987.
15. Dobbins, E. G. and Feldman, J. L., Brainstem network controlling descending drive to phrenic motoneurons in rat, *J. Comp. Neurol.*, 347, 64, 1994.

16. Druz, W. S. and Sharp, J. T., Activity of respiratory muscles in upright and recumbent humans, *J. Appl. Physiol.*, 51, 1552, 1981.
17. Duffin, J., Neural drives to breathing during exercise, *Can. J. Appl. Physiol.*, 19, 289, 1994.
18. Duron, B. and Marlot, D., Intercostal and diaphragmatic electrical activity during wakefulness and sleep in normal unrestrained cats, *Sleep*, 3, 269, 1980.
19. Ellenberger, H. H., Vera, P. L., Haselton, J. R., Haselton, C. L., and Schneiderman, N., Brainstem projections to the phrenic nucleus: an anterograde and retrograde HRP study in the rabbit, *Brain Res. Bull.*, 24, 163, 1990.
20. Elliott, M., Fisher, K., and Ames, E., The effects of rocking on the state and respiration of normal and excessive criers, *Can. J. Psychol.*, 42, 163, 1988.
21. Elmund, J., Bowman, J. P., and Morgan, R. J., Vestibular influence on tongue activity, *Exp. Neurol.*, 81, 126, 1983.
22. Estenne, M., Ninane, V., and De Troyer, A., Triangularis sterni muscle use during eupnea in humans: effect of posture, *Resp. Physiol.*, 74, 151, 1988.
23. Farkas, G. A., Mechanical characteristics and functional length of canine expiratory muscles, *Resp. Physiol.*, 90, 87, 1992.
24. Farkas, G. A., Baer, R. E., Estenne, M., and De Troyer, A., Mechanical role of expiratory muscles during breathing in upright dogs, *J. Appl. Physiol.*, 64, 1060, 1988.
25. Farkas, G. A., Estenne, M., and De Troyer, A., Expiratory muscle contribution to tidal volume in head-up dogs, *J. Appl. Physiol.*, 67, 1438, 1989.
26. Farkas, G. A. and Schroeder, M. A., Mechanical role of expiratory muscles during breathing in prone anesthetized dogs, *J. Appl. Physiol.*, 69, 2137, 1990.
27. Farrimond, T., Sudden infant death syndrome and possible relation to vestibular function, *Percept. Mot. Skills*, 71, 419, 1990.
28. Gernandt, B. E., A comparison between autonomic and somatic motor outflow to vestibular stimulation, *Confin. Neurol.*, 24, 140, 1964.
29. Ghelarducci, B., Pompeiano, O., and Spyer, K. M., Effect of natural stimulation of macular labyrinthine receptors on the discharge pattern of brain stem respiratory neurons, *Arch. Ital. Biol.*, 112, 33, 1974.
30. Ghelarducci, B., Pompeiano, O., and Spyer, K. M., Discharges of medullary respiratory neurones during tilting, *Brain Res.*, 68, 174, 1974.
31. Gilmartin, J. J., Ninane, V., and De Troyer, A., Abdominal muscle use during breathing in the anesthetized dog, *Resp. Physiol.*, 70, 159, 1987.
32. Gorini, M. and Estenne, M., Effect of head-up tilt on neural inspiratory drive in the anesthetized dog, *Resp. Physiol.*, 85, 83, 1991.
33. Green, M., Mead., J., and Sears, T. A., Muscle activity during chest wall restriction and positive pressure breathing in man, *Respir. Physiol.*, 35, 283, 1978.
34. Grillner, S., Nilsson, J., and Thorstensson, A., Intra-abdominal pressure changes during natural movements in man, *Acta Physiol. Scand.*, 103, 275, 1978.
35. Horner, M. and Kummel, H., Topographical representation of shoulder motor nuclei in the cat spinal cord as revealed by retrograde fluorochrome tracers, *J. Comp. Neurol.*, 335, 309, 1993.
36. Huang, Q., Zhou, D., and St. John, W. M., Vestibular and cerebellar modulation of expiratory motor activities in the cat, *J. Physiol.*, 436, 385, 1991.
37. Kinnear, W., Higenbottam, T., Shaw, D., Wallwork, J., and Estenne, M., Ventilatory compensation for changes in posture after human heart-lung transplantation, *Resp. Physiol., 77, 75, 1989.*
38. Korner, A., Kraemer, H., Haffner, M., and Cosper, L., Effects of waterbed flotation on premature infants: a pilot study, *Pediatrics*, 56, 361, 1975.
39. Leevers, A. M. and Road, J. D. Effect of lung inflation and upright posture on diaphragmatic shortening in dogs, *Resp. Physiol.*, 85, 29, 1991.
40. Lloyd, T. C., Effect of inspiration on inferior vena caval blood flow in dogs, *J. Appl. Physiol.*, 55, 1701, 1983.
41. Loring, S. H. and De Troyer, A., Actions of the respiratory muscles, in *The Thorax*, Roussos, C. and Macklem, P. T., Eds., Marcel Dekker, New York, 1985, chap. 10.
42. Mameli, O., Tolu, E., Melis, F., and Caria, M. A., Labyrinthine projection to the hypoglossal nucleus, *Brain Res. Bull.*, 20, 83, 1988.
43. Megirian, D., Vestibular control of laryngeal and phrenic motoneurons of cat, *Arch. Ital. Biol.*, 106, 333, 1968.
44. Miller, A. D., Tan, L. K., and Lakos, S. F., Brainstem projections to cats' upper lumbar spinal cord: implications for abdominal muscle control, *Brain Res.*, 493, 348, 1989.
45. Miller, A. D., Yamaguchi, T., Siniaia, M. S., and Yates, B. J., Ventral respiratory group bulbospinal inspiratory neurons participate in vestibular-respiratory reflexes, *J. Neurophysiol.* 73, 1303, 1995.

46. Mixter, G., Reflex augmentation of inferior venal caval flow demonstrated by a low-resistance phasic flowmeter, *Am. J. Physiol.*, 172, 446, 1953.
47. Moreno, A. H., Burchell, A. R., and Burke, J. H., Respiratory regulation of splanchnic and systemic venous return, *Am. J. Physiol.*, 213, 455, 1967.
48. Mortola, J. and Sant'Ambrogio, G., Reflex responses of respiratory muscles to tilting and positive pressure breathing, *Arch. Fisiol.*, 70, 197, 1973.
49. Newman, S. L., Road, J. D., and Grassino, A., In vivo length and shortening of canine diaphragm with body postural change, *J. Appl. Physiol.*, 60, 661, 1986.
50. Onai, T. and Miura, M., Projections of supraspinal structures to the phrenic motor nucleus in cats studied by a horseradish peroxidase microinjection method, *J. Auton. Nerv. Syst.*, 16, 61, 1986.
51. Onai, T., Saji, M., and Miura, M., Projections of supraspinal structures to the phrenic motor nucleus in rats studied by a horseradish peroxidase microinjection method, *J. Auton. Nerv. Syst.*, 21, 233, 1987.
52. Reid, M. B., Banzett, R. B., Feldman, H. A., and Mead, J., Reflex compensation of spontaneous breathing when immersion changes diaphragm length. *J. Appl. Physiol.*, 58, 1136, 1985.
53. Rikard-Bell, G. C., Bystrzycka, E. K., and Nail, B. S., Brainstem projections to the phrenic nucleus: a HRP study in the cat, *Brain Res. Bull.* 12, 469, 1984.
54. Rikard-Bell, G. C., Bystrzycka, E. K., and Nail, B. S., The identification of brainstem neurones projecting to thoracic respiratory motoneurones in the cat as demonstrated by retrograde transport of HRP, *Brain Res. Bull.*, 14, 25, 1985.
55. Sammon, M. P. and Darnall, R. A., Entrainment of respiration to rocking in premature infants: coherence analysis, *J. Appl. Physiol.*, 77, 1548, 1994.
56. Spiegel, E. A., Respiratory reactions upon vertical movements, *Am. J. Physiol.*, 117, 349, 1936.
57. Takata, M. and Robotham, J. L., Effects of diaphragmatic descent on inferior vena caval venous return, *J. Appl. Physiol.*, 72, 597, 1992.
58. Thorstensson, A., Oddsson, L., and Carlson, H., Motor control of voluntary trunk movements in standing, *Acta Physiol. Scand.*,125, 309, 1985
59. van Lunteren, E., Haxhiu, M. A., Cherniack, N. S., and Goldman, M. D., Differential costal and crural diaphragm compensation for posture changes, *J. Appl. Physiol.*, 58, 1895, 1985.
60. Whitelaw, W. A., Ford, G. T., Rimmer, K. P., and De Troyer, A., Intercostal muscles are used during rotation of the thorax in humans, *J. Appl. Physiol.*, 72, 1940, 1992.
61. Wilson, V. J. and Melvill Jones, G., *Mammalian Vestibular Physiology*, Plenum Press, New York, 1979.
62. Yates, B. J., Balaban, C. D., Miller, A. D., Endo, K., and Yamaguchi, Y., Vestibular inputs to the lateral tegmental field of the cat: potential role in autonomic control, *Brain Res.*, 689, 197, 1995.
63. Yates, B. J., Grélot, L., Kerman, I. A., Balaban, C. D., Jakuš, J., and Miller, A. D., Organization of vestibular inputs to nucleus tractus solitarius and adjacent structures in cat brain stem, *Am. J. Physiol.* (*Reg. Integr. Comp. Physiol.*, 36), 267, R974, 1994.
64. Yates, B. J., Jakuš, J., and Miller, A. D., Vestibular effects on respiratory outflow in the decerebrate cat, *Brain Res.*, 629, 209, 1993.
65. Yates, B. J., Siniaia, M. S., and Miller, A.D., Descending pathways necessary for vestibular influences on sympathetic and inspiratory outflow, *Am. J. Physiol. (Reg. Integr. Comp. Physiol.*, 37), 268, R1381, 1995.
66. Youmans, W. B., Tjioe, D. T., and Tong, E. Y., Control of involuntary activity of abdominal muscles, *Am. J. Physical Med.*, 53, 57, 1974.

8

The Role of the Cerebellum in Vestibular Autonomic Function

Carey D. Balaban

CONTENTS

0-8493-7668-1/96/$0.00+$.50

1 INTRODUCTION

A discussion of the role of the cerebellum in controlling reflexes and voluntary movements of striated muscle is a standard component of basic neuroscience textbooks. However, the parallel role of the cerebellum in autonomic control is appreciated less widely. The relative neglect of the cerebellar role in autonomic function is ironic because one of the earliest hypotheses of functional localization in the brain[82] was based upon this notion. This chapter reviews the historical development of the hypothesis of a cerebellar role in autonomic function. Central to this review is the concept that models of the role of the cerebellum in autonomic function have sought to integrate three areas in brain research: (1) heuristic understanding of the intrinsic operations of the cerebellum, (2) heuristic models of neural control of autonomic (or involuntary) functions and (3) heuristics for understanding the cerebellar influence on central autonomic pathways. Our attempts at approaching these questions have a rich history that began in the 17th century. This history provides a clear illustration of the evolution of ideas in neuroscience, which represents a synthesis of contemporaneous heuristic concepts for understanding physical, biological, and pathological phenomena and philosophical approaches to defining questions regarding the mind, voluntary and involuntary actions, sensation, and perception. Within a modern context, this literature analysis yields five basic hypotheses regarding cerebellar contributions to autonomic control:

1. The cerebellum can adjust the sensitivity of autonomic reflexes (e.g., baroreflexes) by modulating a tonic inhibition of brain stem circuits. Since the cerebellar regions include areas with prominent vestibular inputs, these circuits provide potential substrates for vestibulo-autonomic interactions.
2. The cerebellum contributes a phasic boost to cardiovascular reflexes during postural adjustments. In the case of orthostatic reflexes, it is suggested that both motor corollary discharge and vestibular information may transiently assist in maintaining cerebral blood flow during changes in head position relative to gravity.
3. The cerebellum contributes to adaptive modification of autonomic responses. A key example is provided by the effects of medial cerebellar lesions on classically conditioned cardiovascular responses.[67,76]
4. The cerebellum coordinates both movements of striated muscle and regional blood flow changes associated with the movements.
5. The cerebellum coordinates the components of autonomic "orienting responses," homeostatic responses, and aversive responses to external stimuli.

2 HISTORICAL PERSPECTIVES

Speculations regarding the role of the cerebellum in autonomic functions developed as a corollary of interest in the localization of voluntary and involuntary motor functions in the brain. As a result, the development of hypotheses are intertwined intimately with contemporaneous perspectives of voluntary, involuntary, somatic, and visceral functions. The suggestion of a specific cerebellar role in autonomic function can be attributed to Thomas Willis,[82] who viewed the cerebellum as the site of involuntary functions and the cerebral cortex as the source of voluntary actions (see Dow and Moruzzi[21] and Neuberger[51] for reviews). This was stated explicitly (Reference 81, p.111):

> But the office of the Cerebel seems to be for the animal Spirits to supply some nerves; by which involuntary actions (such as the beating of the Heart, the Concoction of the Aliment, the protrusion of the Chyle, and many others) which are made after a constant manner unknown to us, whether we will or no, are performed. As often we go about voluntary motion, we seem as it were to perceive within us the Spirits residing within the fore-part

of the Head to be stirred up to action, or an influx. But the Spirits inhabiting the Cerebel perform unperceivedly and silently their works of Nature without our knowledge or care.

Despite the explicit nature of the hypothesis, its logical framework is very different from modern ideas of cerebello-autonomic interactions. The main argument supporting this hypothesis was based upon a combination of comparative anatomical arguments and the iatrochemical concept that the "Spirits" for brain function were distilled from the blood by the cerebellar and cerebral cortex. These arguments produced the conclusion that the cerebellum was the site of the involuntary component of the "sensitive (or animal) soul" (Reference 82, p. 114), responsible for both involuntary cardiac, visceral, and respiratory functions and for the modification of those functions during arousal of "... a violent passion, as Joy, Sadness, Anger, Fear, or of any other kind ..." (p. 116). Thus, despite the fact that Willis' reasoning shows insight into the necessity of autonomic control mechanisms for both steady-state behavior and changes during activity or emotional arousal, the supporting evidence was strictly speculative.

The Willis hypothesis was the subject of intense investigation over the succeeding century, serving as a focus for the first major dispute regarding localization. As reviewed by Neuberger,[51] the concept of an important cerebellar role in cardiac and respiratory control was developed further through the mid-19th century, most notably by Jacob Fidelis Ackerman, I. Döllinger, and Adam Karl August Eschenmeyer. One interesting footnote is provided by Gall's[26] clinically based hypothesis that the cerebellum was a center for sexual function. However, as summarized by Ferrier,[23] the results from experimental investigations employing lesions, cocaine and strychnine application, and electrical stimulation of the cerebellum indicated a major role of the cerebellum in the coordination of movement and locomotion. However, Ferrier[23] also noted that a cerebellar role in autonomic function was possible because stimulation of the cerebellar cortex produced pupillary responses.

Research from the latter decades of the 19th century through the first half of the 20th century defined two modern concepts of cerebellar function: (1) the concept that the cerebellum can both modulate reflex performance and coordinate more complex (and hence, voluntary) movements and (2) the concept of functional localization within the cerebellum in relation to afferent and efferent circuitry (e.g., see Fulton and Dow[25] or Dow and Moruzzi[21]). These concepts regarding the cerebellar role in somatic motor function provided a framework for a modern approach to cerebellar influences on autonomic function, or, in the words of Moruzzi,[50] "the cerebellar influence on the autonomic sphere." This approach requires the assumption that the cerebellum influences both autonomic regulatory reflexes and more complex autonomic and automatic/involuntary responses in a manner that is analogous to its influences on somatic motor reflexes. By induction, then, this assumption leads to the general hypothesis that cerebellar circuits assist in the coordination of autonomic and somatic components of motor activity. This perspective was implicit in Moruzzi's[50] division of cerebellar influences on autonomic function into four areas:

1. Cerebellar influence on vasomotor reflexes and blood pressure waves
2. Cerebellar influence on carotid sinus reflexes
3. Cerebellar influence on somatic and autonomic manifestations of sham rage
4. Cerebellar influence on pupils and nictitating membrane

3 GENERAL OVERVIEW: CEREBELLAR ROLE IN MOTOR CONTROL

The assumption that the cerebellum is performing similar functions in somatic and autonomic control provides a systematic framework for developing hypotheses regarding cerebellar influences on autonomic function. First, it is necessary to review basic principles derived from studies of cerebellar contributions to somatic motor control. The applicability of these principles to autonomic control can then be examined. This discussion is not intended to provide a detailed discussion of cerebellar circuitry and contributions to motor control (for review see Reference 37). Rather, it is intended to discuss broad principles that provide an intuitive framework for investigating the role of the cerebellum in autonomic function.

3.1 Cellular Organization of Cerebellar Cortex

The cerebellum consists of two major anatomical divisions, the cerebellar cortex and the cerebellar nuclei. The reader is referred to neurobiology textbooks or monographs (e.g., References 22, 37, 58) for a more detailed description of cerebellar anatomy and neurochemistry. Briefly, the cerebellar cortex consists of the three cellular layers. The innermost gray layer, the granule cell layer, is a dense cellular layer that receives the vast majority of incoming axon terminals, which end as mossy fibers. The granule cell layer contains two classes of cerebellar interneurons, granule cells (excitatory interneurons) and Golgi cells (inhibitory interneurons). The Purkinje cell layer, the second layer, contains Purkinje cells, which are the sole output neurons of the cerebellar cortex. They use GABA as a neurotransmitter and also express a variety of neuropeptides that may also function in neurotransmission. Finally, the molecular layer is the outermost layer of the cerebellar cortex. It contains dendrites of Purkinje cells, axons of granule cells (called parallel fibers, excitatory amino acid transmitter), and two classes of inhibitory interneurons, stellate cells and basket cells (transmitter: GABA). Climbing fiber (excitatory amino acid transmitter), serotonergic and noradrenergic afferents terminate on Purkinje cell dendrites in this layer. The Purkinje cells affect activity in motor pathways via their projections to the cerebellar nuclei, vestibular nuclei, and the parabrachial nucleus (see below).

The understanding of synaptic integration by Purkinje cells has been a major focus of research efforts because Purkinje cells are the final common output pathway for the cerebellar cortex. The interpretation of our present understanding of cerebello-autonomic function, though, requires consideration of the two major inputs to the Purkinje cells, parallel fibers (axons of granule cells) and climbing fibers (axons of inferior olivary cells). The mossy fiber-granule cell/parallel fiber pathway is the major synaptic drive to Purkinje cells. Mossy fibers originate in a variety of sites in the neuraxis; they include primary vestibulo-nodulus afferents, secondary vestibulo-cerebellar afferents, spino-, ponto-, reticulo-, and solitary-cerebellar pathways. These axons end as mossy fiber rosettes in the granule layer, which synapse on dendrites of granule cells. The axons of granule cells then ascend into the molecular layer, where they bifurcate to form parallel fibers, which run perpendicular to the dendritic expansions of the Purkinje cells and terminate on spiny branchlets of Purkinje cell dendrites. In the cat, each Purkinje cell receives on the order of 80,000 parallel fiber synapses, which produce the *simple spikes* that comprise the majority of the discharges of the Purkinje cells. By contrast, each Purkinje cell receives a single climbing fiber afferent, which makes multiple synapses along the soma and smooth dendritic branches. Activation of climbing fibers produces *complex spike* discharges of the Purkinje cell, which have a very slow maximum firing rate (fewer than 4 spikes per second). When a series of complex spikes coincides temporally with simple spike discharges at a particular parallel fiber synapse, the synaptic weight at that synapse appears to decrease via a mechanism termed long-term depression.[38] Hence, climbing fibers provide an important drive for adaptive modification of Purkinje cell responses to parallel fiber inputs, constituting an "error signal" for motor performance.[37,38] This mechanism is believed to contribute to the adaptive organization of motor performance.

Since the advent of modern connection tracing methods in the 1950s, the concept of topographical organization of the cerebellum has been refined to account for anatomical data concerning the parasagittal organization of both climbing fiber afferents to the cerebellar cortex and Purkinje cell connections with cerebellar and vestibular nuclei. These studies have resulted in the concept that there is a zonal pattern of organization of cerebellar circuitry such that different regions of the cerebellar cortex can differentially affect motor output pathways such as the cortico-spinal, rubro-spinal, pontine reticulo-spinal, vestibulo-ocular, and vestibulo-spinal pathways. These zones are distributed sagittally within the flocculonodular lobe, posterior lobe vermis, anterior lobe, and cerebellar hemispheres, and are defined by both the distribution of climbing fiber inputs from the inferior olive to Purkinje cells and the efferent projections of the Purkinje cells to cerebellar and vestibular nuclei. A formal statement of this concept is the microcomplex hypothesis,[4,5,37] which defines cerebellar corticonuclear microcomplexes as a pool of inferior olivary neurons, their strictly crossed climbing fiber projections to both a sagittal group of cerebellar Purkinje cells and a distinct pool of cells in the cerebellar or vestibular nuclei, and the connections of Purkinje cells to cerebellar and vestibular nuclei. This implies that the cerebellar cortical component (a microzone[2,3]) is characterized both by the topographic organization of its efferent pathways to the cerebellar or vestibular nuclei and by the parasagittal topographic projection of inferior olivary afferents. Long-term depression due to climbing fiber activity within each microcomplex is then expected to adaptively modify Purkinje cell activity to maintain accurate motor performance.

3.2 Cerebellar Contributions to Motor Activity

Two global themes for understanding cerebellar contributions to motor control were established by the first half of the 20th century: (1) cerebellar circuitry is organized topographically, and (2) cerebellar circuits contribute to both the coordination of movements and the adaptive modification of reflex and voluntary movements. Anatomical studies of the afferent and efferent connections of the cerebellar cortex, extensive clinical observations, correlative experimental studies of the effects of localized lesions on motor function, and results of electrical-stimulation studies suggested that there is a topographic relationship between cerebellar regions and specific somatic and autonomic motor output pathways or functions.[21,25] Fulton and Dow[25] provided a summary of results from anatomical studies of vestibular, spinal, and pontine mossy fiber afferent projections to cerebellar cortex that justified an initial division of the cerebellar cortex into a vestibulo-cerebellum (flocculo-nodular lobe, uvula, and lingula), a spino-cerebellum (pyramis, paraflocculus and anterior lobe culmen, and lobus centralis), and a neocerebellum (cerebellar hemispheres, tuber, declive, and lobulus simplex). They also ascribed a similar pattern of efferent organization to the cerebellar cortex: the flocculonodular lobe and uvula were believed to project to both the vestibular nuclei and the fastigial nucleus. The anterior lobe, the medial lobulus simplex, and the medial anterior lobe were stated to project to the fastigial nucleus only, the neocerebellum was stated to project to the dentate nucleus, and the nucleus interpositus was believed to receive input from "the no-man's land" between pontocerebellar and spinocerebellar parts. As a result, the cerebellum was divided into a flocculonodular or vestibular region, a paleocerebellar region (the anterior lobe lingula, central, and culmen, and the posterior lobe pyramis, uvula, and paraflocculus) that integrates spinal and vestibular information and a neocerebellar region that was concerned with cortico-pontine processing. This pattern was particularly attractive clinically because it provides an anatomical explanation for the neurologic signs that characterized the neocerebellar, anterior lobe, and flocculo-nodular lobe syndromes.[21,25]

The topographical organization of cerebellar circuitry provides a basis for understanding regional effects of cerebellar stimulation or lesions on motor function. The uniform cytological organization of the cerebellar cortex has produced the hypothesis that the modular regions of cerebellar cortex perform analogous neural processing within different motor control pathways (e.g., References 21, 22, 37). The application of this hypothesis to autonomic control, then, proceeds in three stages: (1) a discussion of the roles of cerebellar circuitry that affect externally directed motor control, such as voluntary movement and reflexes involving striated muscle; (2) an exploration of the analogous motor activities in the internal, or autonomic control domain; (3) a review of evidence of cerebellar influences on these hypothetically analogous autonomic motor activities.

The seminal work of the neurologist Gordon Holmes[32-34] provides one of the most insightful descriptions of the effects of cerebellar injury on motor performance and perception. Although Holmes enumerated 12 clinical features suggesting the roles of the cerebellum in somatic motor control, they may be summarized as four main manifestations: (1) postural hypotonia and resulting impairment of hypotonic muscle, (2) mild asthenia and fatigability, (3) abnormal rate, force, and regularity of voluntary movements, and (4) loss of associated and automatic movements (e.g., the arm that is ipsilateral to a cerebellar lesion does not swing synchronously with the opposite leg during walking). These features indicate a role of the cerebellum in coordinating the timing and magnitude of activation of synergistic and antagonistic muscle groups.

Recent physiologic studies have provided a wealth of information regarding the role of cerebellar circuits in the control of somatic motor activities, including vestibulo-ocular reflexes, saccadic and smooth pursuit eye movements, and voluntary limb movements (e.g., References 15, 37, 40). However, in light of the relative paucity of parametric information regarding cerebellar effects on autonomic functions, only a few general principles are necessary for the present discussion. First, the fairly high resting discharge rate of Purkinje cells is consistent with the existence of a tonic inhibitory influence on their targets in the cerebellar and vestibular nuclei. Second, the temporal relationship between changes in Purkinje cell activity and activity at postsynaptic targets can result in either inhibition or disinhibition at that site. Third, cerebellar output can affect the amplitude of responses of brain stem reflex pathways. Finally, the cerebellar cortex appears to be important for adaptive modification of the performance of reflexes, with climbing fiber inputs serving as a performance error signal. These neural operations, then, contribute to the appropriate performance of movements requiring a coordinated activation of striated muscles.

4 AUTONOMIC MOTOR ACTIVITY: DEFINITION

In the early 20th century, the work of Gaskell[27] and Langley[42] produced the concept that autonomic motor control mechanisms are neural circuits that control the activity of smooth muscle in the cardiovascular system and viscera. These mechanisms may be either synaptic or neuroendocrine, mediated by either neural activity of the parasympathetic and sympathetic nervous systems in the periphery or release of bioactive substances into the systemic circulation by neurons or modified neurons in the hypophysis, epiphysis, or suprarenal glands. These seemingly diverse mechanisms share the common features of (1) sensitivity to visceral sensory input which monitors the status of the internal milieu and (2) a direct effect to control the status of the internal milieu. Hence, by extension from cerebellar influences on somatic motor activity, the aspects of autonomic motor activity that are potentially under cerebellar control may include both visceral motor and neuroendocrine mechanisms that affect smooth muscle to subserve homeostatic reflexes, mechanisms for regulating visceral and respiratory function during voluntary activity, and visceral motor and neuroendocrine components of both emotional responses and orienting responses to internal and external stimulation. Although respiratory activity is mediated largely by actions of striated muscle, it may be viewed analogously because it has an automatic, "involuntary" component that is sensitive to the blood pO_2 and pCO_2. The discussion that follows will focus on the cardiovascular effects of cerebellar stimulation because it has been the subject of most intense investigation.

5 CEREBELLAR EFFECTS ON THE CARDIOVASCULAR SYSTEM

5.1 Overview

Although Willis[82] originally proposed a cerebellar role in cardiovascular control, the first convincing experimental support for this hypothesis did not appear until the late 1930s (an excellent review of the pre-1940 literature is provided by Dow and Moruzzi[21]). Between 1937 and 1940, Moruzzi[48,49] reported four main effects of electrical stimulation of the cerebellar anterior lobe: (1) abolition of third-order blood pressure waves (Meyer waves), (2) inhibition of both pressor and respiratory responses to carotid clamping, (3) attenuation of both vasopressor responses elicited by stimulating the central stump of the superior laryngeal nerve and depressor responses elicited by electrical stimulation of the central stump of the vagus nerve, and (4) a depressor effect when blood pressure was elevated spontaneously; this latter effect was absent when blood pressure was either low or elevated by adrenaline or ephedrine administration. The former two findings were replicated by Wiggers[81] in barbiturate-anesthetized dogs. Moruzzi[48-50] concluded that these effects of anterior lobe stimulation on blood pressure reflect a direct cerebellar action on central vasomotor pathways because they were abolished by cocainization of the anterior lobe, yet were neither affected by bilateral cervical vagotomy nor accompanied by reflex (baroreceptor reflex-evoked) changes in heart rate. This conclusion is significant because it produced two major insights into the cerebellar role in mediating visceral motor activity. First, it provided experimental support for the hypothesis that cerebellar contributions to somatic and visceral motor activity are mediated by a "central inhibition of bulbospinal centers by the paleocerebellar cortex" (Reference 49, p. 30), that is, by similar mechanisms and neuronal circuitry. Second, the effects of cerebellum on centrally mediated cardiovascular reflexes implied that "the aims of paleocerebellar activity should be described as a control more than a simple unidirectional inhibition of somatic and autonomic bulbospinal reflexes" (Reference 49, p. 29). This reflected the view that the cerebellum controls blood pressure by modulating activity in brain stem visceral motor pathways, such that "... the bulbopontine centers impose on the paleocerebellum the nature and intensity of the effort which must be accomplished, both in the somatic and the autonomic sphere" (Reference 49, p. 30). This concept of the role of the cerebellum as a modulator of autonomic reflexes also may extend to cardiovascular responses to sciatic nerve stimulation, which are converted from pressor to depressor responses by cerebellectomy.[60]

Studies of the cardiovascular effects of stimulation of the medial aspect of cerebellar cortex indicate that multiple regions may influence cardiovascular control. Since the original studies by Moruzzi[48-50] and Wiggers,[81] more refined electrical and chemical microstimulation methods have been used to both confirm their results and extend these findings to other sites along the medial aspect of the cerebellar cortex. These studies report that relatively long trains of low-intensity (<100 μA) stimuli are required to elicit cardiovascular responses (e.g., 6 to 10-s trains of 0.1-ms pulses at 200 Hz)[55] from cerebellar sites. These methods have yielded evidence of four regions of cerebellar cortex that may affect cardiovascular control: an anterior lobe region, two posterior lobe regions, and a lateral

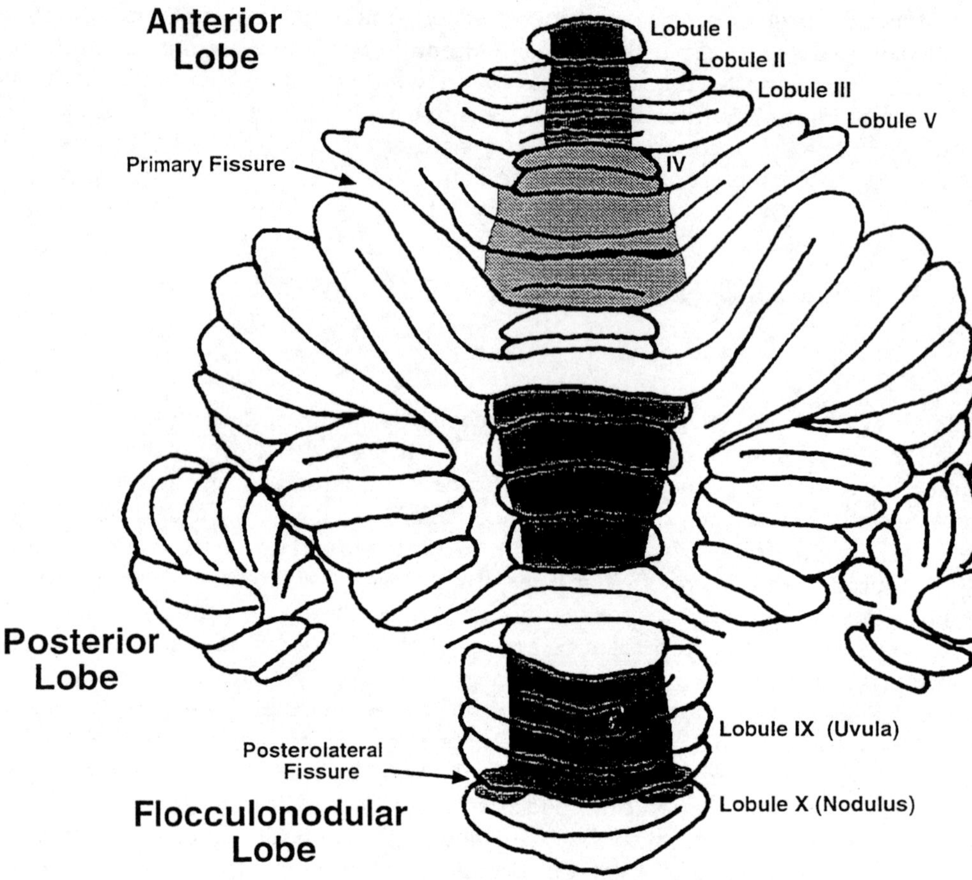

FIGURE 1
The location of cerebellar cortical areas implicated in cardiovascular (dark shading) and respiratory (light shading) control are illustrated on a schematic view of an unfolded rabbit cerebellum. The areas are all located along the medial aspect of cerebellar cortex. The anterior lobe contains two regions, a lobule I-III region that affects cardiovascular reflex control and a lobule IV-V region that appears to affect respiratory reflexes. The posterior lobe also contains two areas, a lobule VI-VIIIa region and a lobule IX-X region. It is hypothesized that the former region participates in classical conditioning of stimulus parameters for triggering orienting responses, while the latter area may modulate vestibulo-autonomic responses.

nodulus-uvula region (Figure 1). Since these cerebellar regions receive vestibular,[37,41] nucleus of the solitary tract,[72] and parabrachial nucleus[71] mossy fiber inputs, they are potential sites for integrated processing of vestibular and autonomic signals. The former three regions are located within the most medial region of the cerebellar cortex, zone A. Although Purkinje cells within this longitudinal region project predominantly to the fastigial nucleus,[79,80] more recent studies have suggested the existence of smaller projections to the vestibular nuclei[4,59] and parabrachial nucleus.[59,74] Purkinje cells of the nodulus-uvula area, on the other hand, appear to affect blood pressure via output to the dorsal aspect of the superior vestibular nucleus and nucleus of the brachium conjunctivum.[28] These output pathways then form potential substrates for influencing autonomic function.

5.2 Anterior Lobe

Both Moruzzi[48-50] and Wiggers[81] reported that the cerebellar sites eliciting effects on blood pressure reflex regulation were located along the vermian (medial) aspect of the cerebellar anterior lobe. Their findings suggested that circuits mediating the reflex control of blood pressure are affected by anterior lobe activation. Subsequent electrical stimulation studies have confirmed the existence of sites that elicit both depressor and pressor responses in this region in cats[31,61] and rabbits.[55] A rebound pressor response after cessation of the stimulus was also observed consistently. Other sites produced an increase in skeletal muscle blood flow without affecting systemic blood pressure.[31] The available evidence indicates that there is no differential topographic distribution of these different response sites across the surface of the anterior lobe cortex. However, Nisimaru et al.[55] reported that anterior lobe sites producing pressor and depressor responses in rabbits are predominantly within 1.5 mm of the midline of lobules I-III. Given the close proximity of these regions to the rostral fastigial nucleus and subfastigial bundle,[46] though, the attribution of blood pressure response to effects at a single site should be interpreted with caution.

The location of stimulation sites in the anterior lobe that produce cardiovascular effects has been described in the greatest detail by Nisimaru et al.,[55] who reported that the sites were located predominantly within zone A of the rabbit. Zone A is divided into two parasagittal regions that receive different climbing fiber inputs from the caudal aspect of the medial accessory olive. The medial band, termed zone A_1, has been defined as a region that receives spino-olivocerebellar input from the hindlimb, while the more lateral zone A_2 receives spino-olivocerebellar input from the forelimb.[14] The location of these neurons in the central third of the caudal medial accessory olive is consistent with the distribution of stimulation sites that either (1) inhibit the depressor component of the carotid sinus reflex[69] or (2) elicit depression of renal sympathetic activity.[52] There is less information, though, regarding the organization of efferent pathways mediating cardiovascular effects of anterior lobe stimulation. Anatomical studies have identified projections of zone A Purkinje cells to the rostral fastigial nucleus (e.g., References 37, 79, 80), rostral medial vestibular nucleus,[4] and parabrachial region (from scattered zone A and zone B Purkinje cells[75]). Of these regions, the fastigial nucleus[37] and parabrachial region connections seem to be likely candidates for these pathways, since they are connected reciprocally and they contribute projections to the nucleus of the solitary tract.

The mossy fiber inputs to the anterior lobe are also consistent with a role in integrating vestibular, proprioceptive, and autonomic information. In addition to spinocerebellar (dorsal and ventral spinocerebellar tract), cuneocerebellar, and vestibular nuclear mossy fiber projections (for review see Reference 37), anatomical studies have revealed potential autonomic mossy fiber inputs to this region from the nucleus ambiguus/para-ambiguus region,[41,62] parabrachial region,[62,71] and caudal nucleus tractus solitarius.[72] Anatomic[19] and physiologic[74] data also indicate that the anterior lobe cortex receives hypothalamic influences via non-climbing fiber pathways and that Purkinje cell responses are sensitive to Pavlovian fear-conditioning paradigms.[75] However, Saint Cyr and Woodward[63,64] were unable to locate fornix-evoked responses in the vermis of lobules I-III, which suggests that fornix inputs may not directly influence Purkinje cells in the autonomic response sites in the anterior lobe.

The tonic effects of anterior lobe stimulation on elevated blood pressure were noted by Moruzzi[48-50] as evidence for a tonic inhibitory effect of cerebellar cortex on vascular tone. More recent studies indicate that the systemic pressor, systemic depressor, and skeletal muscle flow changes during anterior lobe stimulation reflect a cerebellar influence on multiple mechanisms influencing tonic sympathetic vasomotor drive. The independence of central mechanisms for the skeletal muscle blood flow changes is suggested by their occurrence in paralyzed animals in the absence of either renal blood flow changes or system blood flow effects.[31] Further, the pressor responses can be converted into depressor responses by either deep anesthesia,[61] electrolytic lesions of the posterior hypothalamus, or decerebration,[31] without affecting the subsequent rebound excitation. These effects have suggested that the pressor mechanisms require an intact hypothalamus (and/or telencephalic limbic circuits), while brain stem mechanisms alone are sufficient to elicit depressor effects.[31] Furthermore, two main findings are consistent with Moruzzi's hypothesis[49,50] that the cerebellar anterior lobe inhibits central sympathetic control circuits. First, anterior lobe-evoked depressor and pressor responses are correlated with respective decreases or increases in renal nerve discharge rates.[55] Second, the pressor responses are accompanied by both decreased renal arterial flow and a cutaneous vasoconstriction, all of which are abolished by systemic phenotolamine administration.[31] Thus, they suggest that these blood pressure and blood flow responses reflect central effects on outflow to peripheral vascular sympathetic responses mediated by α-adrenoreceptors.

Since the initial observations that anterior lobe stimulation suppressed Meyer waves and carotid sinus reflexes,[48-50,81] exploration of a potential role of the anterior lobe vermis in modulating the cardiovascular reflexes has been modest. Ramu and Bergmann[60] reported that Meyer waves are enhanced by anterior lobe ablation and abolished by complete cerebellectomy, while Hering-Traube waves (blood pressure variations at respiratory frequency) were preserved. These effects on spontaneous oscillations in blood pressure raise the hypothesis that the anterior lobe may contribute to control of blood pressure *lability*, such that increased Purkinje cell activity depresses Meyer waves. The lack of heart rate changes during stimulation-evoked blood pressure changes,[48,49] on the other hand, suggests that activation of some sites in the anterior lobe may depress baroreflex sensitivity, which would tend to decrease the magnitude of compensatory heart rate fluctuations in response to blood pressure fluctuations. These findings, then, suggest that the anterior lobe cortex may participate in regulation of the degree of precision of blood pressure control during motor activity.

Summary and Hypothesis — The midline areas of the anterior lobe display characteristics that are consistent with a role in regulating cardiovascular reflex performance during postural adjustments. This region receives mossy fiber inputs from vestibular, somatic sensory (including proprioceptive), and autonomic pathways. Climbing fiber inputs, on the other hand, reflect somatic sensory spino-olivary pathways, which is consistent with a role of the anterior lobe in the control of posture.[14,21,37] The available physiologic data suggest that zone A of lobules I-III can both produce transient effects on blood pressure by central inhibition of sympathetic control pathways and inhibit cardiovascular reflexes and Meyer waves. These findings suggested that the anterior lobe may participate in controlling the lability of blood pressure (i.e., precision of reflex control of blood pressure) during static posture and postural shifts, which imposes an orthostatic load on the cardiovascular system. However, this linkage between postural changes and blood pooling carries an additional implication: posture adjustments may be viewed as another mechanism for cardiovascular control (e.g., adoption of a head-down posture for light-headedness). This argument leads to the hypothesis that the circuitry involving the lobule I-III region may coordinate or match the cardiovascular consequences of movement, cardiac, and baroreceptor reflex sensitivity with sympathetic activity during postural changes.

5.3 Rostral Posterior Lobe: Lobules VIIa-VIIIa

A region eliciting both a depressor response and inhibition of sympathetic activity was reported in posterior lobe vermal lobules VI-VIIIb[54,55] of chloralose-urethane-anesthetized rabbits. By contrast, Bradley et al.[11-13] reported that they could not elicit responses from this region with shorter pulse train stimuli (5 vs. 10 s). The lowest-threshold sites were located predominantly in the medial aspect (zone A) of lobules VIIa-VIIIa.[55] Purkinje cells in this region receive climbing fiber inputs from the medial aspect of the medial accessory olive,[52] which can produce vagally driven complex spikes[57] that may also reflect a convergence of somatic-renal and renal afferent activity.[78] This area contributes efferent projections to the fastigial nucleus.[36,37,79,80]

Like the anterior lobe region, the lobule VII-VIIIa region receives mossy fiber inputs from the superior, medial and inferior vestibular, and parabrachial nuclei.[10,41,71] However, the afferent organization appears to differ from the lobule I-III autonomic region in three basic respects. First, Purkinje cell responses in this region reflect a wealth of different sensory and motor-related signals. Lobule VII is included in the classic auditory and visual areas of cerebellar cortex.[21,70] Lobule VII has also been termed the oculomotor area because stimulation can evoke eye movements and Purkinje cells reflect both saccades and smooth pursuit eye movements and stretch receptor information from extraocular muscles.[24,40,56,77,83] Second, Purkinje cells in the vermis of lobules VI-VIIb respond prominently to electrical stimulation of the fornix,[63-64] indicating a limbic input that is absent in the other cardiovascular regions of cerebellar cortex. Finally, the lobule VII-VIIIa region appears to receive only a minor direct input from the nucleus of the solitary tract.[10,72] The existence of limbic inputs and the absence of an effect of lobule VI-VIII ablation on the cardioinhibitory component of the baroreflex[43] suggest that the rostral posterior lobe cardiovascular region may serve a different role than either the anterior lobe or the caudal posterior lobe (uvula) in control of cardiovascular activity. This has been confirmed by the finding that ablation of the vermis of lobules IV-VIII in rats severely impairs the development of classically conditioned bradycardic responses, without a significant effect on either the baseline heart rate during the initial session or initial heart rate responses to the conditioned stimulus (6-s, 92-dB, 1-kHz tone) alone.[76] This attenuation of conditioned bradycardic responses was confirmed by Sebastiani et al.[67] in the rabbit, and they further reported that the development of these classically conditioned responses was unaffected by lesions of the uvula (lobule IX) alone. Since the bradycardic component of orienting responses was unaffected by lobule VII lesions, these results imply that the lobule VIIa-VIIIa cardiovascular region contributes to modification of the sensory conditions that elicit a bradycardic response (classical conditioning in an experimental sense). This line of reasoning raises the hypothesis that lobule VIIa-VIIIa participates in modifying the criteria for activation of bradycardic orienting responses.

Summary and Hypothesis — The lobule VIIa-VIIIa region displays characteristics that are consistent with a role in selecting or adjusting stimulus conditions for autonomic components of orienting responses. Anatomic and physiologic data indicate that the lobule VIIa-VIIIa region receives mossy fiber afferents that may reflect the magnitude and location of stimuli in exteroceptive space: secondary vestibular, extraocular proprioceptive, oculomotor, and auditory inputs have been documented. Purkinje cells in this region also respond to fornix stimulation, which may reflect spatial and/or affective information. Consistent with a role in regulation of visceral motor responses, climbing fiber inputs reflect renal and convergent somatic-renal afferent activity. However, it is important to note that autonomic components of orienting responses are part of a coordinated motor program that includes gaze adjustments, which are mediated by both eye movements (saccades, vergence, and smooth pursuit), lens accommodation, pupillary responses, and body movements. Hence, it is noteworthy that electrical stimulation of lobule VII produces lens accommodation in cats[35] and pupillary responses[61] (either mydriasis or meiosis). These considerations lead to the hypothesis that the lobule VIIa-VIIIa cardiovascular region participates in the regulation of orienting responses by (1) contributing to the selection of stimulus criteria for triggering the responses and (2) assisting in coordination of the visceral and somatic motor components of orienting responses.

5.4 Caudal Posterior Lobe

5.4.1 Lobule IX

Rasheed et al.[61] reported that electrical stimulation of unspecified sites in the uvula produces a depressor response in anesthetized cats, but a pressor response in unanesthetized, paralyzed preparations. This region was explored in greater detail by Bradley et al.,[11,12] who reported a slight bradycardia, a depressor response, inhibition of renal nerve activity, and vasodilitation of the hindlimb vascular bed after electrical or chemical stimulation of the medial aspect of the uvula (lobule IX) in anesthetized rabbits and cats. The opposite responses (tachycardia, a pressor response, increased renal nerve activity, and vasoconstriction of renal and femoral artery beds) were observed in decerebrate preparations. Paton et al.[59] demonstrated that these responses represent two distinct cerebellar output circuits: lesions of the superior cerebellar peduncle abolished depressor responses in anesthetized animals, without affecting the pressor response patterns after decerebration, while lesions of the inferior cerebellar peduncle produced the opposite result (selective abolition of pressor responses in decerebrate preparations). This suggested that the depressor/bradycardic responses in anesthetized animals are mediated via ascending projections to the lateral parabrachial nucleus, but that pressor/tachycardic responses are mediated via projection to either the vestibular nuclei or medial parabrachial nucleus.[59]

The sites eliciting these responses appear to correspond with zones A and B of lobule IX. The efferent projections of the Purkinje cells in this region to the fastigial nucleus and the superior, rostral medial and inferior vestibular nuclei have been described repeatedly in the literature with a variety of anatomic tracing methods (e.g., References 4, 59, 79). In addition, Paton et al.[59] reported that lobule IXb Purkinje cells also project to the caudal aspect of the lateral and medial parabrachial nucleus. It is perhaps significant that these output relations to the fastigial, vestibular, and parabrachial nuclei are similar to the pattern for zone A of lobules I-III, particularly since both regions (1) receive autonomic-related inputs (presumably mossy fibers) from both the caudal nucleus tractus solitarius[62,72] and the parabrachial nucleus,[71] (2) receive mossy fiber projections from the superior, medial, and inferior vestibular nuclei,[41] and (3) lie outside the region that responds to fornix stimulation.[63,64] These superficial similarities, then, are consistent with the fact that stimulation of the two areas produces similar cardiovascular effects.

Despite these connectional similarities with the anterior lobe area, the climbing fiber inputs to the uvula differ markedly because they arise from inferior olivary divisions that respond both to vertical semicircular canal and otolith inputs.[8,9,65] A sagittal band along the medial aspect of the uvula receives climbing fiber inputs from the caudal half of nucleus β of the medial accessory olive,[65] a region that responds to natural rotation in the plane of the ipsilateral anterior/contralateral posterior semicircular canals.[9] The rostral half of nucleus β projects to an adjacent, lateral band[65] and responds to stimuli in the plane of the ipsilateral posterior/contralateral anterior semicircular canal pair.[9] This is flanked by a narrow band that receives input from the caudolateral medial accessory olive,[65] a region that receives a projection from nucleus tractus solitarius[44] and also projects to zone A of lobule VII.[52] Finally, this region is flanked laterally by a zone innervated by the dorsal half of the dorsomedial cell column,[8,65] which contains neurons that show prominent modulation that reflects utricular input.[8] These inputs, then, suggest that parasagittal zones in the uvula may serve an important role cerebellar modulation of vestibulo-autonomic responses.

The existing database gives relatively few clues regarding the role of the uvula in autonomic or vestibulo-autonomic control. A role in baroreceptor reflex modulation is suggested by the finding that, after uvula ablation,[43] both the absolute magnitude and the gain of baroreceptor reflex bradycardic responses were greater than prelesion control values in 14/19 rabbits. By contrast, lesions of this region failed to affect classical conditioning of bradycardia. Although these lesions also encroached on the lateral nodulus-uvula region, this finding is consistent with the hypothesis that the caudal posterior lobe may contribute to control of vestibular interactions with baroreflex mechanisms.

5.4.2 *Lateral Nodulus-Uvula Border Region*

Cardiovascular responses are also elicited from a site along the border of the nodulus and uvula of the rabbit, centered approximately 3 mm lateral to the midline.[30,53] Low-threshold, pulse-train (10-s trains) electrical stimulation of this site produces a depressor response, accompanied by a depression of renal nerve activity[53] but no change in heart rate.[30] This depressor effect was attenuated by both α-adrenergic (phentolamine) and sympathetic ganglionic blockade, but was unaffected by either β-adrenergic blockade (propranolol) or parasympatholytic (atropine methylnitrate) treatment.[30] The attenuation of the depressor response by phentolamine treatment was not confounded by the hypotensive actions of the drug because (1) the effects of phentolamine remained when blood pressure was raised to the pretreatment level by angiotensin II infusion and (2) nitroprusside-induced hypotension did not produce an equivalent change in the nodulus-uvula depressor response in untreated animals.[30] Thus, it was suggested that this region inhibits brain stem circuits that provide sympathetic drive to vasculature via alpha-adrenergic mechanisms.

Henry et al.[30] presented two lines of evidence that the nodulus-uvula depressor response is mediated via Purkinje cell projections to a region that includes the dorsal aspect of the superior vestibular nucleus and the adjacent nucleus of the brachium conjunctivum, a small cell group that caps the superior cerebellar peduncle. First, anterograde tracing experiments revealed direct projections to these regions. Second, microinjection of bicuculline methiodide into these terminal regions either significantly depressed or abolished the nodulus-uvula evoked response. Since the depressor response was unchanged after microinjections of hydrastine (a very-low-potency analog of bicuculline) in the same regions, it was suggested that the depressor effect reflects direct Purkinje cell inhibition of dorsal superior vestibular nucleus and nucleus of the brachium conjunctivum neurons, via actions at GABA-A receptors. Thus, it seems likely that this region is involved in coordinating vestibular-autonomic responses.

The present data do not elucidate whether the medial uvula and nodulus-uvula sites should be regarded as separate regions or as two specialized components of a single cerebellar mechanism that influences vestibulo-autonomic interactions. For example, anatomic studies suggest that the climbing fiber inputs to this region arise from the dorsomedial cell column and/or the rostrolateral medial accessory olive.[6,39] Since these olivary regions also innervate the lateral aspect of lobule IX,[65] the nodulus-uvula site may be considered a caudolateral extension of the lobule IX region. The close proximity of these sites further complicates differentiation of these regions. For example, bilateral lidocaine infusions into this region increased the sensitivity of baroreceptor reflex responses to pressor doses of angiotensin II,[28] which is consistent with the effects of lobule IX lesions on the cardioinhibitory baroreceptor reflex.[43] However, it is not possible to determine whether this evidence for baroreceptor reflex modulation reflects a contribution of one or both of the regions.

The nodulus-uvula depressor response has several characteristic properties that may shed light upon its potential functions. These properties were described by Henry et al.[30] as follows. First, the response is transient, consisting of a 7 to 10 s decrease in blood pressure followed by a recovery. This transient response profile is constant for stimulus trains lasting at least 1 min. Second, the response is refractory. After the response is elicited, a second response of equal magnitude can be evoked from the same site only after a 40 to 60 s pause. This refractoriness appears to be a property of a lateralized brain stem or spinal mechanism because activation of one site does not affect the magnitude of the response to a contralateral stimulus. Third, responses to simultaneous, bilateral stimulation are not additive. These properties suggest that this nodulus-uvula region may be engaged to provide a transient and intermittent effect on blood pressure control during head movements. For example, Doba and Reis[20] demonstrated that vestibular nerve section increases the depth and prolongs the recovery phase of orthostatic hypotension. The lateral nodulus-uvula circuit provides one potential vestibular-related mechanism for transiently modulating α-sympathetic drive to resistance vessels during orthostatic responses and adjusting the sensitivity of baroreceptor reflexes for longer-term control.

Summary and Hypothesis — The medial uvula (lobule IX) and lateral uvula-nodulus cardiovascular regions display properties that are consistent with modulation of vestibulo-autonomic interactions. These regions receive vestibular and autonomic mossy fiber inputs and appear to correspond to regions that receive climbing fiber inputs that respond to otolith and vertical semicircular canal activation. These responses are likely mediated by direct projections to vestibulo-autonomic pathways. For example, the nodulus-uvula response is mediated by projections to the dorsal aspect of the superior vestibular nucleus, a region that gives rise to projections to the parabrachial nucleus.[7] On the other hand, the medial uvula bradycardia/depressor response is mediated by a direct projection to the lateral parabrachial nucleus, a structure that receives direct vestibular nucleus inputs.[7] Thus, it may provide modulation to vestibular-related brain stem circuits that influence sympathetic outflow.

5.5 Fastigial Nucleus

Since many vermis zone A Purkinje cells send axons to the fastigial nucleus, there is an expectation that fastigial nucleus neurons will contribute to cardiovascular control. However, the nature of these responses has been the source of some controversy. Although electrical stimulation of the rostral fastigial nucleus produces a tachycardic-pressor response,[1,46] more recent evidence indicates that this response is due to fiber-of-passage activation of commissural projections of the lateral parabrachial nucleus in a subfastigial fiber bundle.[47] This conclusion is consistent with the finding that the pressor response is unaffected by kainic acid ablation of the fastigial nucleus.[29] Further, although chemical stimulation of the fastigial nucleus revealed a depressor site[16] in the rat, this finding has not been replicable in the cat.[47]

6 CEREBELLAR CONTRIBUTIONS TO NON-CARDIOVASCULAR AUTONOMIC RESPONSES

Although evidence has been presented of non-cardiovascular autonomic responses after electrical or chemical stimulation of the cerebellum,[21,37] the results are primarily anecdotal. The findings for stimulation of the anterior lobe present a case in point. For example, McDonald[45] reported piloerection and forceful urination after stimulation of the medial aspect of lobule II of cats, while Sawyer et al.[66] and Rasheed et al.[61] reported that pupillary responses are elicited by stimulation of sites within lobules I-III (lingula and lobulus centralis). In addition, the latter paper described lingula-evoked penile erection in male and intrauterine pressure increase in female cats during direct stimulation of anterior lobe sites. Similar findings have been reported for the other "cardiovascular regions" of the vermis. The fact that these other stimulus-evoked autonomic responses appear grossly co-localized with cardiovascular responses is consistent with the general hypothesis that, like its role in coordinating and calibrating somatic motor activity, the cerebellum contributes to the generation of coordinated autonomic and autonomic-somatic motor activity.

7 CEREBELLAR CONTRIBUTIONS TO RESPIRATORY CONTROL

Respiratory effects have been reported in association with cardiovascular effects of stimulation of the lobule I-III[21] and lobule IX[11,12] regions, suggesting that respiratory circuitry is another component of the somatic and autonomic motor programs that are coordinated by these regions. However, there is also evidence suggesting the existence of a non-cardiovascular region that influences respiratory function. Decima and von Euler[18] reported that tetanic stimulation of anterior lobe lobules IV-V produced a predominantly ipsilateral depression of phrenic nerve activity. Shorter-duration stimuli produce both a reduction in inspiratory-related discharges and augmented phrenic nerve reflex responses to intercostal nerve stimulation. The finding that ventilatory responses to carbon dioxide inhalation are depressed by anterior lobe ablation, but unaffected by posterior lobe ablation[68] is also consistent with the implication that lobule IV-V participates in some aspect of respiratory reflex control.

The zonal location of lobule IV-V sites that affect respiratory function has not been described. However, lobules IV-V receive afferents from nucleus tractus solitarius[72] and the parabrachial nucleus,[71] and at the lateral aspect the region responds to intercostal nerve stimulation.[17] Since the region that responds to electrical stimulation of the fornix spans lobules IV-V,[63,64] it is suggested that this region integrates limbic, visceral, and respiratory-related information for control of inspiratory-related muscle activity. In particular, its location in the anterior lobe raises the hypothesis that this region may participate in the coordination of respiratory and postural activity in trunk muscles.

8 SUMMARY AND CONCLUSIONS

1. Three sites for cerebellar influences on cardiovascular function have been identified in the medial cerebellum: anterior lobe region (zone A of lobules I-III), rostral posterior lobe region (zone A of lobules VIIa-VIIIa), and caudal posterior lobe-nodulus region (zone A of lobule IX and a lateral site on the border of lobule IX and the nodulus).
2. These regions can produce both pressor and depressor responses through inhibition of brainstem circuits influencing sympathetic outflow. The anterior lobe and caudal posterior lobe regions may also influence the magnitude of cardiovascular reflexes.
3. The connectivity and the physiological effects of stimulation and lesions of cerebellar cortex raise the hypothesis that these areas are specialized for coordinating somatic and autonomic activity. It is suggested that the rostral posterior lobe region participates in the regulation of orienting responses by (1) contributing to the selection of stimulus criteria for triggering the responses and (2) assisting in coordination of the visceral and somatic motor components of orienting responses. The anterior lobe region, by contrast, is hypothesized to participate in coordinating the anticipated cardiovascular consequences of movement, cardiac and baroreceptor reflex sensitivity, and sympathetic activity during postural changes. Finally, it is proposed that the caudal posterior lobe-nodulus area may participate in the control of vestibulo-autonomic reactions.

REFERENCES

1. Achari, N. K. and Downman, C. B. B., Autonomic responses elicited by stimulation of fastigial nuclei in the cat, *J. Physiol. (London)*, 204, 130, 1969.
2. Andersson, G. and Oscarsson, O., Projections to the lateral vestibular nucleus from cerebellar climbing fiber zones, *Exp. Brain Res.*, 32, 549, 1978.

3. Andersson, G. and Oscarsson, O., Climbing fiber microzones in cerebellar vermis and their projection to different groups of cells in the lateral vestibular nucleus, *Exp. Brain Res.*, 32, 565, 1978.
4. Balaban, C. D., Olivo-vestibular and cerebello-vestibular connections in albino rabbits, *Neuroscience*, 12, 129, 1984.
5. Balaban, C. D., Distribution of inferior olivary projections to the vestibular nuclei of albino rabbits, *Neuroscience*, 24, 119, 1988.
6. Balaban, C. D. and Henry, R. T., Zonal organization of olivo-nodulus projections in albino rabbits, *Neurosci. Res.*, 5, 409, 1988.
7. Balaban, C. D., Vestibular nucleus projections to the parabrachial nucleus in rabbits: Implications for vestibulo-autonomic interactions, in Abstracts, 1995 Midwinter Res. Meet. Assoc. Res. Otolaryngol., No. 402, 1995.
8. Barmack, N. H. and Fagerson, M. H., Vestibularly evoked activity of single neurons in the dorsomedial cell column of the vestibular nuclei in rabbit, *Neurosci. Abstr.*, 20, 1190, 1994.
9. Barmack, N. H., Fagerson, M., Fredette, B. J., Mugnaini, E., and Shojaku, H., Activity of neurons in the beta nucleus of the inferior olive of the rabbit evoked by natural vestibular stimulation, *Exp. Brain Res.*, 94, 203, 1993.
10. Batini, C., Buisseret-Delmas, C., Corvisier, J., Hardy, O., and Jassik-Gerschenfeld, D., Brain stem nuclei giving fibers to lobules VI and VII of the cerebellar vermis, *Brain Res.*, 153, 241, 1978.
11. Bradley, D. J., Ghelarducci, B., Paton, J. F. R., and Spyer, K. M., The cardiovascular responses elicited from the posterior cerebellar cortex in the anaesthetized and decerebrate rabbit, *J. Physiol.*, 383, 537, 1987.
12. Bradley, D. J., Pascoe, J. P., Paton, J. F. R., and Spyer, K. M., Cardiovascular and respiratory responses evoked from the posterior cerebellar cortex and fastigial nucleus in the cat, *J. Physiol.*, 383, 107, 1987.
13. Bradley, D. J., Ghelarducci, B., and Spyer, K. M., The role of the posterior cerebellar vermis in cardiovascular control, *Neurosci. Res.*, 12, 45, 1991.
14. Brodal, A. and Kawamura, K., *Olivocerebellar Projection: A Review*, Springer-Verlag, Berlin, 1980.
15. Brooks, V. and Thach, T., Cerebellar control of movement, in *Handbook of Physiology, The Nervous System*, Vol. II: American Physiological Society, Bethesda, MD, 1982, chap. 10.
16. Chida, K., Iadeccola, C., Underwood, M., and Reis, D. J., A novel vasodepressor response elicited from the rat cerebellar fastigial nucleus: the fastigial depressor response, *Brain Res.*, 370, 378, 1986.
17. Coffey, G. L., Godwin-Austen, R. B., Macgillivray, B. B., and Sears, T. A., The form and distribution of the surface evoked responses in cerebellar cortex from intercostal nerves in the cat, *J. Physiol.*, 212, 129, 1971.
18. Decima, E. E. and von Euler, C., Intercostal and cerebellar influences on efferent phrenic activity in the decerebrate cat, *Acta Physiol. Scand.*, 76, 148, 1969.
19. Dietrichs, E. and Haines, D. F., Interconnections between hypothalamus and cerebellum, *Anat. Embryol.*, 179, 207, 1989.
20. Doba, N., and Reis, D. J., Role of the cerebellum and vestibular apparatus in regulation of orthostatic reflexes in the cat, *Circ. Res.*, 34, 9, 1974.
21. Dow, R. S. and Moruzzi, G., *The Physiology and Pathology of the Cerebellum*, University of Minnesota Press, Minneapolis, 1958.
22. Eccles, J. C., Ito, M., and Szentagothai, J., *The Cerebellum as a Neuronal Machine*, Springer-Verlag, New York, 1967.
23. Ferrier, D., *The Functions of the Brain*, Smith & Elder, London, 1876.
24. Fuchs, A. F. and Kornhuber, H. H., Extraocular muscle afferents to the cerebellum of the cat, *J. Physiol.*, 200, 713, 1969.
25. Fulton, J. F. and Dow, R. S., The cerebellum: a summary of functional localization, *Yale J. Biol. Med.*, 10, 89, 1937.
26. Gall, F. J., *On the Functions of the Cerebellum by Drs. Gall, Vimont and Broussais*, Comb, G., Transl., Machlachlan & Stewart, Edinburgh, 1838.
27. Gaskell, W. H., *The Involuntary Nervous System* (Monographs on Physiology), Longmans, Green and Co., New York, 1916.
28. Henry, R. T., Balaban, C. D., and Connor, J. D., Phasic and tonic influences of the cerebellar nodulus-uvula on cardiovascular tone, *FASEB J.*, 2, A1325, 1988.
29. Henry, R. T. and Connor, J. D., Evidence that the fastigial pressor response may be due to stimulation of axons of passage, *Soc. Neurosci. Abstr.*, 12, 536, 1986.
30. Henry, R. T., Connor, J. D., and Balaban, C. D., Nodulus-uvula depressor response: central GABA-mediated inhibition of α-adrenergic outflow, *Am. J. Physiol.*, 256 (*Heart Circ. Physiol.*, 25), H1601, 1989.
31. Hoffer, B. J., Mitra, J., and Snider, R. S., Cerebellar influences on the cardiovascular system, in *Limbic System Mechanisms and Autonomic Function*, Hockman, C. H., Ed., Charles C. Thomas, Springfield, IL, 1972, 91.

32. Holmes, G., The Croonian lectures on the clinical symptoms of cerebellar disease and their interpretation, I, *Lancet*, 100, 1177, 1922.
33. Holmes, G., The Croonian lectures on the clinical symptoms of cerebellar disease and their interpretation, II, *Lancet*, 100, 1231, 1922.
34. Holmes, G., The Croonian lectures on the clinical symptoms of cerebellar disease and their interpretation, III, *Lancet*, 100, 56, 1922.
35. Hosoba, M., Bando, T., and Tsukahara, N., The cerebellar control of accomodation of the eye in the cat, *Brain Res.*, 153, 495, 1978.
36. Ikeda, Y., Noda, H., and Sugita, S., Olivocerebellar and cerebelloolivary connections of the oculomotor region of the fastigial nucleus in the macaque monkey, *J. Comp. Neurol.*, 284, 463, 1989.
37. Ito, M., *The Cerebellum and Neural Control*, Raven Press, New York, 1984.
38. Ito, M., Long-term depression, *Annu. Rev. Neurosci.*, 12, 85, 1989.
39. Katayama, S. and Nisimaru, N., Parasagittal zonal pattern of olivonodular projections in rabbit cerebellum, *Neurosci. Res.*, 5, 424, 1988.
40. Keller, E. L., The cerebellum, in *The Neurobiology of Saccadic Eye Movements*, Wurtz, R. H. and Goldberg, M. E., Eds., Elsevier, New York, 391, 1989.
41. Kotchabhakdi, N. and Walberg, F., Cerebellar afferent projections from the vestibular nuclei in the cat: an experimental study with the method of retrograde axonal transport of horseradish peroxidase, *Exp. Brain Res.*, 31, 591, 1978.
42. Langley, N. N., *The Autonomic Nervous System, Part I*, W. Heffer, Cambridge, 1921.
43. LaNoce, A., Bradley, D. J., Goring, M. A., and Spyer, K. M., The influence of lobule IX of the cerebellar posterior vermis on the varoreceptor reflex in the decerebrate rabbit, *J. Autonom. Nerv. Sys.*, 32, 31, 1991.
44. Loewy, A. D. and Burton, H., Nuclei of the solitary tract: efferent projections to the lower brain stem and spinal cord of the cat, *J. Comp. Neurol.*, 181, 421, 1978.
45. McDonald, J. V., Responses following electrical stimulation of anterior lobe of cerebellum in cat, *J. Neurophysiol.*, 16, 69, 1953.
46. Miura, M. and Reis, D. J., Cerebellum: a pressor response elicited from the fastigial nucleus and its efferent pathways in the cerebellum, *Brain Res.*, 12, 595, 1969.
47. Miura, M. and Takayama, K., The site of the origin of the so-called fastigial pressor response, *Brain Res.*, 473, 352, 1988.
48. Moruzzi, G., Action inhibitrice du paleocervelet sur les reflexes circulatoires et respiratoires d'origine sino-carotidienne, *Comptes Rendus. Soc. Belge Biol.*, 128, 533, 1938.
49. Moruzzi, G., Paleocerebellar inhibition of vasomotor and respiratory carotid sinus reflexes, *J. Neurophysiol.*, 20, 1940.
50. Moruzzi, G., *Problems in Cerebellar Physiology*, Charles C. Thomas, Springfield, IL, 1950.
51. Neuberger, M., *The Historical Development of Experimental Brain and Spinal Cord Physiology before Fluorens*, Clarke, E., Transl., Johns Hopkins University Press, Baltimore. Translation of *Die Historische Entwicklung*, 1897.
52. Nisimaru, N., Okahara, K., and Nagao, S., Olivocerebellar projection to the cardiovascular zone of rabbit cerebellum, *Neurosci. Res.*, 12, 240, 1991.
53. Nisimaru, N. and Watanabe, Y., A depressant area in the lateral nodulus-uvula of the cerebellum for renal sympathetic nerve activity and systemic blood pressure in the rabbit, *Neurosci. Res.*, 3, 177, 1985.
54. Nisimaru, N. and Yamamoto, M., Depressant action of the posterior lobe of the cerebellum upon renal sympathetic nerve activity, *Brain Res.*, 133, 371, 1977.
55. Nisimaru, N., Yamamoto, M., and Shimoyama, I., Inhibitory effects of cerebellar cortical stimulation on sympathetic nerve activity in rabbits, *Jpn. J. Physiol.*, 34, 539, 1984.
56. Noda, H. and Fujikado, T., Topography of the oculomotor area of the cerebellar vermis in macaques as determined by microstimulation, *J. Neurophysiol.*, 58, 359, 1987.
57. Okahara, K. and Nisimaru, N., Climbing fiber responses evoked in lobule VII of the posterior cerebellum from a vagal nerve in rabbits, *Neurosci. Res.*, 12, 231, 1991.
58. Palay, S. L. and Palay, V. C., *Cerebellar Cortex Cytology and Organization*, Springer-Verlag, New York, 1974.
59. Paton, J. F. R., LaNoce, A., Sykes, R. M., Sebastiani, L., Bagnoli, P., Ghelarducci, B., and Bradley, D. J., Efferent connections of lobule IX of the posterior cerebellar cortex in the rabbit: some functional considerations, *J. Autonom. Nerv. Sys.*, 36, 209, 1991.
60. Ramu, A. and Bergmann, F., The role of the cerebellum in blood pressure regulation, *Experientia*, 23, 383, 1966.
61. Rasheed, B. M. A., Manchanda, S. K., and Anand, B. K., Effects of the stimulation of paleocerebellum on certain vegetative functions in the cat, *Brain Res.*, 20, 293, 1970.

62. Saigal, R. P., Karamanlidis, A. N., Voogd, J., Michaloudi, H., and Mangana, O., Cerebellar afferents from motor nuclei of cranial nerves, the nucleus of the solitary tract, and nuclei coeruleus and parabrachialis in sheep, demonstrated with retrograde transport of horseradish peroxidase, *Brain Res.*, 197, 200, 1980.
63. Saint-Cyr, J. A. and Woodward, D. J., Activation of mossy and climbing fiber pathways in the cerebellar cortex by stimulation of the fornix in the rat, *Exp. Brain Res.*, 40, 112, 1980.
64. Saint-Cyr, J. A. and Woodward, D. J., A topographic analysis of limbic and somatic inputs to the cerebellar cortex in the rat, *Exp. Brain Res.*, 40, 13, 1980.
65. Sato, Y. and Barmack, N. H., Zonal organization of olivocerebellar projections to the uvula in rabbits, *Brain Res.*, 359, 281, 1985.
66. Sawyer, C. H., Hilliard, J., and Ban, T., Autonomic and EEG responses to cerebellar stimulation in rabbits, *Am. J. Physiol.*, 200, 405, 1961.
67. Sebastiani, L., La Noce, A., Paton, J. F., and Gherlarducci, B., Influence of the cerebellar posterior vermis on the acquisition of the classically conditioned bradycardic response in the rabbit, *Exp. Brain Res.*, 88, 193, 1992.
68. Senapati, J. M., Jain, S. K., Parida, B., Panda, A., and Fahim, M., The influence of cerebellum on carbon dioxide responses in the dog, *Jpn. J. Physiol.*, 40, 471, 1990.
69. Smith, O. A., Jr. and Nathan, M. A., Inhibition of the carotid sinus reflex by stimulation of the inferior olive, *Science*, 154, 674, 1966.
70. Snider, R. S. and Stowell, A., Receiving areas of the tactile, auditory and visual systems in the cerebellum, *J. Neurophysiol.*, 7, 331, 1944.
71. Somana, R. and Walberg, F., The cerebellar projection from the parabrachial nucleus in the cat, *Brain Res.*, 172, 144, 1979.
72. Somana, R. and Walberg, F., Cerebellar afferents from the nucleus of the solitary tract, *Neurosci. Lett.*, 11, 41, 1979.
73. Supple, W. F., Jr., Purkinje cell responses in the anterior cerebellar vermis during Pavlovian fear conditioning in the rabbit, *NeuroReport*, 4, 975, 1993.
74. Supple, W. F., Jr., Hypothalamic modulation of Purkinje cell activity in the anterior cerebellar vermis, *NeuroReport*, 4, 979, 1993.
75. Supple, W. F., Jr. and Kapp, B. S., Anatomical and physiological relationships between the anterior cerebellar vermis and the pontine parabrachial nucleus in the rabbit, *Brain Res. Bull.*, 33, 561, 1994.
76. Supple, W. F., Jr. and Leaton, R. N., Cerebellar vermis: essential for classically conditioned bradycardia in the rat, *Brain Res.*, 509, 17, 1990.
77. Suzuki, D. A., Noda, H., and Kase, M., Visual and pursuit eye movement-related activity in posterior vermis of monkey cerebellum, *J. Neurophysiol.*, 46, 1120, 1981.
78. Tong, G., Robertson, L. T., and Brons, J., Climbing fiber representation of the renal afferent nerve in the vermal cortex of the cat cerebellum, *Brain Res.*, 601, 65, 1993.
79. Van Rossum, J., *Cortioconuclear and Cortico-Vestibular Projections of the Cerebellum*, Van Gorcum, Assen, 1969.
80. Voogd, J. and Bigare, F., Topographical distribution of olivary and corticonuclear fibers in the cerebellum: a review, in *The Inferior Olivary Nucleus: Anatomy and Physiology*, Courville, J. et al., Eds., Raven Press, New York, 1980, 207–234.
81. Wiggers, K., The influence of the cerebellum on the heart and the circulation of the blood II. *Arch. Physiol.*, TOME, 27, 301, 1943.
82. Willis, T., *The Anatomy of the Brain and Nerves* (Facsimile reprint of *The Anatomy of the Brain and the Description of the Use of the Nerves, from the Remaining Medical Works of Dr. Thomas Willis.* Englished by S[amuel].P[ordage]. Printed for T. Dring, C. Harper, and J. Leigh, 1681), Feindel, W., Ed., Classics of Medicine Library, Alabama, 1978.
83. Wolfe, J. W., Relationship of cerebellar potentials to saccadic eye movements, *Brain Res.*, 204, 1986.

Section III

CLINICAL ASPECTS OF VESTIBULAR AUTONOMIC REGULATION

9

The Autonomic Nervous System and Motion Sickness

Kenneth E. Money, James R. Lackner, and Robert S. K. Cheung

CONTENTS

0-8493-7668-1/96/$0.00+$.50

1 INTRODUCTION

An elementary description of the autonomic nervous system is presented, followed by a brief description of motion sickness. Finally, changes in the autonomic nervous system with motion sickness are described. More details concerning the autonomic nervous system are found in Chapter 3, and Chapter 5 provides further information about the neural basis of nausea and vomiting.

2 BASIC DESCRIPTION OF THE AUTONOMIC NERVOUS SYSTEM

2.1 Definition of the Autonomic Nervous System

A vertebrate "nervous system" is usually understood to include peripheral sensory (afferent) nerves, a brain and spinal cord, and peripheral motor (efferent) nerves. The term *central nervous system* (CNS) refers to the brain and spinal cord, and *peripheral nervous system* refers to all the neural structures outside the CNS. The term *autonomic nervous system* is normally used to refer discretely to certain specific neural pathways that are only peripheral in location and only motor in function. These autonomic pathways are efferent from the brain and spinal cord to three specific kinds of effectors: smooth muscle, cardiac muscle, and glands. The autonomic nervous system usually refers specifically and exclusively to these peripheral motor pathways to smooth muscle, cardiac muscle, and glands. Some authorities define the autonomic nervous system in this way,[6,120,130] and others do not provide definitions but characterize the autonomic nervous system in terms of its parasympathetic and sympathetic parts, describing both of the parts as exclusively peripheral and efferent and visceral.[17,27,50] The autonomic nervous system (ANS) could logically be renamed the *peripheral motor pathways to the viscera* (PMPV) or *efferent autonomic pathways* (EAP), but reviewers, like dictionary writers, should be historians rather than law givers and we shall refrain from inventing new terms. "In German one speaks of *das viszerale Nervensystem*, in French of *le system vegetatif*."[17] Nevertheless, the autonomic nervous system is not a nervous system.

Some authorities use the term *autonomic nervous system* to include also certain afferent pathways from the viscera and certain central mechanisms,[83] but most do not. Other authorities define the ANS as comprising "all the neurons lying completely outside the central nervous system,"[38] a definition that is consistent with the original classification proposed by Langley in 1921. This classification excludes any afferent pathways and any central mechanisms, but includes[128] the enteric neurons which lie completely within the walls of the gut. We shall exclude the enteric neurons from the definition because it is probably best to use the term ANS in the way it is actually used most of the time, to refer only to the parasympathetic and sympathetic pathways (peripheral visceral efferent pathways). We shall refer to other mechanisms specifically as required.

These peripheral efferent pathways to smooth muscle, cardiac muscle, and glands do deserve a special name, because they have special characteristics. Of all the peripheral motor pathways, only the autonomic pathways have a synapse located between the CNS and the effector, and only the autonomic pathways include neurons that are located completely outside the CNS. Unlike other motor fibers, the motor fibers that contact smooth muscle, cardiac muscle, and glands belong to neurons the cell bodies and processes of which lie entirely outside the CNS; the cell bodies of these neurons are located in autonomic ganglia (outside the CNS). "Neither somatic motor neurons, whose cell bodies lie within the spinal cord, nor primary sensory neurons, whose axons project into the spinal cord or brain stem, can be considered autonomic."[38]

2.2 Automaticity of the ANS

Autonomic pathways usually function with autonomy, with independence from voluntary control, and hence the origin of the term. It should be understood, however, that special feedback training can confer the ability to exert some voluntary control of activity in some of the autonomic pathways.[18,65,84,118] It is also important to note that, although the autonomic nervous system functions with considerable independence from voluntary control, it does not function with autonomy from the rest of the body. The autonomic nervous system is fully integrated with the rest of the body, and is controlled by the brain and spinal cord. Control of the autonomic nervous system is based on information about the environment and the body, including information provided by afferent nerves from the viscera.

The body has, of course, some motor pathways that are *not* autonomic but are nevertheless independent of conscious volition most of the time. Major respiratory muscles, such as the diaphragm and the intercostal muscles, are striated muscles innervated by somatic (not autonomic) nerves and they function independently of voluntary control most of the time. The moving of air in and out of the lungs is not an autonomic function even though it is usually done without participation of conscious volition. Similarly, the external eye muscles are striated muscles that are innervated by somatic motor fibers in cranial nerves 3, 4, and 6, and these fibers are influenced constantly by "automatic" neural mechanisms that receive inertial and gravitational information from the vestibular apparatus of the inner ear, and visual information from the retina. These nerve fibers can of course be controlled readily by conscious volition, as in making saccadic eye movements. The *torsional* eye movements (rotational displacements of the eye about the optic axis) that are routinely produced reflexly during tilt of the head in roll[15,30] cannot be produced voluntarily except by people who have received special training. The tensor tympani muscles and the stapedius muscles of the middle ears are also fast striated muscles that are innervated by somatic, not autonomic, nerves; these nerves also are controlled independently of conscious volition, by the acoustic reflex (although some people can contract and relax the middle ear muscles voluntarily).

Probably all the muscles that are normally controlled voluntarily (striated muscles innervated by somatic neurons) are occasionally controlled automatically, i.e. reflexly, as in postural reflexes, flexion reflexes, and shivering. In general, striated muscle is controlled both by conscious volition and automatically. In contrast, smooth muscle, cardiac muscle, and glands receive only autonomic motor nerve supplies and are controlled automatically almost always. There are, of course, those rare people who have had biofeedback training and have learned how to change some autonomic activity voluntarily.[18,21,119] Also, autonomic activity is routinely influenced by voluntary activity. For example, when visual fixation is voluntarily changed from far to near, changes occur in the autonomic control of two smooth muscles in the eye, the ciliary muscle and the iris, so that clearer vision is achieved.

2.3 Parasympathetic and Sympathetic Divisions of the ANS

Understanding the autonomic changes associated with motion sickness requires a basic understanding of the differences between the parasympathetic and the sympathetic divisions of the ANS. The parasympathetic pathways are also called the cranio-sacral pathways because they exit the CNS in the cranium and in the sacral part of the spinal cord (in cranial nerves 3, 7, 9, and 10, and from spinal cord sacral segments 2, 3, and 4); the sympathetic pathways are also called thoracico-lumbar pathways because they exit the CNS from spinal cord segments thoracic 1 to lumbar 3.

All the junctions of the parasympathetic fibers with their effectors are "cholinergic," i.e., one of the major neurotransmitters at the effector is acetylcholine; most, but not all, of the junctions of the sympathetic fibers with their effectors are "adrenergic," i.e., one of the major neurotransmitters at the effector is norepinephrine. Three of the sympathetic junctions are exceptional; the sympathetic junctions at the sweat glands and at the piloerector muscles, and also the sympathetic junctions that *dilate* blood vessels with sympathetic stimulation, are actually cholinergic.

Autonomic nerve endings at effectors may secrete many other neurotransmitters in addition to acetylcholine (ACh) and norepinephrine (NA). Other such neurotransmitters include "adenosine 5′-triphosphate (ATP), 5-hydroxytryptamine (5-HT), dopamine, and a number of peptides, notably neuropeptide Y (NPY), vasoactive intestinal polypeptide (VIP), enkephalin, somatostatin, cholecystokinen, substance P (SP), and calcitonin gene-related peptide (CGRP)."[11] At most, if not all, autonomic effector junctions a *mixture* of two or more (reportedly, up to six) transmitters is secreted. Certain mixtures or "codings" are commonly found. "The predominant 'codings' are: sympathetic nerves, NA, ATP, and NPY; parasympathetic nerves, ACh and VIP."[11] Also, at some autonomic effectors there are two or more different kinds of receptors for a single neurotransmitter, so that blockers that are effective for some receptors do not necessarily block receptivity in other receptors.

Most autonomic effectors are supplied by fibers of both the parasympathetic system and the sympathetic system. In such dually innervated effectors, the influence of the two systems are often opposite. For example, an *increase* in action potential frequencies in the *parasympathetic* fibers to the heart tends to slow the heart rate, and a *decrease* tends to speed up the heart rate; an *increase* in action potential frequencies in the *sympathetic* fibers to the heart tends to speed up the heart rate, and a *decrease* tends to slow the heart rate.

Another important difference between the parasympathetic and the sympathetic systems is that the sympathetic system controls secretions by the adrenal medulla, which secretes into the bloodstream the major transmitter of the sympathetic system, norepinephrine, together with epinephrine; the secretion is roughly 20% norepinephrine and 80% epinephrine but the proportions are different in different circumstances.[39,50] These secretions have the effect of mimicking, all at once, increases in activity in most of the sympathetic pathways. The parasympathetic system has no comparable endocrine gland.

A large number of the sympathetic pathways can be activated all at the same time as part of a "stress response," whereas large numbers of the parasympathetic pathways cannot be activated all at the same time. With alarm or stress (often because of a conscious perception), the "stress" or "fight-or-flight" responses become evident. These responses include *conscious sensations* (sometimes fear or anger), *endocrine responses*, and *the autonomic response*. The autonomic response is an en bloc activation of a large part of the sympathetic nervous system, an increase in action potential frequencies in a large number of the sympathetic neural pathways. This increase in sympathetic activity causes, among other things, a rise in the heart rate, an increase in the strength of contractions of the heart, a rise of blood pressure, an increase in circulating red blood cells as they are released from the spleen (not quantitatively important in humans), a blood shift out of the skin and abdominal organs and into skeletal and cardiac muscle, decreased digestive activity, increased secretions by sweat glands, increased diameters of bronchioles, increased diameters of the pupils, and an increase in circulating epinephrine and norepinephrine; the increase in circulating epinephrine has a variety of effects, including increases in blood glucose and blood fatty acids.

2.4 Other Distinguishing Characteristics of the ANS

Autonomic pathways are different from somatic motor pathways not only in their automaticity but also in the *function* of the basal or "tonic" activity: at rest, there is a basal frequency of action potentials along the autonomic pathways to smooth muscle, cardiac muscle, and glands, and some of the important adjustments in autonomic control are made by *decreasing* the frequencies of action potentials from that basal level. There is also tonic activity, at rest recumbent, in the somatic motor nerves to striated muscle, but this is maintenance activity rather than controlling activity, and important adjustments in the functional control of these muscles do not involve decreases from the resting levels of activity (with the occasional, and arguable, exception of the external anal sphincter).

Another difference between autonomic motor pathways and somatic pathways is that somatic pathways innervate only one kind of effector (striated muscle) using only one kind of neurotransmitter (acetylcholine) at the effector, whereas autonomic pathways innervate a variety of effectors using a variety of neurotransmitters. Autonomic fibers are also smaller in diameter, conduct at slower speeds, and use generally lower frequencies of action potentials.

Parts of the autonomic nervous system will react to an isolated small stimulus such as a puff of wind, a light touch, a flash of light, or a sound. Stimuli that consist of specific meaningful patterns can provoke in the autonomic nervous system certain specific and meaningful responses. A flash of light might cause changes in the autonomic nervous system that result only in a small, brief increase of heart rate, but an appropriate *pattern* of optical stimuli can cause changes in the autonomic nervous system with the result that the diameter of the pupil and the focus of the eye are altered to produce clearer vision. A pattern of optical stimuli that reveals a large, angry grizzly bear running at the observer can provoke autonomic responses that include large increases in heart rate and blood pressure, increases of blood flow to skeletal muscles, increased diameter of the bronchioles, etc.

Similarly, a brief vestibular stimulation can also cause small discrete autonomic changes.[123] However, stimulating the vestibular system over a period of time can provoke motion sickness, a syndrome that includes a variety of large changes in the autonomic nervous system. The autonomic changes in motion sickness can be seen as appropriate to a poisoning situation that involves vomiting. The autonomic responses to single, small vestibular stimuli have no particular relationship to the autonomic changes that occur when motion sickness is provoked, just as the autonomic responses to single flashes of light have no particular relationship to the autonomic changes that occur when the eye reports a charging grizzly bear.

3 MOTION SICKNESS

3.1 Definition and Essential Nature

Motion sickness is a sickness caused by motion. Motion sickness is: the signs and symptoms of the sickness caused by motion. Motion sickness appears to be a poison response caused by unnatural motion acting on the vestibular system.[101,103] Motions of the body can act on the vestibular apparatus and thereby on the central vestibular mechanisms to generate motion sickness. Motions of visual fields apparently cause motion sickness[26,69,90,113,124] "by stimulating the vestibular system,"[100] by acting through central connections to the central vestibular mechanisms.[13] In many motion sickness situations there is probably a combined action of several different kinds of information about orientation and motion: information from the vestibular apparatus,[63,68] from the eyes,[90,124] from proprioceptors,[76] and from memory.[110,116]

Various aspects of motion sickness have been reviewed recently.[7,8,12,23,28,29,94,96,122] Motion sickness never occurs in people or animals who lack vestibular function,[63,68] and susceptible animals become absolutely nonsusceptible to all forms of motion sickness after bilateral surgical removal of the vestibular apparatus.[105,121] People lacking vestibular function are nonsusceptible even to the motion sickness caused by purely visual stimuli.[13] The vestibular system is necessary for motion sickness. Also, discrete stimulation of the vestibular system is sufficient to cause motion sickness in normal people who are at rest with the eyes closed.[100] Discrete vestibular stimulation can give rise to a group of signs and symptoms that looks like a poison response.

Motion sickness is not a sickness in the normal sense of the word; it is the usual response of normal people to certain kinds of abnormal motion stimuli. In motion sickness the problem is the response to the stimulus, and failure to respond will completely eliminate the problem. This is unlike trauma or low temperatures or low oxygen pressures or bacterial infection, where the problem is not the response and failure of the body to respond would *not* eliminate the problem. Motion sickness is an inappropriate response to a stimulus.

Essentially all normal people can be made motion sick if the stimulus is appropriate and lasts for a long enough time, but there are important differences in individual susceptibility. Differences in individual susceptibility are determined by a variety of characteristics,[49] some of them genetic,[1,4] and the degree of motion sickness experienced in a given situation can be determined by special circumstances,[40] by the activities skills or strategies of an individual in the motion environment,[147] by perceptual style,[49] by the functional capacity of the vestibular apparatus of the inner ear,[63,68] by habituation,[24,45,88,125] medications,[86,129,144,145] psychological influences,[31,49,67,99,114] or special biofeedback training.[5,18,19,65,134]

The vomiting of motion sickness is one of several known kinds of inappropriate vomiting, "inappropriate" because the stomach is emptied even though there is no poison in it. Inappropriate vomiting can occur in the morning sickness of pregnancy,[140] in response to emotional stress,[10,44] severe physical pain,[10] radiation,[8] mechanical irritation of abdominal viscera,[10] or physical exercise that is extremely vigorous and prolonged.

In many cases there is little understanding of the essential nature of inappropriate vomiting, but in the case of motion sickness it is now understood[16,135] that the vestibular system plays an important role in the vomiting response to certain poisons as well as in the vomiting response to motion. In animals after bilateral surgical removal of the vestibular apparatus, the emetic response to motion is lacking[121] and also the emetic response to certain poisons is grossly defective.[104] Since the vestibular system functions in both the vomiting response to certain poisons and the vomiting response to certain motions, it is reasonable to think (1) that the vestibular mechanism that functions in the vomiting response to poisons arose as a biological protective mechanism, and (2) that certain unnatural motions can stimulate that protective mechanism to produce the poison response that is called motion sickness. This is much more reasonable than assuming that there are two different emetic mechanisms in the vestibular system (one for poisons, and a separate one for motions). The essence of motion sickness seems to be that certain unnatural motions stimulate, inappropriately, a protective mechanism.

3.2 Motion Sickness as a Poison Response

If there is only one emetic mechanism in the vestibular system, then the sensitivity of that mechanism in a given individual might be expected to determine the individual's sensitivity both to emetic poisons and to motion. The individuals who are more susceptible to emetic poisons might be expected also to be more susceptible to motion sickness. In one study,[107] positive evidence of this was found: persons who were more susceptible to motion sickness were also more susceptible to the nausea and vomiting of drugs that have emetic side effects. Seventy-seven cancer patients who were found by questionnaire to be susceptible to motion sickness were "matched to cancer patient controls without previous motion sickness by sex, age, type and dose of chemotherapeutic drug received, and antiemetic medication." It was found that the cancer chemotherapy produced, in the patients with a history of motion sickness, significantly "more frequent, severe, and longer-lasting nausea and vomiting than controls."[107] This evidence suggests that there is only one emetic mechanism in the vestibular system.

The sensitivity to the nausea and vomiting side effects of drugs and the sensitivity to motion are possibly both determined by the same protective mechanism. The results from the experiment with chemotherapeutic drugs[107] resemble the results from the experiment that tested experimental animals before and after labyrinthectomy:[104] the animals that were more susceptible to the vomiting of drugs (the preoperative animals) were also more sensitive to motion sickness, and the animals less sensitive to the vomiting of drugs (the postoperative animals) were also less sensitive to motion sickness.[104] It has been found also in rats, using kaolin ingestion (pica) as the index of sickness, that sensitivity to motion correlates with sensitivity to emetic poisons.[53] The failure of labyrinthectomized rats to develop conditioned taste aversions 19 days postoperatively[112] is another indication of the involvement of the vestibular system in poison responses (although they recovered the ability to acquire conditioned taste aversions 29–30 days postoperatively).

Authorities in the field of poisoning[2,115] apparently do not make lists of the "basic" or "physiological" responses to poisoning, perhaps because of their primary concern about the differential determination of which poison is present and how to counteract it. There are different direct effects of different poisons and also different bodily responses to different poisons. However, poisoning authorities do recognize that some "non-specific" reactions to poisoning occur. "Nausea and vomiting are probably the most common non-specific alimentary symptoms of poisoning."[115] A list of the signs and symptoms of motion sickness is perhaps a list of nonspecific reactions to poisoning.

"Motion sickness is possibly the purest and simplest poison response available for study, because it can be produced for study without the complicating presence of a poison. It is produced by a motion stimulus that masquerades as a poison."[103] Unnatural motion seems to activate a poison response mechanism, and as a result the body inappropriately inflicts on itself malaise, anxiety, nausea, and other conscious sensations, and also pallor, cold sweating, vomiting, endocrine responses, and autonomic responses (not necessarily in that order). It is of course possible that an otherwise-harmless dose of an emetic such as apomorphine could provoke the same collection of signs and symptoms, not by provoking each sign or symptom individually but by simply turning on a poison-response mechanism so that the body then inflicts on itself the whole poison-response syndrome. Such a poison response to a harmless dose of apomorphine would be just as "inappropriate" as the poison response in a motion sickness episode and, as in motion sickness, failure to respond would completely eliminate the problem.

3.3 The "Two Phenomena" of Motion Sickness

Motion sickness appears to be a poison response that includes two major phenomena: (1) emptying of the stomach, and (2) a stress response. This "two-phenomena" concept can accommodate all the known signs and symptoms of motion sickness. The emptying of the stomach includes the vomiting and some accompanying conscious sensations. The stress response includes most of the endocrine responses and autonomic responses and some accompanying conscious sensations. It is reasonable to describe motion sickness as including a stress response because the endocrine phenomena of motion sickness and the autonomic phenomena of motion sickness, together, do comprise an easily recognized stress response.[18,34,70-75,97,113] Presumably, all or most emetic poison responses include a stress response as well as the emptying of the stomach; perhaps the stress response improves the chances of surviving some kinds of poisons. It is known that experimental animals lacking stress response mechanisms are much more likely to die when exposed to any one of a variety of stresses.

Presumably motion sickness is stressful *only* because the poison response includes a stress response; that is, if the poison response that is motion sickness included only a program for producing stomach emptying and no program for producing a stress response, the stomach would be emptied without any stress being experienced. It is only because the poison response includes a stress response that motion sickness is experienced as stressful.

In motion sickness situations both parts of the poison response, the emptying of the stomach and the stress response, are "inappropriate": the stomach is emptied in the absence of a poison, and a stress response occurs in the absence of a stress. No one suggests that there is a poison present in the stomach with motion sickness, but sometimes it is suggested that there is a stress present. There have been many attempts to see something noxious or stressful in the motions that elicit motion sickness, but such attempts have been remarkably unsuccessful.[100] This is not surprising, since most such motions are not stressful; failure to respond to the motions completely eliminates the problem. It is a new and different use of the word *stress* to argue that a physical stimulus such as a motion is a sickness-causing stress when simple failure of the nervous system to respond would completely eliminate the problem. Other kinds of physical stresses cause problems that cannot be eliminated by simple failure to respond. For example, the problems caused by the physical stresses of cold, heat, dehydration, bacterial infection, radiation, anoxia, etc., are not eliminated by simple failure to respond.

The problem in motion sickness is not a stress; the problem is the inappropriate response to the motion. In fact, some provocative motions are clearly and obviously not stressful, except for causing motion sickness. The most common provocative motions, the motions of ships in moderate seas, are not noxious, not injurious, not a threat to the calibration of the vestibulo-ocular reflex, not frightening, not stressful in any way to adults who do not respond inappropriately, and not stressful even for babies before the age of mobility;[100] such babies eat and sleep and smile happily in these environments. To vomit and to generate a stress response in reaction to these motions is remarkably inappropriate.

A poison response includes a stress response, and therefore vomiting as part of a poison response is a stressful experience, but vomiting that is not part of a poison response does not have to be accompanied by stress. As reported by a psychiatrist specialist in eating disorders,[148] vomiting that is not part of a poison response is sometimes not a stressful experience. Some bulimics who vomit voluntarily do not find that the vomiting itself is stressful, although many feel guilty about it afterwards. Some bulimics look forward to vomiting and consider it something positive, and some learn to vomit spontaneously, without mechanical inducement. There seems to be no reason to assume that emptying of the stomach per se *has* to be stressful.

On rare occasions the stress response seems to be missing from motion sickness. Some unusual test subjects exposed to provocative motion vomit after brief exposure and are ready to continue the testing because they do not find it stressful to vomit.

One common view of motion sickness is that the unnatural motion produces the stomach-emptying phenomena (nausea and vomiting) and that the stomach-emptying phenomena are stressful, causing the stress response. This seems unlikely since, as suggested above, there is no reason to assume that a simple emptying of the stomach has to be stressful. Also, in some people the parts of motion sickness that comprise a stress response are well developed before vomiting or nausea occurs. It seems unreasonable to say that the stress response is caused by the nausea in those individuals who have the stress response before the nausea.

The reverse view of motion sickness is also common; this view suggests that the motion causes a stress response (an "unbalance" of the autonomic nervous system) and that the stress response then causes the stomach-emptying phenomena. This also seems unlikely since, as suggested above, the motion is not usually stressful except for its ability to cause motion sickness. Moreover, most stress responses do not cause nausea or stomach emptying: sky divers and sprinters, for example, experience stress and have extremely high heart rates but they seldom, if ever, vomit or experience nausea from the stress. Anecdotally, several hundred sprinters and several hundred skydivers have been observed before, during, and after the events, and not a single one was seen vomiting.[149] Furthermore, in some people the nausea appears before any signs of a stress response. It seems especially unreasonable to say that the nausea is caused by the stress response in those individuals who have the nausea before the stress response.

Specific evidence of the sequence of events in motion sickness might be of interest. Although conscious perception of abdominal changes is often the first indication of motion sickness in strongly provocative test situations,[48,82] parts of the stress response in fact appear before the start of nausea in some people. One important part of the stress response, the two- or threefold increase in blood flow in skeletal muscle,[126] occurs in some subjects *before* any report of nausea[127] and is therefore unlikely to be caused by the nausea or other stomach-emptying phenomena in these subjects. On the other hand, in some other subjects the nausea is well developed before the increase in muscle blood flow occurs and it is therefore unlikely that the change in muscle blood flow causes the nausea in these subjects. Two other parts of the stress response, pallor and cold sweating, also occur in some subjects before any perceived nausea and in other subjects only after nausea is perceived.[52,55,64,126] One major study found a "remarkable consistency of pallor detected prior to the report of nausea,"[22] and another major study[82] found that autonomic [stress response] effects "preceded [nausea] onset in very few individuals."

Overall, there is no preponderance of evidence for thinking either that the stomach-emptying phenomena occur before the stress response or that the stress response occurs before the stomach-emptying phenomena. It seems that the stomach-emptying phenomena and the stress response are produced at about the same time; there is no evidence that one causes the other.

The autonomic phenomena associated with emptying of the stomach cannot be used as indicators of whether the stress response precedes or follows or occurs at the same time as the stomach emptying, because these autonomic phenomena are themselves "stomach-emptying" phenomena rather than "stress response" phenomena. While considering the sequencing of events in motion sickness, however, it should be mentioned that the change of gastric motility with motion sickness, the gastric tachyarrhythmia, "typically precedes the first reports of nausea by 2 to 3 minutes."[70]

3.4 The Signs and Symptoms of Motion Sickness

The major signs and symptoms of motion sickness are listed below. Autonomic signs are included here but some of them are described more fully in Section 4.

3.4.1 Stomach-Emptying Phenomena

3.4.1.1 Conscious Sensations

- Abdominal sensations such as awareness and discomfort[47,52]
- Malaise, drowsiness[47]
- Nausea[70,100,124]
- Sensations of vomiting taking place (visual, auditory, proprioceptive, and tactile sensations of vomiting taking place)

The origin of nausea is not known, and nausea is sometimes assumed to be the conscious awareness of unusual activity in the central mechanisms of vomiting.[100] It is now known, however, that vomiting can occur without consciousness of nausea, in bulimics[148] and astronauts in space,[133] and it seems that the assumption about awareness of unusual activity in the central mechanisms is not a valid one; that is, the central mechanisms of vomiting seem to be capable of functioning without provoking the conscious sensation called nausea. One caution: it might be that the culture of astronauts makes it difficult for them to experience or report nausea,[102] or to acknowledge on-going symptoms of motion sickness.

When the stomach is present and innervated, certain stimuli to it can elicit nausea. Nevertheless, it is clear that the stomach is not the only elicitor of nausea, because "human subjects with total gastrectomy not infrequently experience nausea."[25] Nothing that the stomach might do is necessary for nausea, and since motion sickness does occur in gut-denervated animals it seems unlikely that the sensation nausea in motion sickness is triggered primarily from the stomach. It seems possible that the sensation of nausea in motion sickness is generated in the brain as part of a poison response. It is also possible that the nausea of motion sickness results from the increased plasma levels of the antidiuretic hormone AVP[54,132] that occur with motion sickness. The vomiting of motion sickness in dogs is not impeded or delayed by the surgical removal of the pituitary gland, however, and this procedure removes a major source of AVP and produces polyuria.[106] Furthermore, nausea in humans can be provoked without any increase of plasma AVP, by giving ipecacuanha.[108]

3.4.1.2 Vomiting

- Increased Plasma Antidiuretic Hormone, i.e., AVP[23,32,36,70,71,131,146] — Plasma levels of AVP are increased up to 20 or 30 times the pre-motion levels.
- Salivation — In most people, measured salivation flow is decreased with motion sickness.[41-43] In most squirrel monkeys, the measured salivation flow is increased.[59-61] In some people, just before vomiting, there is a brief but strong increase in salivation.
- Decreased motility of the stomach and changes in the electrogastrogram.[77,79,80,124,133]
- Decreased motility of the upper small intestine.[80,93,95,133]
- Giant Retrograde Contraction — A single, strong retrograde peristaltic wave from the mid-small intestine to the gastric antrum.[77-79]
- Phasic contractions of pylorus and small intestine.[80]

- Elevation of cardia of stomach into thorax, and relaxation of lower esophagus.
- Adjustments of body posture.
- Retching, i.e., special coordinated contractions of the respiratory muscles and muscles of the abdominal wall.[95]
- Closure of the glottis, elevation of the soft palate.
- Compression of stomach by abdominal wall and diaphragm; relaxation of diaphragmatic hiatus.[78,95,141]
- Opening of the mouth.
- Expulsion of vomitus.
- Repetitive swallowing.

Motion sickness in dogs often includes, before vomiting, defecation, urination, and obvious drooling.[105] In squirrel monkeys there is marked foaming at the mouth, chewing motions of the jaw,[142,143] licking,[14] and a motionless trance-like state that occurs invariably in some experimental conditions but not[142,143] in others. There is also visible salivation in cats, both in motion sickness and in other poison responses.[87]

The massive increases of plasma antidiuretic hormone (AVP) are likely related to the stomach-emptying phenomena, rather than to the stress response. The evidence is convincing. Increased levels of AVP are thought to be closely associated with nausea.[150] When large doses of AVP are injected in humans, "Increased intestinal activity is likely to cause nausea, belching, cramps, and an urge to defecate."[54] Vomiting as well as nausea is very common after administration of AVP to humans.[132] Increases of AVP are known to be absent in some vomiting situations but present in some other vomiting situations,[108] and present also in "anaesthetized man or experimental animals" in which "traction on the intestines consistently resulted in very large increases in plasma antidiuretic activity."[117]

Increases of plasma AVP probably play some role in the nausea and vomiting of motion sickness[32,36,70,71] although it is difficult to say how important this role might be. The tendencies of AVP to cause cramps and the urge to defecate are possibly counteracted in human motion sickness by the stress response phenomena, but in dogs motion sickness often does include defecation.

Also, "in the course of many studies in both rats and human volunteers, we were impressed with how rarely pain or other obvious stresses ... had any discernible effect on plasma AVP."[117] In rats, the researchers[117] were surprised to find that "none of the three stresses employed — light ether anesthesia, water immersion, or pain — had any detectable effect at any time on plasma AVP even though all three resulted in very large increases in plasma corticosterone ..." It seems clear that the increases in plasma AVP with motion sickness are *not* part of the stress response and that they *are* part of the vomiting response; these AVP increases would help to prevent dehydration and loss of blood pressure[39] in prolonged vomiting situations.

There are several possible influences of AVP in motion sickness. These include the antidiuresis of motion sickness, the sensation of nausea, the vomiting, the facial pallor, and the increased blood flow in skeletal muscle. It is also possible that AVP increases the rate of selective transfer of water from saliva into the blood vessels of the tissues of the mouth. Several AVP antagonist medications have been tested for their influence on motion sickness in squirrel monkeys, and one of them abolished emesis in all six of the animals tested.[14]

The central and peripheral mechanisms of vomiting have been well reviewed recently (see References 9, 77, 80, 81, 93, 95, 98, 141; see also Chapter 5 in this book). A few minutes before vomiting, the motility of the stomach and duodenum decreases markedly and following this decrease of motility some contents of the upper half of the small intestine are emptied back into the stomach by a single powerful retrograde peristaltic contraction called the RGC, retrograde giant contraction.[80] The RGC "begins at mid small intestine and propagates retrograde to the gastric antrum."[79] This giant contraction consists of a long, continuous portion of the gut that is contracted to a small diameter, a length of some 70 cm, and the 70-cm area of small diameter travels along the gut from mid small intestine to the gastric antrum of the stomach.[151] The emptying, or partial emptying, of the upper small intestine into the stomach would tend to move any poison to the stomach where vomiting could remove it, and also the acidic stomach contents would tend to be neutralized by the intestinal contents so that damage to the esophagus during vomiting would be less likely.[77]

Following the RGC there are "post-RGC phasic contractions occurring in the stomach and small intestine."[79] These phasic contractions are especially prominent in the lower small intestine, and serve to "move the contents of the lower small intestine into the colon in a stripping fashion."[77] Of course, absorption of a poison from the colon into the blood stream is much less likely than from the small intestine, and this movement into the colon can be considered protective.

Overall, it seems that the RGC empties the upper half of the small intestine into the stomach, and the phasic contractions empty the lower half of the small intestine into the colon. These two actions would do much to stop the absorption of the contents of the gut and would therefore be protective against poisons in the gut.

The lower esophageal sphincter relaxes and the lower half of the esophagus relaxes, and the cardia of the stomach is drawn through the diaphragmatic hiatus into the thorax by contraction of the upper esophagus. Apparently the "acute angle" between the stomach and the esophagus is thereby altered and the "fundic reservoir above the gastroesophageal junction is eliminated."[77] Movement of stomach contents into the esophagus is facilitated. The lower half of the esophagus passively accepts vomitus, which moves in and out of the lower half of the esophagus with retching.

The glottis closes,[9] and the soft palate rises. "Vomitus expulsion occurs at the end of retching as a single maximal [and prolonged] contraction of the rectus abdominis and diaphragmatic dome, while the diaphragmatic hiatus relaxes."[78] "The external (inspiratory) muscles also co-contract with the diaphragm and abdominal muscles during vomiting."[95] Apparently, the positive pressure in the abdomen and the negative pressure in the thorax moves the vomitus from the stomach to as high as the top third or so of the esophagus (the body of the stomach is a passive sac that is squeezed by the striated muscles of the diaphragm and the abdominal wall). The upper esophageal sphincter (UES) relaxes.

The top half, or more,[81] of the esophagus in dogs (top quarter in humans) is striated muscle and is an active participant in ejecting the vomitus. The "cervical esophagus about 5–10 cm from the UES contracts. This cervical esophageal contraction propagates orad at about 10 cm/s,...The cervical esophageal retrograde contraction may represent one of the primary differences between retching and vomiting and may be responsible for projectile vomiting."[78] Another crucial difference between retching and vomiting is, apparently, the relaxation of the hiatal area of the diaphragm.[77] The striated muscle of the pharynx is also active in ejecting the vomitus.

The head is lowered, the mouth opens, the tongue protrudes, and the vomitus is expelled. After expulsion of vomitus the pharynx is cleared by repetitive swallowing. The stomach slowly returns to its normal position.

The very active role of the esophagus in vomiting is noteworthy. Rupture of the esophagus while vomiting, with fatal outcome, has been reported.[66]

The peripheral neural mechanisms of vomiting are further described in Section 4.1.2 below. The central mechanisms of vomiting are described in Chapter 5 (by Miller and Grélot).

3.4.2 Stress Response

3.4.2.1 Conscious Sensations

- Anxiety, or distress, or dread[70,82,124]

Motion sickness produces no sensation of anger, but there is a dread or anxiety that is perhaps related to fear, and there is a tendency to contemplate *exclusively* the motion sickness, and to contemplate it with apprehension. There is a striking absence of inclination to daydream or to think of business, politics, sports, romance, or anything except the sickness. People suffering severe motion sickness in a laboratory just do not spontaneously think about baseball games or the stock market, and they do not smile or laugh. It is reasonable to think that in operational situations a motion-sick person might be delinquent in maintaining attention and vigilance to tasks, especially "background" tasks such as eliciting crew cooperation, checking fuel levels, and making radio reports.

The anxiety of motion sickness is distinguishable from the nausea of motion sickness.[82] The changes in consciousness and behavior that indicate anxiety or dread are not simply nausea and not simply an awareness that vomiting is impending. These changes are utterly unlike the changes in a bulimic who has no nausea but knows that voluntary vomiting is impending (and who looks forward to it). The victim of motion sickness is possessed by a remarkable anxiety/distress/dread.

3.4.2.2 Endocrine Responses

- Increased plasma ACTH[72,73]
- Increased plasma cortisol[32,70,71]
- Increased plasma epinephrine[22,23,70-72,74,75,146]
- Increased plasma norepinephrine[70-72,74,75]
- Increased plasma thyroid hormones, a tiny increase indirectly detected[51]
- Decreased plasma thyroid stimulating hormone[51]
- Increased plasma growth hormone[32]
- Increased plasma prolactin[32]
- Increased plasma beta endorphin, anticipatory and after vomiting[70,71]

The plasma levels of the classic stress response hormones, epinephrine, norepinephrine, ACTH, and cortisol, are all elevated with motion sickness. Plasma levels of growth hormone, prolactin (in both males and females), and beta endorphin also rise in response to various stresses in humans,[97] and they do rise also in motion sickness. Plasma levels of thyroid stimulating hormone are reported to be decreased by stress in animal studies and to be unchanged or increased by stress in humans,[97] but they were found to be decreased in humans by motion sickness.

Note also that the massive increase of plasma antidiuretic hormone (AVP) with motion sickness is *not* listed here as part of the stress response. The increased plasma AVP is better considered part of the vomiting response and is listed as such, above.

The nausea of motion sickness is what concerns the motion-sick person, and it is probably for that reason that there is a popular tendency to consider the nausea as the real motion sickness and the other phenomena as accompanying changes that might or might not *correlate* well with "the motion sickness." It should be noted that the stress response phenomena are real and are part of motion sickness regardless of the correlation with the nausea.

3.4.2.3 Autonomic Responses

The autonomic parts of the stress response are part of Section 4.

4 CHANGES IN THE AUTONOMIC NERVOUS SYSTEM WITH MOTION SICKNESS

There are autonomic changes associated both with the emptying of the stomach and with the stress response of motion sickness. The autonomic changes associated with emptying the stomach include important increases in the level of activity in the parasympathetic pathways, as well as possible increases in the sympathetic pathways.

The autonomic changes associated with the stress response appear to be, predominantly, increases in activity in the sympathetic pathways and/or decreases in activity in the parasympathetic system. The two exceptions, both indications of increased parasympathetic activity, are (1) an increase in variability of the intervals between heart beats in squirrel monkeys,[58,62] although a different measure of variability in these intervals in humans decreases,[70] and (2) a small decrease in the dark field focus distance with motion sickness in humans.[35] The more prominent autonomic changes with the stress response of motion sickness are indicative of increased sympathetic activity: pallor, cold sweating, and increased blood flow to skeletal muscle.

The changes in heart rate and blood pressure with motion sickness are of small magnitude, even with severe motion sickness. Many stress responses have peculiarities, however, and the small magnitudes of these changes are perhaps just the idiosyncrasies of the stress response of motion sickness. In general, the stress responses from different kinds of stress resemble each other[34,50] but they are not identical.[3,97,113,138] It would be reasonable to expect some differences between the responses to heat stress, cold stress, psychological stress, exercise stress, fear, anger, revulsion, and poisoning stress. One experimental study found, for example, that heart rate increased with motion sickness but decreased with the revulsion evoked by viewing a film showing battlefield surgical operations.[113] The reason for the modest cardiovascular part of the stress response of motion sickness could be the high plasma level of AVP with motion sickness. High levels of AVP can substantially reduce both heart rate and blood pressure.[108,132]

The changes described below are probably caused primarily by the autonomic nervous system, but it has not been ruled out that some of the changes are related to other, possibly hormonal, influences. Although the remarkable pallor of motion sickness is usually ascribed to increases in activity in the sympathetic neural pathways, a role of the antidiuretic hormone AVP in the production of the pallor has not been ruled out.

4.1 ANS Changes Associated with Emptying of the Stomach

4.1.1 Salivation

There seems to be no compelling evidence on which to base the assignment of salivation changes to either the "emptying of the stomach" part or the "stress response" part of motion sickness. We shall locate it here as part of the emptying of the stomach, for the reason that salivation is closely related to gastric function.

Investigation has revealed that in humans there is a *perceived increase* of salivation with motion sickness in most subjects[42] but actually a *measured decrease* of salivary flow rate in most subjects.[42,43] There is also a surprising increase in the sodium concentration of saliva with motion sickness,[42,43] surprising because an increased concentration of sodium is an indication of an increased rate of secretion of saliva.[50] The concentration of salivary sodium in motion sickness has been found to increase also in squirrel monkeys, but in this species the measured flow rate of saliva does tend to increase with motion sickness, as would be expected when the concentration of sodium increases.[58,59]

It has been suggested that the measured *decrease* of flow rate in motion-sick humans[42,43] could be the result of an increased activity in the *sympathetic* division of the autonomic nervous system[41,43,50] or a decreased activity in the parasympathetic division[39,43,50] or both. The measured *increase* of flow rate in motion-sick squirrel monkeys is probably the result of increased activity in the *para*sympathetic division of the autonomic nervous system.[58,59] In dogs the increase of salivation immediately before apomorphine vomiting has been shown to be driven by the parasympathetic system.[37]

The decreased salivary flow rates with motion sickness in humans seem to be unlike the situation in cats, dogs, and monkeys who drool visibly or foam at the mouth with motion sickness. It is possible, however, that with motion sickness in these animals the salivary flow rate remains unchanged but the animals decrease the rate of swallowing. The *perceived* increase of salivation (in the face of a measured decrease) in humans could also be caused by a decrease in the normal rate of swallowing[42] with motion sickness. A decreased rate of swallowing in a poisoning situation might have protective value, since it could reduce the amount of poison swallowed.

It seems possible that the rate of swallowing is decreased and the rate of absorption of water into the tissues of the mouth is increased in motion sickness. If so, then the measured concentrations of salivary constituents in the mouth would presumably increase. In fact, it has been reported that the concentrations of the salivary constituents measured, sodium, potassium, and proteins, do increase with motion sickness.[43,58,59-61] In prolonged vomiting situations an increased rate of absorption of water into the tissues of the mouth, possibly under the influence of AVP, would perhaps tend to protect against dehydration: more of the water in saliva would be returned to the blood, instead of being either lost out the mouth directly (if swallowing is inhibited), or swallowed and then lost in vomitus.

In some human subjects there is a brief but strong increase in salivation, with swallowing, and sometimes with "pressure" sensations localized to the salivary glands themselves, immediately before vomiting; this latter phenomenon seems closely related to the vomiting, but it is seldom reported in motion sickness and it is perhaps separate from the changes in salivation that occur earlier, before vomiting is imminent. The earlier changes in salivation are the ones reported in the motion sickness literature. In dogs given apomorphine it has been reported that there is increased salivation[37] and increased frequency of swallowing[81] starting 2 min or so before vomiting.

4.1.2 Gastrointestinal Motility

The major control of functional gastrointestinal motility is by the parasympathetic part of the ANS. The sympathetic part of the ANS appears to influence gastrointestinal motility only moderately and in a general way, with an overall inhibiting and relaxing effect.

Surface electrodes on the abdomen of humans can record the electrical activity of the wall of the stomach, the electrogastrogram or EGG.[70,71,124] The EGG changes with motion sickness. The basic electrical rhythm of the EGG is 3 cycles per minute, but with the onset of motion sickness a "tachyarrhythmia" occurs: the frequency of the EGG increases to 4-9 cycles per minute,[71,146] and also the magnitude of the EGG voltage decreases. The muscular activity of the stomach decouples from the electrical activity when the frequency increases in motion sickness, and the increased frequency of the EGG is actually accompanied by a marked *decrease* of the tonus and motility of the stomach.

The initial inhibition of motility of the stomach and small intestine that occurs with motion sickness is imposed by inhibitory parasympathetic fibers from the vagus nerve.[141] These inhibitory fibers are thought to be the same fibers that provide the inhibitory phases of normal motility.[98] The inhibition/relaxation is mainly in the proximal part of the gastrointestinal tract, i.e., in the antrum of the stomach and in the duodenum and "proximal jejunum."[98] The inhibitory fibers that provide this initial inhibition are parasympathetic but their neurotransmitter is thought not to be acetylcholine. The fibers have been referred to as "non-adrenergic, non-cholinergic inhibitory vagal system."[141] Transection of the vagus nerve at the diaphragm prevents this initial inhibition, except possibly in the antrum of the stomach.[79,80,141]

The giant retrograde contraction that follows the initial inhibition is also driven by parasympathetic fibers from the vagus nerve. This GRC is eliminated by cutting the vagus nerve[81,98] and is also eliminated by atropine,[79,80,98] although one report suggests that it is not eliminated by atropine.[141]

The phasic contractions that follow the GRC are also driven by parasympathetic fibers from the vagus nerve. These contractions are also eliminated by transection of the vagus nerve.[78]

Presumably the elimination of the prodromal gastrointestinal changes would prevent the emptying of the top half of the small intestine into the stomach before vomiting, so that animals lacking these prodromal activities would vomit without removing any ingested material that had gained access to that part of the small intestine. With these prodromal changes eliminated, vomiting nevertheless occurs apparently normally; the stomach contents are expelled.[78-80] In fact, dogs with the entire sympathetic and parasympathetic nerve supplies to the gut severed are able to vomit in an apparently normal way after exposure to motion.[137]

4.1.3 The Esophagus

Neither the pharyngoesophageal actions in vomiting, nor the expulsion of vomitus, is altered by vagal blockade at the cervical level. "Therefore, vagal reflexes of thoracic or abdominal origin or efferent fibers of the recurrent laryngeal nerves do not control these pharyngoesophageal motor responses."[78] Since the pharynx and the upper part of the esophagus are *striated* muscle, the ANS is not involved in their motor control. No doubt the pharyngoesophageal motor contributions to the expulsion of vomitus are controlled by somatic nerves. This is what would be expected of a fast strong response that needs precise timing to be effective (striated muscle, somatic nerves). The pharyngoesophageal contractions are presumably timed to coordinate with the other contractions controlled by somatic nerves (the contractions of the diaphragmatic dome and the abdominal wall). It has been suggested[80] that the central mechanism of vomiting "may consist of two functionally distinct parts that are activated sequentially: one controlling the gastrointestinal responses and the other the somatomotor responses." In fact, the striated part of the esophagus, and the pharynx, as well as the abdominal wall and the diaphragm, would be controlled by the "somatomotor" part of the central mechanism (not via the ANS).

4.2 ANS Changes Associated with the Stress Response

4.2.1 Heart Rate

The cardiovascular changes in motion sickness[18-22,46,82] are generally in the same directions as in well-known stress responses, but in motion sickness some of these changes are surprisingly small[57] and some are transient.[126,127] In some studies the heart rate increased with motion sickness and in some studies it did not; in some studies a low heart rate or an unchanged heart rate is reported.[100] Transient changes in heart rate could be missed in studies[56] that stop the motion and measure the heart rate by feeling it with the experimenter's fingers. The required head movements of some motion sickness experiments, when performed in a motionless environment, are actually sufficient to produce increases in heart rate although of course no motion sickness is provoked.[82] Such increases should not be attributed to motion sickness.

The investigators who do observe increases in heart rates with motion sickness[20,22,57,82] appear to be using more severe levels of motion sickness than investigators who observe no increases,[46,82] and it seems likely that increases in heart rate occur but only with the more severe levels of the sickness. Investigators who stop the motion stimulus before the appearance of the more severe sicknesses do not see any increases in heart rates.[82] The increases of heart rate with more severe motion sickness, as recorded during the motion by electrocardiograph (not afterwards using fingers and a stopwatch), appear to be real.[20,82,113]

With moderate or severe levels of nausea, the observed increases of heart rate are statistically significant in some studies, although they do not occur in all the subjects, and the magnitudes of the increases are only 5 to 15 or so beats per minute.[20,82] This modest increase of heart rate with motion sickness, 5 to 20%, is much less than the 100% or so found in swimmers just before a competitive swim sprint,[89] or the 150 to 199% increase produced by skydiving or fast running in non-experts, or the 40 to 99% or so increase in astronauts prior to rocket launch into orbit.[82] Considering the level of misery and anxiety being suffered by victims of severe motion sickness, it is surprising that the increase of heart rate is not greater. Apparently there have been no comprehensive studies of heart rate during the frank vomiting of motion sickness, although there is some indication that it does increase.[56]

It is possible that the heart rate with motion sickness would be more typical of stress response situations (higher) if high plasma levels of AVP were not generated by motion sickness. It is not clear that the levels of AVP in motion sickness are sufficiently high to influence heart rate, but it is known that high levels of plasma AVP cause substantial decreases of heart rate and coronary blood flow.[132] The decreases of heart rate after injection of large doses of AVP in humans are roughly 15%. Perhaps it is relevant that stimulation of the area postrema inhibits heart rate in some species.[33] Also, when apomorphine was administered to normal humans,[108] nausea and vomiting was provoked and plasma AVP rose dramatically and mean arterial pressure *fell* significantly; in people with diabetes insipidus however, the apomorphine provoked the nausea and vomiting but of course no rise in plasma AVP, and in these people the fall in blood pressure was smaller and not statistically significant. When ipecacuanha was given to the subjects,[108] nausea and vomiting were provoked but no rise in plasma AVP occurred, and with ipecacuanha the blood pressure and the pulse rate both *rose* significantly.

Increased activity in the sympathetic division of the ANS can cause increased heart rates, and the modestly increased heart rates with motion sickness can be considered part of a stress response.

4.2.2 R-R Interval

The measure of interest here is the variability in the interval between heart beats. The R-R interval (RRI), the duration of the interval between heart beats, is more variable when the *para*sympathetic system is more active in controlling the heart, and less variable when the sympathetic system is more active in controlling the heart.[136] In a study of 52 human subjects exposed to a rotating optokinetic drum that provoked motion sickness, it was found that the RRI variability *decreased* (that is, the pulse became steadier) with motion sickness.[57,70] Specifically, "the mean successive differences of R-R intervals" decreased with motion sickness. Considering this result together with the observed gastric tachyarrhythmia and increased skin conductance levels with motion sickness, a small mean increase in pulse rate in the subjects, and other evidence, it was suggested that "high sympathetic nervous system activity plays an important role in the appearance of ... symptoms of motion sickness."[57]

On the other hand, in motion sickness tests of five normal squirrel monkeys it was found that the variability of the RRI *increased* with motion sickness.[62] The measure of variability of the RRI in the monkeys was "the coefficient of variance (CV) of mean R-R intervals," not the same measure of variability as in the study of humans,[57] but the finding must be given careful attention for the reason that it is tempting to ignore it because it is an exception to the generalization.

It is an exception to the generalization that the stress response of motion sickness consists only of increases in the sympathetic influences and/or decreases in the parasympathetic influences. However, the measure of variability employed, CV of RRI, is a sensitive measure that is possibly more appropriate for detecting variability in RRI than mean successive differences of RRI intervals which could be confounded by beat frequency. The CV of RRI was also validated pharmacologically in 11 monkeys: a drug that reduces the parasympathetic influence, atropine, produced a decrease in the CV, and a drug that increases beta-adrenergic (sympathetic) influence, isoproterenal, similarly produced a decrease in the CV; both of these results suggest that the CV measure is a valid one. Two drugs that might be expected to *increase* the CV, carbachol ("stimulates the parasympathetic nerve system") and propanolol (produces "beta-adrenergic blockade") did not produce significant changes in the CV,[62] but significance is often technically difficult to achieve with small test populations. The motion sickness tests in the five monkeys, with increased CV of mean RRI, seem to be valid indicators of increased *para*sympathetic influence on the heart with motion sickness in this species.

Surprisingly, in the same study[62] three monkeys with unilateral labyrinthectomy were exposed to the motion sickness stimulation and all three showed *decreased* CV of the RRI, the *same* sort of result as found in the study[57] of humans. Unilaterally labyrinthectomized animals might be expected to be somewhat less susceptible to motion sickness than normals, but if they do get motion sickness, presumably the sickness should be similar to that of normal animals although perhaps less severe. The findings in the three unilaterally labyrinthectomized monkeys therefore appears to contradict the findings in the five normal monkeys. In this monkey study there was, unfortunately, no mention of the behavior of the heart *rate* with the motion sickness. A later study by some of the same investigators[58] confirmed the increased CV of RRI in squirrel monkeys with motion sickness but, as in the earlier study, no information was provided about heart rates. The later article also confirms their view that there is an increased influence of the parasympathetic system on the salivary glands with motion sickness in squirrel monkeys. Salivary responses are presumably part of the stomach-emptying part of motion sickness and do not suggest anything about the stress response of motion sickness.

The stress response of motion sickness in squirrel monkeys, in its cardiac RRI manifestation, is exceptional. In these monkeys the increase of RRI variability with motion sickness is not characteristic of a typical stress response. Increased activity in the sympathetic division of the ANS can cause decreased variability of the RRI, and in humans the decreased variability of the RRI with motion sickness can be considered part of a stress response.

4.2.3 Blood Pressure

Several studies of blood pressure in motion sickness have found no large changes.[64,82,126,127] Blood pressure tends to vary with motion sickness, but it "can rise or fall in response to motion in both susceptible and nonsusceptible individuals."[100] Even two subjects who had their blood pressure measured "through a needle inserted into the brachial artery"[64] did not show clear directions of changes in blood pressure with motion sickness.

Since the blood flow in skeletal muscle can increase markedly in motion sickness, and since no corresponding loss of blood pressure is observed, therefore either the cardiac output increases (unlikely since the pulse rate increases are small) or there is a decreased blood flow elsewhere in the body (probably in skin and, more importantly, in viscera).[126,132] There is presumably an increase of peripheral resistance in skin and viscera that balances nicely the decrease of peripheral resistance in skeletal muscle.

Absence of a large increase in blood pressure is not typical of a stress response. The enormous increase of blood flow in skeletal muscle with motion sickness can be expected to be a powerful reducer of blood pressure in motion sickness, and could explain the failure of blood pressure to rise with motion sickness. Also, the large increases of plasma AVP with motion sickness could explain the failure of blood pressure to rise. It is known that large doses of AVP injected into humans cause substantial decreases in cardiac output, decreases in coronary blood flow, and substantial decreases in heart rate.[132] Since AVP does decrease coronary blood flow,[132] it would be prudent for people with heart disease to avoid conditions in which motion sickness could occur. Plasma AVP in motion sickness increases to 20–30 times the normal levels.

4.2.4 *Blood Flow in Skeletal Muscle*

With moderate-to-severe motion sickness there is a two- to threefold increase in the blood flow through the forearm.[126,127] There is also an increase in the total limb volume with motion sickness. This was measured in 46 human subjects exposed to head movements while rotating at 30 rpm, and it was found that the rate of flow of blood through the forearm increased two to three times in susceptible subjects.[126,127] The blood flow was measured by venous occlusion plethysmography (the venous return from the arm is blocked with a pressure cuff and the rate of swelling of the arm is measured). The measurements were done repeatedly without interrupting the motion stimulus, and the subjects called for cessation of the stimulus when they thought that emesis was imminent.

Thirty-two of the subjects stopped the stimulus before five minutes had elapsed, and in all of these 32 subjects all of the measurements showed at least some increase in forearm blood flow, compared to the pre-motion flow. "Although there were some exceptions, the magnitude of the change in forearm blood flow correlated with the severity of the symptoms, and it was often possible, by observing the blood flow recordings, to predict when the subject would signal that he would vomit if stimulation were not discontinued."[126] Since the human forearm is 85% muscle by volume, the results are taken to be an indication of increases in skeletal muscle blood flow generally. The blood flow in the calf of the leg of one subject was found to increase similarly. One study did not find the increase in forearm blood flow, possibly because a lesser degree of motion sickness was provoked.[82]

Increased activity in the sympathetic division of the ANS can cause increased blood flow in skeletal muscle, and the increased blood flow in skeletal muscle with motion sickness can be considered part of a stress response.

4.2.5 *Blood Flow in Skin and Blood Content of Skin*

The hallmark facial pallor, the pasty or greenish face of people suffering from motion sickness, is presumably caused by a lack of red color in the face, a decreased presence of red blood in the skin. The pallor is apparently caused by a reduction in the amount of blood present in the skin rather than a reduction in the rate of flow of blood through the skin, and measurements of blood flow in human fingers did not reveal any decrease during motion sickness,[82] although earlier studies found decreases of finger pulse volume with motion sickness.[22,113] With motion sickness onset there is actually a brief flushing of the face in some people,[52,111] but the flushing is usually replaced by pallor as the sickness continues. In most people the pallor then persists for as long as the sickness lasts. In one study the "onset of pallor, as detected on the color films, occurred before the report of nausea in all but 3 of 14 cases."[22]

The spectacular decrease in the amount of blood in the skin of the face, the remarkable pallor, is exactly what would be expected from a very strong increase in activity in the sympathetic nervous system, and the shunting of blood out of the skin is thought to be beneficial in stressful situations because more blood would be available to skeletal muscle and less blood would be lost if the skin were broken. The extreme degree of pallor seen in motion sickness, however, is seldom seen in response to other stresses, seldom in athletes even during extremely vigorous exercise. People who exercise to the point of exhaustion and nausea will sometimes exhibit facial pallor, but sprinters and skydivers just do not look like people suffering from motion sickness, even though they have heart rates that suggest a strong stress response.

It is puzzling that motion sickness seems to produce such an extreme sympathetic effect on the blood vessels of the skin of the face (constricting them) and on the blood vessels of skeletal muscle (dilating them), but such a modest effect on the heart rate and blood pressure. A possible explanation of the extreme change in the skin is that the massive increase in plasma AVP also constricts the blood vessels in skin. At higher levels of AVP, "Circulation in the skin and the gastrointestinal tract is markedly reduced"[39] and "After the injection of large doses of vasopressin, marked facial pallor as a result of cutaneous vasoconstriction is commonly observed."[54]

In motion sickness the plasma levels of AVP can reach 20 or even 30 times normal levels. If 20 or 30 times normal levels can be called a "large dose," then the hallmark pallor of motion sickness might be considered partly the result of the body's release of AVP into the bloodstream. At this time it is reasonable to think that *both* the sympathetic nervous system and the plasma AVP play roles in causing the remarkable pallor of motion sickness.

With motion sickness in spaceflight it has been reported that the pallor is much less than with motion sickness on Earth, possibly because the fluid shift caused by weightlessness keeps blood in the skin of the face.[102] In colder environmental temperatures there is probably more pallor with motion sickness than in warmer environmental temperatures. "Environmental temperature affects the degree of pallor but not the time of its onset."[22]

Increased activity in the sympathetic division of the ANS can cause pallor, and the pallor of motion sickness can be considered part of a stress response.

4.2.6 Cold Sweating or Increased Skin Conductance

"Cold sweating is the sweating that occurs without an adequate thermal stimulus,"[55] and this kind of sweating is a prominent sign of motion sickness. In an experiment in which motion sickness is provoked by head movements on a rotating platform, "A sweat response can be obtained from the palmar aspect of the hand during the initial few head movements...This response is maximal on the first one or two head movements and quickly declines with continuation of the stimulus. At a varying time interval after the commencement of head movements, a distinct sweat response occurs on the dorsal aspect of the hand and arm, independent of the palmar response."[92] The initial brief palmar sweating is "arousal" sweating, and the more extensive later sweating is "thermal" (but inappropriate) sweating.[92]

The sweating with motion sickness is sweating without an adequate thermal stimulus, but the environmental temperature retains an important influence. With motion sickness "more sweating took place at the 94° than at the 84° environmental temperature."[22] It was further found that as "the temperature was lowered increased amounts of vestibular stimulation were required to evoke the sweat response,"[91] and with environmental temperatures of 65°F or below there was little or no sweating in most subjects even though the nausea criterion was reached. Sweating or increased skin conductance with motion sickness is a consistent finding.[20-22,55,57,91,113,114]

Increased activity in the sympathetic division of the ANS can cause sweating, and the sweating of motion sickness can be considered part of a stress response.

4.2.7 Decrease in the Dark Field Focus Distance

Motion sickness has been found to increase the refractive power of the lens and thereby decrease the dark field focus distance of the eye.[35] The increase of refractive power was less than 0.5 diopters, a small fraction of the 60 diopters that is the total refractive power of the normal relaxed eye and a small fraction of the 10 diopters that is the maximum possible *change* in focus that can be achieved (in young people) by voluntary accommodation. In a normal eye, focused on infinity when relaxed, a maximum effort at near vision can increase the refractive power of the lens by roughly 10 diopters in a 20-year-old (and zero diopters in a 50-year-old: approximately 1/3 of a diopter of accommodative power is lost per year, with aging, starting from childhood). An increase of 0.5 diopters, and the corresponding decrease in the focus distance, is not insignificant.

An increase in the accommodative power of the lens of the eye is achieved by an increase in the activity of the parasympathetic fibers to the ciliary muscle and a resulting increase of contraction of the ciliary muscle.[50] The observed decrease in the dark field focus distance with motion sickness is therefore an indication of an increase in parasympathetic activity, and such an increase is exceptional in the stress response of motion sickness. The autonomic parts of the stress response of motion sickness are predominantly increases in sympathetic activity and/or decreases of parasympathetic activity. In the context of a stress response, the increase in parasympathetic activity to the ciliary muscle of the eye, and the resulting decrease in dark focus distance, must be considered paradoxical.

It has been pointed out[35] that motion sickness is not the only stress that provokes a decrease in the dark field focus of humans. "Two subjects were made angry by insulting remarks directed to them while their accommodative level was being tracked in complete darkness. In both cases a rise in accommodative level, lasting for several minutes, occurred; in one instance the rise exceeded 1 [diopter]."[139] It might be that the peculiarity of an increase of the parasympathetic activity to the ciliary muscle, in the stress response of motion sickness, is not so much a peculiarity of motion sickness as it is a peculiarity of the optical accommodative responses of humans to stresses in general. The parasympathetic fibers to the ciliary muscle exert the major control of focus distance; the sympathetic fibers to the ciliary muscle are in fact incapable of exerting any strong influence on focus distance since "they supply very weak innervation to the ciliary muscle."[50]

The optical accommodative response of a human to a near object usually occurs as part of a "near response" or "triad response" that includes, in addition to a decrease in the focus distance, a convergence of the eyes (directions of gaze move together more), and a decrease in the size of the pupils. The sympathetic fibers to the iris are much more influential than the sympathetic fibers to the ciliary muscle, and it would be interesting to find out whether the pupil shows an increase of sympathetic activity in motion sickness. Whether the other two parts of the human "near response" (pupillary constriction, and convergence) occur with motion sickness has apparently not been investigated.

In a different species, rabbits, exposed to a different stress, the expected *decreases* of accommodative power (and increases of focus distances) in response to stress were found; in this species the expected decrease of parasympathetic activity with stress seems to have occurred. When, in dim light, "rabbits were startled by a sudden tap on the nose (the most convenient stimulus), by having their heads sharply raised and lowered or turned side to side, or by being suddenly jostled about, or upon hearing a sudden noise such as the slamming of the door, they immediately"[109] showed *decreases* in accommodative power of the lens (increases of focus distance). These changes lasted 2 to 5 min and on average were 1 diopter.

5 SUMMARY AND CONCLUSIONS

Motion sickness can be seen as a poison response that occurs, inappropriately, in response to certain kinds of motion. The signs and symptoms form two groups, (1) stomach-emptying phenomena and (2) stress response phenomena. It is sometimes postulated that the stomach emptying phenomena of motion sickness are stressful and cause the stress response; it is also postulated sometimes that the stress response of motion sickness causes the stomach-emptying phenomena. In fact, the available evidence does not support either of these (mutually exclusive) etiological postulates. The evidence suggests that motion acts on the vestibular system to trigger a poison response, and that the poison response provides both the stomach emptying and the stress response.

The stress response of motion sickness includes changes of conscious sensations (including distress or dread), endocrinological changes (increased plasma levels of norepinephrine, epinephrine, ACTH, cortisol, etc.), and autonomic changes. The autonomic changes of the stress response of motion sickness seem to be the result of increases of activity in the sympathetic pathways: pallor, cold sweating, increased blood flow in skeletal muscle, and a modest increase in pulse rate.

The stomach-emptying phenomena include a suppression of upper gastrointestinal motility, followed by the emptying of the upper half of the small intestine into the stomach and the lower half into the colon (thereby preventing the further absorption of most substances). Then retching and vomiting empty the stomach. Conscious sensations of nausea, and sensations of vomiting, also occur. Another prominent stomach-emptying phenomenon of motion sickness is an increase in the concentration of the antidiuretic hormone AVP in the circulating blood, an increase up to 30 times the normal circulating concentration. There is evidence that the high levels of AVP in motion sickness prevent the strong increase of arterial blood pressure that typically does occur with stress responses but does not occur in motion sickness. The AVP would also give some protection against water loss in prolonged vomiting situations.

The autonomic phenomena of stomach emptying are, of course, controlled mostly by the parasympathetic pathways. Controversies about whether sympathetic or parasympathetic responses are dominant in motion sickness can probably be resolved by noting that the stress response of motion sickness involves increases in sympathetic activity whereas the stomach-emptying phenomena of motion sickness involve increases in parasympathetic activity.

REFERENCES

1. Abe, K., Amatomi, M., and Kajiyama, S., Genetical and developmental aspects of susceptibility to motion sickness and frost-bite, *Human Heredity*, 30, 507, 1970.
2. Arena, J. M., *Poisoning — Toxicology, Symptoms, Treatments*, 4th ed., Charles C. Thomas, Springfield, IL, 1979.
3. Ax, A. F., The physiological differentiation between fear and anger in humans, *Psychsom. Med.*,15, 433, 1953.
4. Bakwin, H., Car-sickness in twins, *Dev. Med. Child Neurol.*, 13, 310, 1971.
5. Banks, R. D., Salisbury, D. A., and Ceresia, P. J., The Canadian Forces airsickness rehabilitation program, 1981-1991. *Aviat. Space Environ. Med.*, 63, 1098, 1992.
6. Bard, P., Autonomic nervous system or efferent pathways to visceral effectors, in *Medical Physiology*, 12th ed., Mountcastle, V. B., Ed., C.V. Mosby, St. Louis, 1968, chap. 79.
7. Benson, A. J., Ed., *Motion Sickness: Significance in Aerospace Operations and Prophylaxis*, NATO/AGARD Lecture Ser. 175, 1991.
8. Bianchi, A. L., Grélot, L., Miller, A. D., and King, G. L., *Mechanisms and Control of Emesis, Colloques INSERM*/John Libbey Eurotext, Montrouge, France, 1992.

9. Borison, A. L., Anatomy and physiology of the chemoreceptor trigger zone and area postrema, in *Nausea and Vomiting: Mechanisms and Treatment,* Davis, C. J., Lake-Bakaar, G. V., and Grahame-Smith, D. G., Eds., Springer-Verlag, New York, 1986, 10.
10. Brizee, K. R. and Mehler, W. R., The central connections involved in the vomiting reflex, in *Nausea and Vomiting: Mechanisms and Treatment,* Davis, C. J., Lake-Bakaar, G. V., and Grahame-Smith, D. G., Eds., Springer-Verlag, New York, 1986, 31.
11. Burnstock, G., Plasticity of autonomic transmission, *J. Auton. Nerv. Syst.,* 33, 137, 1991.
12. Case, C., Cheung, B., Fry, M., Garshnek, V., Money, K., Smith, G., Uri, J., and Winisdoerffer, F., Human factors and physiological aspects of the Mars mission, *Acta Astronautica,* 31, 73, 1993.
13. Cheung, B. S. K., Howard, I. P., and Money, K. E., Visually-induced sickness in normal and bilateral labyrinthine-defective subjects, *Aviat. Space Environ. Med.,* 62, 527, 1991.
14. Cheung, B. S. K., Kohl, R. L., Money, K. E., and Kinter, L. B., Etiologic significance of arginine vasopressin in motion sickness, *J. Clin. Parmacol.,* 34, 664, 1994.
15. Cheung, B. S. K., Money, K. E., and Howard, I. P., Dynamics of torsional optokinetic nystagmus under altered gravitoinertial forces, *Exp. Brain Res.,* 102, 511, 1995.
16. Claremont, C. A., The psychology of sea-sickness, *Psyche,* 11, 86, 1930.
17. Considine, D. M., Ed., *Van Nostrand's Scientific Encyclopedia,* 7th ed., Van Nostrand Reinhold, New York. 1989.
18. Cowings, P. S., Autogenic-feedback training: a treatment for motion and space sickness, in *Motion and Space Sickness,* Crampton, G. H., Ed., CRC Press, Boca Raton, FL, 1990, chap. 17.
19. Cowings, P. S. and Toscano, W. B., The relationship of motion sickness susceptibility to learned autonomic control for symptom suppression, *Aviat. Space Environ. Med.,* 53, 570, 1982.
20. Cowings, P. S., Naifeh, K. H., and Toscano, W. B., The stability of individual patterns of autonomic responses to motion sickness stimulation, *Aviat. Space Environ. Med.,* 61, 399, 1990.
21. Cowings, P. S., Suter, S., Toscano, W. B., Kamiya, J., and Naifeh, K., General autonomic components of motion sickness, *Psychophysiology,* 23, 542, 1986.
22. Crampton, G. H., Studies of motion sickness. XVII. Physiological changes accompanying sickness in man, *J. Appl. Physiol.,* 7, 501, 1955.
23. Crampton, G. H., Ed., *Motion and Space Sickness,* CRC Press, Boca Raton, FL, 1990.
24. Crampton, G. H. and Lucot, J. B., Habituation of motion sickness in the cat, *Aviat. Space Environ. Med.,* 62, 212, 1991.
25. Cummins, A. J., The physiology of symptoms. III. Nausea and vomiting, *Am. J. Dig. Dis.,* 3, 710, 1958.
26. Daunton, N. G., Fox, R. A., and Crampton, G. H., Susceptibility of cat and squirrel monkey to motion sickness induced by visual stimulation: correlation with susceptibility to vestibular stimulation, in *Motion Sickness: Mechanisms, Prediction, Prevention and Treatment,* NATO/AGARD-CP-372, Advisory Group for Aerospace Res. Dev. Conf. Proc. No. 372, 1984, pp. 31-1 to 31-5.
27. Davis, B. O., Holtz, N., and Davis, J. C., *Conceptual Human Physiology,* Charles E. Merrill, Toronto, 1985.
28. Davis, C. J., Harding, R. K., Leslie, R. A., and Andrews, P. L. R., The organization of vomiting as a protective reflex, in *Nausea and Vomiting: Mechanisms and Treatment,* Davis, C. J, Lake-Bakaar, G. V., and Grahame-Smith D. G., Eds., Springer-Verlag, New York, 1986, 65.
29. Davis, C. J., Lake-Bakaar, G. V., and Grahame-Smith, D. G., Eds., *Nausea and Vomiting: Mechanisms and Treatment,* Springer-Verlag, New York, 1986.
30. Diamond, S. G. and Markham, C. H., Ocular torsion in upright and tilted positions during hypo- and hypergravity of parabolic flight, *Aviat. Space Environ. Med.,* 59, 1158, 1988.
31. Dobie, T. G. and May, J. G., Cognitive-behavioral management of motion sickness, *Aviat. Space Environ. Med.,* 65 (10, Suppl), C1-C10, 1994.
32. Eversmann, T., Gottsmann, M., Uhlich, E., Ulbrecht, G., von Werder, K., and Scriba, P. C., Increased secretion of growth hormone, prolactin, antidiuretic hormone, and cortisol induced by the stress of motion sickness, *Aviat. Space Environ. Med.,* 49, 53, 1978.
33. Ferguson, A. V. and Smith, P., Autonomic mechanisms underlying area postrema stimulation-induced cardiovascular responses in rats, *Am. J. Physiol.,* 261 (*Reg. Integr. Comp. Physiol.,* 30), R1-R8, 1991.
34. Fischer, L. A. and Brown, M. R., Central regulation of stress: regulation of the autonomic nervous system and visceral function by corticotrophin releasing factor -41, *Bailler's Clin. Endocrinol. Metab.,* 5, 35, 1991.
35. Fowlkes, J. E., Kennedy, R. S., Hettinger, L. J., and Harm, D. L., Changes in the dark focus of accommodation associated with simulator sickness, *Aviat. Space Environ. Med.,* 64, 612, 1993.
36. Fox, R. A., Keil, L. C., Daunton, N. G., Crampton, G. H., and Lucot, J. B., Vasopressin and motion sickness in cats, *Aviat. Space Environ. Med.,* 58 (9, Suppl), A143-A147, 1987.
37. Furukawa, N. and Okada, H., Canine secretion from submaxillary glands before and during retching, *Am. J. Physiol.,* 267 *(Gastrointest. Liver Physiol.,* 30*),* G810, 1994.

38. Gibbins, I., Peripheral autonomic nervous system, in *The Human Nervous System*, Paxinos, G., Ed., Academic Press, San Diego, 1990, chap. 5.
39. Gilman, A. G., Rall, T. W., Nies, A. S., and Taylor, P., *Goodman and Gilman's The Pharmacological Basis of Therapeutics*, 8th ed., Pergamon Press, New York, 1990.
40. Golding, J. F. and Stott, J. R. R., *Objective and Subjective Time-Course of Recovery from Motion Sickness Assessed by Repeated Motion Challenges*, Royal Air Force Institute of Aviation Medicine Rep. No. 763, AST 45, Project No. S/82/01, February 1994.
41. Gordon, C. R., Ben-Aryeh, H., Spitzer, O., Doweck, I., Gonen, M. A., Melamed, Y., and Shupak, A., Seasickness susceptibility, personality factors, and salivation, *Aviat. Space Environ. Med.*, 65, 610, 1994.
42. Gordon, C. R., Ben-Aryeh, H., Szargel, R., and Attias, J., Salivary changes associated with seasickness, *J. Auton. Nerv. Syst.*, 26, 37, 1989.
43. Gordon, C. R., Ben-Aryeh, H., Szargel, R., Attias, J., Rolnick, A., and Laufer, D., Salivary changes associated with experimental motion sickness condition in man, *J. Auton. Nerv. Syst.*, 22, 91, 1988.
44. Grahame-Smith, D. G., The multiple causes of vomiting: is there a common mechanism? in *Nausea and Vomiting: Mechanisms and Treatment*, Davis, C. J., Lake-Bakaar, G. V., and Grahame-Smith, D. G., Eds., Springer-Verlag, Berlin, 1986, 1.
45. Graybiel, A., Deane, F. R., and Colehour, J. K., Prevention of overt motion sickness by incremental exposure to otherwise highly stressful Coriolis accelerations, *Aerospace Med.*, 40, 142, 1969.
46. Graybiel, A. and Lackner, J. R., Evaluation of the relationship between motion sickness symptomatology and blood pressure, heart rate, and body temperature, *Aviat. Space Environ. Med.*, 51, 211, 1980.
47. Graybiel, A., Wood, C. D., Miller, E. F., and Cramer, D. B., Diagnostic criteria for grading the severity of acute motion sickness, *Aerospace Med.*, 39, 453, 1968.
48. Guedry, F. E., Motion Sickness and Its Relation to Some Forms of Spatial Disorientation, NATO/AGARD Lecture Ser. 175, 1991, pp. 2-1 to 2-30.
49. Guedry, F. E., Factors Influencing Susceptibility: Individual Differences and Human Factors, NATO/AGARD Lecture Ser. 175, 1991, pp. 5-1 to 5-18.
50. Guyton, A. C., *Textbook of Medical Physiology*, 8th ed., W.B. Saunders, Philadelphia, 1991.
51. Habermann, J., Eversmann, T., Erhardt, F., Gottsmann, M., Ulbrecht, G., and Scriba, P. C., Increased urinary excretion of Triiodothyronine (T3) and Thyroxine (T4) and decreased thyreotropic hormone (TSH) induced by motion sickness, *Aviat. Space Environ. Med.*, 49, 58, 1978.
52. Harm, D. L., Physiology of motion sickness symptoms, in *Motion and Space Sickness*, Crampton, G. H., Ed., CRC Press, Boca Raton, FL, 1990, chap. 10.
53. Hasegawa, S., Takeda, N., Morita, M., Horii, A., Koizuka, I., Kubo, T., and Matsunaga, T., Vestibular, central, and gastric triggering of emesis, *Acta Otolaryngol. (Stockholm)*, 112, 927, 1992.
54. Hays, R. M., Agents affecting the renal conservation of water, in *Goodman and Gilman's The Pharmacological Basis of Therapeutics*, 8th ed., Gilman, A. G., Rall, T. W., Nies, A. S., and Taylor, P., Eds., Pergamon Press, New York, 1990, chap. 29.
55. Hemingway, A., Cold sweating in motion sickness, *Am. J. Physiol.*, 141, 172, 1944.
56. Hemingway, A., Cardiovascular changes in motion sickness, *Aviat. Med.*, 16, 417, 1945.
57. Hu, S., Grant, W., Stern, R. M., and Koch, K. L., Motion sickness severity and physiological correlates of repeated exposure to a rotating optokinetic drum, *Aviat. Space Environ. Med.*, 62, 308, 1991.
58. Igarashi, M., Chae, S., MacDonald, S., Himi, T., and Takeda, N., Autonomic indexes during the vestibular-visual conflict exposure: a squirrel monkey study, *Auris. Nasus. Larynx (Tokyo)*, 17, 69, 1990.
59. Igarashi, M., Macdonald, S., Chae., S.-Y., Plishker, G. A., and Kohl, R. L., Sodium concentration in saliva along the time course of experimental Coriolis sickness, *Acta Otolaryngol. (Stockholm)*, 107, 485, 1989.
60. Igarashi, M., Reschke, M. F., Henley, C., Macdonald, S., Kohl, R., and Kanemasa, M., Salivary total protein and experimental Coriolis sickness, *Acta Otolaryngol. (Stockholm) Suppl.*, 504, 38, 1993.
61. Igarshi, M., Reschke, M. F., Henley, C., MacDonald, S., Kohl, R., and Mizukoshi, K., Salivary total protein and experimental Coriolis sickness, *Acta Otolaryngol. (Stockholm) Suppl.*, 504, 38,1993.
62. Ishii, M., Igarashi, M., Patel, S., Himi, T., and Kulecz, W., Autonomic effects on R-R variations in the squirrel monkey: an indicator of autonomic imbalance in conflict sickness, *Am. J. Otolaryngol.*, 3, 144, 1987.
63. James, W., The sense of dizziness in deaf mutes, *Am. J. Otol.*, 4, 239, 1882.
64. Johnson, W. H., Motion sickness part 1. Aetiology and autonomic effects, in *Handbook of Sensory Physiology, Vol. VI/2*, Kornhuber, H. H., Ed., Springer-Verlag, Berlin, 1974, 389.
65. Jones, D. R. and Hartman, B. O., Biofeedback Treatment of Airsickness: A Review, NATO/AGARD Conf. Proc. No. 372, 1984, pp. 42-1 to 42-4.

66. Karabinis, A., Rigas, K., Mega, A. M., Thomopoulos, G., Klouva, F., and Bilalis, D., Spontanée de l'oesophage masquée par la prise d'un anti-inflammatoire non-steroidien (lettre), Spontaneous rupture of the oesophagus masked by the ingestion of a non-steroidal anti-inflammatory, *La Presse Med.*, 22, 1974, 1993.
67. Kemmler, R. W., *Psychological Components in the Development and Prevention of Air Sickness*, NATO/AGARD Conf. Proc. No. 372, 1984, pp. 41-1 to 41-8.
68. Kennedy, R. S., Graybiel, A., McDonough, R. C., and Beckwith, F. D., Symptomatology under storm conditions in the north Atlantic in control subjects and bilateral labyrinthine defects, *Acta Oto-Laryngol.*, 66, 533, 1968.
69. Kennedy, R. S., Lilienthal, M. G., Berbaum, K. S., Baltzley, D. R., and McCauley, M. E., Simulator sickness in U.S. Navy flight simulators, *Aviat. Space Environ. Med.*, 60, 10, 1989.
70. Koch, K. L., Motion sickness, in *The Handbook of Nausea and Vomiting*, Sleisenger, M. H., Ed., Parthenon, New York, 1993, 43.
71. Koch, K. L., Stern, R. R., Vasey, M. W., Seaton, J. F., Demers, L. M., and Harrison, T. S., Neuroendocrine and gastric myoelectric responses to illusory self-motion in men, *Am. J. Physiol.*, 258 (*Endocrinol. Metab.*, 21), E304-E310, 1990.
72. Kohl, R. L., Endocrine correlates of susceptibility to motion sickness, *Aviat. Space Environ. Med.*, 56, 1158, 1985.
73. Kohl, R. L., Hormonal responses of metoclopramide-treated subjects experiencing nausea or emesis during parabolic flight, *Aviat. Space Environ. Med.*, 58 (9, Suppl), A266, 1987.
74. Kohl, R. L., Endocrinology of space/motion sickness, in *Motion and Space Sickness*, Crampton, G. H., Ed., CRC Press, Boca Raton, FL, 1990, 65.
75. Kohl, R. L., Autonomic function and plasma catecholamines following stressful sensory stimuli, *Aviat. Space Environ. Med.*, 64, 921, 1993.
76. Lackner, J. and Di Zio, P., Some efferent and somatosensory influences on body orientation and oculomotor control, in *Sensory Experience, Adaptation and Perception*, Spillman, L. and Wooten, B. R., Eds., Erlbaum and Associates, Clifton, NJ, 1984, 281.
77. Lang, I. M., Digestive tract motor correlates of vomiting and nausea, *Can. J. Physiol. Pharmacol.*, 68, 242, 1990.
78. Lang, I. M., New perspectives on the mechanisms controlling vomitus expulsion, in *Mechanisms and Control of Emesis*, Vol. 223, Bianchi, A. L., Grélot, L., Miller, A. D., and King, G. L., Eds., Colloque INSERM/John Libbey Eurotext Montrouge, France, 1992, 71.
79. Lang, I. M., Marvig, J., Sarna, S. K., and Condon, R. E., Gastrointestinal myoelectric correlates of vomiting in the dog, *Am. J. Physiol.*, 251 (*Gastrointest. Liver Physiol.*, 14), G830, 1986.
80. Lang, I. M., Sarna, S. K., and Condon, R. E., Gastrointestinal motor correlates of vomiting in the dog: quantification and characterization as an independent phenomenon, *Gastroenterology*, 90, 40, 1986.
81. Lang, I. M., Sarna, S. K., and Dodds, W. J., Pharyngeal, esophageal, and proximal gastric responses associated with vomiting, *Am. J. Physiol.*, 265 (*Gastrointest. Liver Physiol.*, 28), G963-G972, 1993.
82. Lawson, B. D., *Human Physiological and Subjective Responses During Motion Sickness Induced by Unusual Visual And Vestibular Stimulation*, Ph.D. dissertation, Department of Psychology, Brandeis University, Waltham, MA, 5 April 1993, 1-109.
83. Lefkowitz, J. R., Hoffman, B. B., and Taylor, P., Neurohumeral transmission: the autonomic and somatic motor nervous systems, in *Goodman and Gilman's The Pharmacological Basis of Therapeutics*, 8th ed., Gilman, A. G., Rall, T. W., Nies, A. S., and Taylor, P., Eds., Pergamon Press, New York, 1990, chap. 5.
84. Levy, R. A., Jones, D. R., and Carlson, E. H., Biofeedback rehabilitation of airsick aircrew, *Aviat. Space Environ. Med.*, 52, 118, 1981.
85. Low, P. A., Autonomic nervous system, *J. Clin. Neurophysiol.*, 10, 14, 1993.
86. Lucot, J. B., Neurochemistry and pharmacology of motion sickness in nonhuman species, in *Motion and Space Sickness*, Crampton, G. H., Ed., CRC Press, Boca Raton, FL, 1990, chap. 5.
87. Lucot, J. B. and Crampton, G. H., Xylazine emesis, Yohimbine and motion sickness susceptibility in the cat, *J. Pharmacol. Exp. Ther.*, 237, 450, 1986.
88. Lucot, J. B. and Crampton, G. H., 8-OH-DPAT does not interfere with habituation to motion-induced emesis in cats, *Brain Res. Bull.*, 26, 919, 1991.
89. Magel, J. R., McArdle, W. D., and Glaser, R. M., Telemetered heart rate response to selected competitive swimming events, *J. Appl. Physiol.*, 26, 764, 1969.
90. McCauley, M. E., Ed., *Research Issues in Simulator Sickness: Proceedings of a Workshop*, National Academy Press, Washington, D.C., 1984.
91. McClure, J. A. and Fregly, A. R., Effect of environmental temperature on sweat onset during motion sickness, *Aerospace Med.*, 43, 959, 1972.
92. McClure, J. A., Fregly, A. R., Molina, E., and Graybiel, A., *Response from Arousal and Thermal Sweat Areas During Motion Sickness*, NAMRL-1142, Naval Aerospace Med. Res. Lab., Pensacola, FL, 1971, 1-8.

93. Miller, A. D. and Kucharczyk, J., Mechanisms of nausea and emesis, in *Nausea and Vomiting: Recent Research and Clinical Advances*, Kucharczyk, J., Stewart, D. J., and Miller, A. D., Eds., CRC Press, Boca Raton, FL, 1991, chap. 1.
94. Miller, A. D., Motion-induced nausea and vomiting, in *Nausea and Vomiting: Recent Research and Clinical Advances*, Kucharczyk, J., Stewart, D. J., and Miller, A. D., Eds., CRC Press, Boca Raton, FL, 1991, chap. 2.
95. Miller, A. D., Neuroanatomy and physiology, in *The Handbook of Nausea and Vomiting*, Sleisenger, M. H., Ed., Parthenon Press, New York, 1993, chap. 1.
96. Miller, A. D. and Leslie, R. A., The area postrema and vomiting, *Frontiers Neuroendocrinol.*, 15, 301, 1994.
97. Mills, F. J., The endocrinology of stress. *Aviat. Space Environ. Med.*, 56, 642, 1985.
98. Miolan, J. P., Neural control of the distal digestive tract responses associated with vomiting, in *Mechanisms and Control of Emesis*, Vol. 223, Bianchi, A. L., Grélot, L., Miller, A. D., and King, G. L., Eds., Colloque INSERM/John Libbey Eurotext, Montrouge, France, 1992, 59.
99. Mirabile, C. S. and Glueck, B. C., Separation of affective disorder into seasonal and nonseasonal types using motion sickness susceptibility as a marker, *J. Neuropsychiat. Clin. Neurosci.*, 5, 330, 1993.
100. Money, K. E., Motion sickness, *Physiol. Rev.*, 50, 1, 1970.
101. Money, K. E., Motion sickness and evolution, in *Motion and Space Sickness*, Crampton, G. H., Ed., CRC Press, Boca Raton, FL, 1990, chap. 1.
102. Money, K. E., Space sickness, in *Motion Sickness: Significance in Aerospace Operations and Prophylaxis*, Benson, A. J., Ed., NATO/AGARD Lecture Ser. 175, pp. 6A-1 to 6A-5, 1991.
103. Money, K. E., Signs and symptoms of motion sickness and its basic nature, in *Mechanisms and Control of Emesis*, Vol. 223, Bianchi, A. L., Grélot, L., Miller, A. D., and King, G. L., Eds., Colloque INSERM/John Libbey Eurotext, Montrouge, France, 1992, 177.
104. Money, K. E. and Cheung, B. S., Another function of the inner ear: facilitation of the emetic response to poisons, *Aviat. Space Environ. Med.*, 54, 208, 1983.
105. Money, K. E. and Friedberg, J., The role of the semicircular canals in causation of motion sickness and nystagmus in the dog, *Can. J. Physiol. Pharmacol.*, 42, 793, 1964.
106. Money, K. E. and Wood, J. D., Neural mechanisms underlying the symptomatology of motion sickness, *4th Symp. Role of the Vestibular Organs in Space Exploration*, NASA SP-187, 35-44, 1968.
107. Morrow, G. R., The effect of susceptibility to motion sickness on the side effects of cancer chemotherapy, *Cancer*, 55, 2766,1985.
108. Nussey, S. S., Hawthorn, J., Page, S. R., Ang, V. T. Y., and Jenkins, J. S., Response of plasma oxytocin and arginine vasopressin to nausea induced by apomorphine and ipecacuanha, *Clin. Endocrinol.*, 28, 297, 1988.
109. Olmsted, J. M. D. and Morgan, M. W., Refraction of the rabbit's eye in the unexcited and excited state, *Am. J. Physiol.*, 127, 602,1939.
110. Oman, C. M., A heuristic mathematical model for the dynamics of sensory conflict and motion sickness, *Acta Otolaryngol. Suppl.*, 392, 1, 1982.
111. Oman, C. M., Spacelab experiments on space motion sickness, *Acta Astronautica*, 15, 55, 1987.
112. Ossenkopp, K.-P., Rabbi, Y. J., and Eckle, L. A., Alcohol-induced conditioned taste aversions in chemically labyrinthectomized rats, *Aviat. Space Environ. Med.*, 65, 824, 1994.
113. Parker, D. M., *An Investigation of Autonomic Function in Motion Sickness and Revulsion*, Doctoral dissertation, Department of Psychology, UCLA, 1964.
114. Parker, D. M., A psychophysiological test for motion sickness susceptibility, *J. Gen. Psychol.*, 85, 87, 1971.
115. Proudfoot, A. T., *Acute Poisoning: Diagnosis and Management*, Butterworth Heinemann, Oxford, 1993.
116. Reason, J. T. and Brand, J. J., *Motion Sickness*, Academic Press, London, 1975.
117. Robertson, G. L., The regulation of vasopressin function in health and disease, *Recent Prog. Hormonal Res.*, 33, 333, 1977.
118. Schwartz, G. E., Biofeedback, self-regulation, and the patterning of physiological processes, *Am. Sci.*, 63, 314, 1975
119. Shapiro, D. and Schwartz, G. E., Biofeedback and visceral learning: clinical applications, *Semin. Psychiatr.*, 4, 171, 1972.
120. Sherwood, L., *Human Physiology from Cells to Systems*, 2nd ed., West Publishing Co., Minneapolis/St. Paul, 1993.
121. Sjoberg, A. A., Experimental studies of the eliciting mechanism of seasickness, *Acta Oto-Laryngol.*, 13, 343, 1929.
122. Sleisenger, M. H., Ed., *The Handbook of Nausea and Vomiting*, Parthenon Press, New York, 1993.
123. Spiegel, E. A., Effect of labyrinthine reflexes on vegetative nervous system, *Arch. Otolaryngol.*, 44, 61, 1946.

124. Stern, R. M., Koch, K. L,, Leibowitz, H. W., Inger, M., Lindblad, I. M., Shupert, C. L., and Stewart, M. S., Tachygastria and motion sickness, *Aviat. Space Environ. Med.*, 56, 1074, 1985.
125. Stott, J. R. R., Adaptation to nauseogenic motion stimuli and its application in the treatment of airsickness, in *Motion and Space Sickness*, Crampton, G. H., Ed., CRC Press, Boca Raton, FL, 1990, 373.
126. Sunahara, F. A., Johnson, W. H., and Taylor, N. B. G., Vestibular stimulation and forearm blood flow, *Can. J. Physiol. Pharmacol.*, 42, 199, 1964.
127. Sunahara, F. A., Farewell, J., Mintz, L., and Johnson, W. H., Pharmacological interventions for motion sickness: cardiovascular effects, *Aviat. Space Environ. Med.*, 58 (9, Suppl), A270-A276, 1987.
128. Surprenant, A., Control of the gastrointestinal tract by enteric neurons, *Annu. Rev. Physiol.*, 56, 117, 1994.
129. Takeda, N., Morita, M., Hasegawa, S., Horii, A., Kubo, T., and Matsunaga, T., Neuropharmacology of motion sickness and emesis: a review, *Acta Otolaryngol.*, 501, 10, 1993.
130. Taylor, E. J., Ed., *Dorland's Illustrated Medical Dictionary*, 27th ed., W.B. Saunders, Philadelphia, 1988.
131. Taylor, N. B. G., Hunter, J., and Johnson, W. H., Antidiuresis as a measurement of motion sickness, *Can. J. Biochem. Physiol.*, 35, 1017, 1957.
132. Thomford, N. R. and Sirinek, K. R., Intravenous vasopressin in patients with portal hypertension: advantages of continuous infusion, *J. Surg. Res.*, 18, 113, 1975.
133. Thornton, W. E., Linder, B. J., Moore, T. P., and Pool, S. L., Gastrointestinal motility in space motion sickness, *Aviat. Space Environ. Med.*, 58 (9, Suppl.), A16-A21, 1987.
134. Toscano, W. B. and Cowings, P. S., Reducing motion sickness: a comparison of autogenic-feedback training and an alternative cognitive task, *Aviat. Space Environ. Med.*, 53, 449, 1982.
135. Treisman, M., Motion sickness: an evolutionary hypothesis, *Science*, 197, 493, 1977.
136. Vybiral, T., Bryg, R. J., Maddeis, M.E., and Boden, W. E., Effect of passive tilt on sympathetic and parasympathetic components of heart rate variability in normal subjects, *Am. J. Cardiol.*, 63, 1117, 1989.
137. Wang, S. C., Chinn, H. I., and Renzi, A. A., Experimental motion sickness in dogs: role of abdominal visceral afferents, *Am. J. Physiol.*, 190, 578, 1957.
138. Wenger, M. A. and Cullen, T. D., Fourteen ANS response patterns in man, Abstract, *Am. Psychol.*, 13, 423, 1958.
139. Westheimer, G., Accommodation measurements in empty visual fields, *J. Opt. Soc. Am.*, 47, 714, 1957.
140. Whitehead, S. A., Holden, W. A., and Andrews, P. L. R., Pregnancy sickness, in *Mechanisms and Control of Emesis*, Vol. 223, Bianchi, A. L., Grélot, L., Miller, A. D., and King, G. L., Eds., Colloque INSERM/John Libbey Eurotext, Montrouge, France, 1992, 297.
141. Willems, J. L. and Lefebvre, R. A., Peripheral nervous pathways involved in nausea and vomiting, in *Nausea and Vomiting: Mechanisms and Treatment*, Davis, C. J., Lake-Bakaar, G. V., and Grahame-Smith, D. G., Eds., Springer-Verlag, Berlin, 1986, 56.
142. Wilpizeski, C. R. and Lowry, L. D., A two-factor model of rotation-induced motion sickness in squirrel monkeys, *Am. J. Otolaryngol.*, 8, 7, 1987.
143. Wilpizeski, C. R., Lowry, L. D., Contrucci, R. B., Green, S. G., and Goldman, W. S., Effects of head and body restraint on experimental motion-induced sickness in squirrel monkeys, *Aviat. Space Environ. Med.*, 56, 1070, 1985.
144. Wood, C. D., Pharmacological countermeasures against motion sickness, in *Motion and Space Sickness*, Crampton, G. H., Ed., CRC Press, Boca Raton, FL, 1990, 343.
145. Wood, C. D., Stewart, J. J., Wood, M. J., and Mims, M., Effectiveness and duration of intramuscular antimotion sickness medications, *J. Clin. Pharmacol.*, 32, 1008, 1992.
146. Xu, L. H., Koch, K. L., Summy-Long, J., Stern, R. M., Seaton, J. F., Harrison, T. S., Demers, S. M., and Bingaman, S., Hypothalamic and gastric myoelectrical responses during vection-induced nausea in healthy Chinese subjects, *Am. J. Physiol.*, 265 (*Endocrinol. Metab.*, 28), E578-E584, 1993.
147. Yardley, L., Motion sickness and perception: a reappraisal of the sensory conflict approach, *Br. J. Psychol.*, 83, 449, 1992.
148. Kroft, F., personal communication.
149. Money, K. E., unpublished observations.
150. Stern, R. M., personal communication.
151. Lang, I. M., personal communication.

10

Autonomic Disturbances Resulting from Exposure to Altered Gravitational Environments

Fred H. Previc

CONTENTS

0-8493-7668-1/96/$0.00+$.50

1 INTRODUCTION

Exposure to unusual gravitational environments — principally, those in which the gravitational forces are either greater than 1-G (hyper-G) or less than 1-G (hypo-G) — has been shown to have widespread autonomic consequences. The most prominent of these effects are the physiological symptoms associated with motion sickness and cardiovascular adaptation. Many different physiological systems besides the labyrinthine ones are directly affected by unusual gravitational conditions, including the cardiovascular, gastrointestinal, renal, and musculoskeletal systems. However, vestibular inputs — particularly those emanating from the otolith organs, which are responsible for graviception — are believed to underlie many of the autonomic effects of unusual gravitational environments, as discussed in recent reviews.[85,113] After presenting some general issues, I will highlight the specific involvement of the vestibular system in the autonomic consequences of altered gravitational exposure. I will then conclude by discussing the theoretical basis for vestibulo-autonomic interactions in humans and its implications for long-term spaceflight and normal and pathological aging.

1.1 Autonomic Measures Studied in Unusual Gravitational Environments

A wide variety of autonomic measures have been used in studying the effects of altered gravitational environments. The most prominent of these are the symptoms of motion and space sickness — nausea and vomiting (emesis) and, to a lesser extent, skin pallor, salivation, sweating, and a variety of central nervous signs (dizziness, drowsiness, and headache). Activation of both the parasympathetic and sympathetic divisions of the autonomic nervous system occurs during motion sickness, but these systems appear to mediate different aspects of the physiological response to motion: whereas parasympathetic activation generally accounts for most of the gastrointestinal changes leading to the vomiting response, sympathetic activation may be part of a general stress response manifested in such symptoms as pallor and sweating (see Money, Lackner, and Cheung, Chapter 9).

Measures of cardiovascular functioning represent a second set of autonomic indices that have been widely studied under altered gravitational conditions. Transient acceleration loading by means of $+G_z$ (head-to-foot force) and, to a lesser extent, $+G_x$ (front-to-back force), produce an increase in heart rate to partly counteract the direct and indirect forces limiting cardiac output (see Reference 12), whereas prolonged exposure to a weightless environment leads to a "cardiac deconditioning" syndrome, in which reduced plasma and blood volume, lowered venous return, an attenuated baroreflex response, reduced stroke volume, and orthostatic intolerance all develop.[15,27] The role of the different autonomic pathways is more clearcut in the case of cardiovascular functioning, with heightened sympathetic activity and/or parasympathetic withdrawal ordinarily leading to increased cardiac output and heightened parasympathetic and/or reduced sympathetic activation producing the opposite change (see Reference 38).

Endocrine functioning — particularly as it pertains to adrenal functioning — is another widely studied measure of autonomic activity in altered gravitational environments (see References 49,54,55,105). The increase in adrenally regulated catecholamine production and other stress hormones associated with terrestrial motion sickness (TMS) has already been noted, and this contrasts with the striking decline in catecholaminergic levels (and severe hypovolemia) that apparently occurs during weightlessness.[55] Many, if not most, of the endocrine changes during spaceflight represent an adaptation to the altered fluid distribution and cardiovascular demands of that environment,[54,55] whereas the increased fluid retention during TMS may help to maintain and shunt cardiovascular output to support the stress response that develops consequent to the motion stimulus.[49]

Finally, several other indirect measures of autonomic functioning have been studied in connection with altered gravitational environments, and these will be mentioned only briefly throughout the remainder of this chapter. As noted in several reviews,[12,15,91,95] altered gravity produces changes in thermoregulation, glucose metabolism, insulin production, red-blood cell production, and leukocyte activity, all of which are affected by sympathetic levels. The directional change in many of these measures appears to distinguish prolonged hypo- and hyper-gravity exposures, which produce blood pooling in different parts of the body and opposite cardiovascular stress and metabolic demands (increased in prolonged hyper-G and decreased in weightlessness, due to the relationship between G-sensitive body weight and metabolic output; see Reference 95). Another measure of autonomic functioning that will receive little mention in the remainder of this chapter is intrinsic oculomotor balance (e.g., pupil size and accommodative state), which may be differentially affected by the canalicular and otolithic portions of the vestibular system.[85] The only study that investigated these variables in an altered gravitational state (weightlessness) found no differences in accommodative state in 0-G relative to the 1-G terrestrial environment,[104] despite previous anecdotal evidence to the contrary.

1.2 Mechanisms Underlying the Autonomic Effects of Altered Gravity

1.2.1 Nonvestibular Mechanisms

The most prominent nonvestibularly based reason why unusual gravitational environments so dramatically effect autonomic changes pertains to the redistribution of fluids caused by the altered gravitoinertial forces. This redistribution primarily requires cardiovascular and adrenal adaptations that, in the case of weightlessness, reduce cardiac output and lower overall blood volume to minimize the cephalic shift in fluids. An illustration of many of the autonomic changes required to compensate for the increased thoracic fluid pressures is shown in Figure 1. Perhaps the best terrestrial model for simulating prolonged 0-G fluid shifts in humans is head-down bedrest (or analogously, hindlimb suspension in animals), which mimics the declines in catecholamine levels, musculoskeletal tone (e.g., extensor reflexes), and cardiovascular conditioning that occur in the 0-G environment.[15,24,56,91,105] Another terrestrial simulation involves water immersion, which perhaps even more convincingly leads to a noradrenergic/sympathetic deficit.[7,30] What neither of these models can account for are the autonomic correlates of space motion sickness (SMS), as prolonged head-down tilt in particular has repeatedly been shown *not* to lead to nausea and vomiting.[33,102] Nor is it clear whether the bedrest model can completely mimic all other 0-G autonomic changes either, given that (1) catecholamine levels and sympathetic activity are not as markedly altered as in actual spaceflight, and (2) hypovolemia does not occur as rapidly and to the same extent as in weightlessness.[101]

Changes in musculoskeletal activity — principally as regards the mass and tonus of the lower-extremity muscles and bones because of the altered antigravity extension required — may also help account for some of the cardiovascular changes observed in unusual gravitational conditions, given the importance of the "skeletal muscle pump"[97] in maintaining normal circulation. Indeed, increased venous compliance — which leads to increased pooling of blood in the lower extremities upon return to a 1-G environment and contributes to orthostatic intolerance — has been linked to the atrophy of the lower-limb (e.g., antigravity) muscles following prolonged 0-G exposure.[10,15] Altered musculoskeletal activity may also contribute to SMS, particularly in the case of abnormal neck proprioception;[50-52] this factor may explain why head movements relative to the Z-axis of the subject are highly nauseogenic in both 0-G and 2-G environments.[52]

Finally, it has recently been speculated that a visceral graviceptor organ(s) may contribute to motion sickness whenever its output is phase-offset relative to that of the

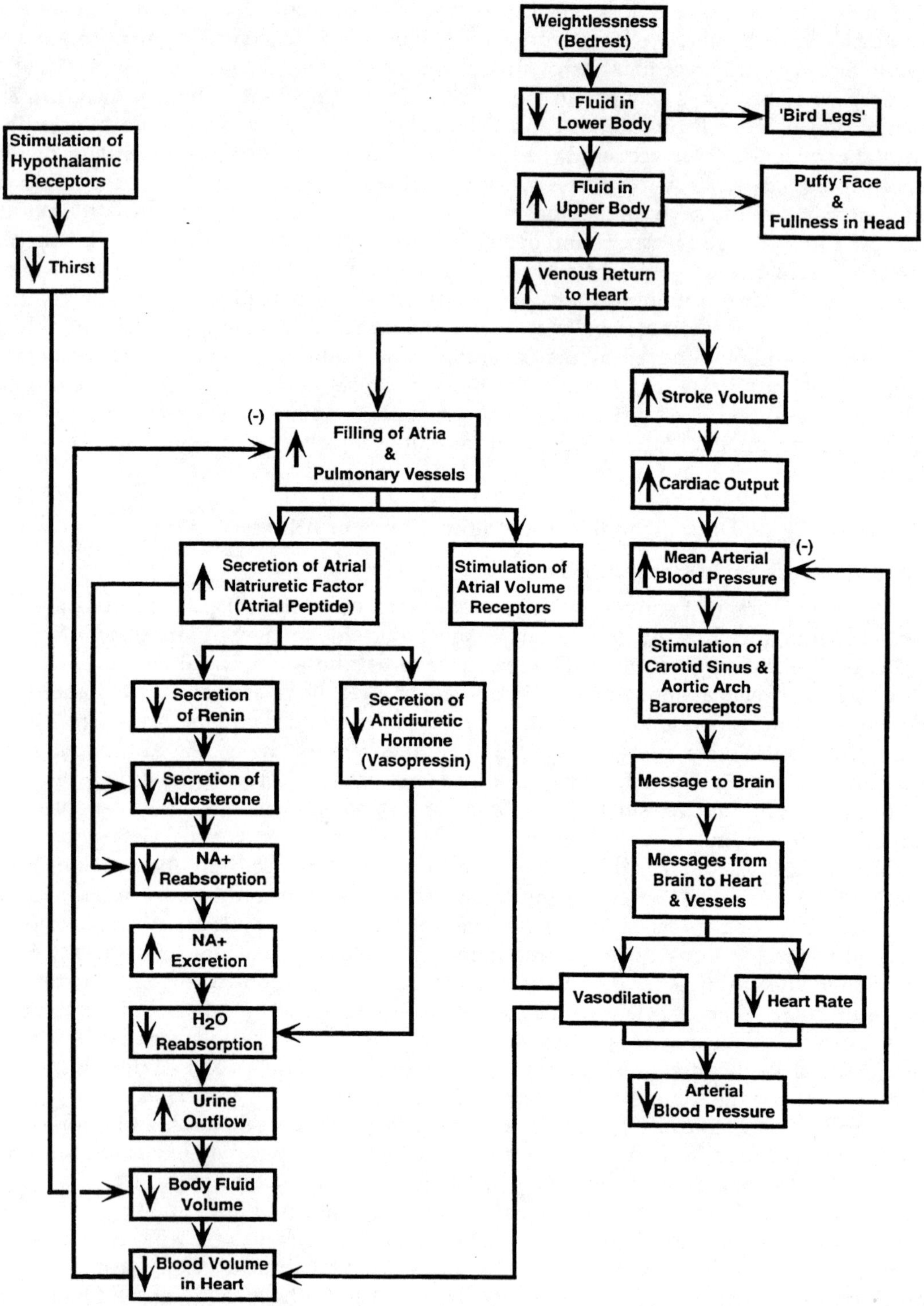

FIGURE 1

A schematic illustration of the cardiovascular and renal adaptations to the effects of weightlessness or simulated weightlessness (top portion of figure). Note the absence of vestibular inputs in this model. (From Frey, M. A. B. et al., in *Circulatory Response to the Upright Posture*, Smith, J. J., Ed., CRC Press, Boca Raton, FL, 1990, 65, Figure 5. With permission.)

vestibular organs.[108] Virtually no direct empirical data concerning this theory have been generated, however, and the mechanisms underlying the visceral graviceptor's role in the genesis of TMS have not been specified. At the very least, the crucial role of the vestibular organs in mediating motion-induced emesis[68] — in contrast to the failure of abdominal deafferentation to eliminate motion sickness[109] — requires that the role of visceral graviception (as well as that of neck proprioception) be supplemental to that of the labyrinth in the induction of motion sickness. Any putative role of altered visceral graviception in the induction of SMS is even more problematic in that gravity itself exerts only a very small direct effect on the stomach and its motility.[4]

1.2.2 Role of Specific Vestibular Systems in Autonomic Control

It has long been recognized that the vestibular system plays an important role in autonomic control,[113] although the exact nature of this control has remained elusive. In a recent theoretical review,[85] I argued that the semicircular canals (and possibly the saccules) exhibit a primarily parasympatho-excitatory effect whereas the utricles exert a sympatho-excitatory effect. My review focused primarily on three areas of autonomic control: motion sickness, cardiovascular tone, and oculomotor behavior.

A long controversy has surrounded the relative contributions of the canals versus otolith organs in the development of motion sickness (see References 85, 87, Miller and Grélot, Chapter 5). Although a great deal of evidence indicates that the semicircular canals are important (if not critical) for the occurrence of motion sickness, the role of the otoliths in motion sickness is less clear. Although lesions of the saccules apparently prevent the nauseogenic effects of vertical motion,[42] no research has shown that *utricular lesions per se* reduce susceptibility to motion sickness. Indeed, the general relationships between altered gravity and motion sickness to be discussed in later sections provide evidence for a generally non-nauseogenic role of utricular stimulation in motion and space sickness.

Another set of findings suggesting opposite roles of the utricles versus other vestibular end-organ systems in autonomic control is derived from studies of cardiovascular function. It has been shown that rotational (and more disputably caloric) stimulation of the horizontal canals produces a vasodepressor response;[58,72,80,99] furthermore, damage to the medial vestibular nucleus (which receives its heaviest inputs from the horizontal canals) abolishes this response,[103] whereas otolithic lesions do not.[58] Conversely, graviceptor stimulation either by sustained head tilt or linear acceleration (which are sensed by the otolith organs) can lead to increased cardiac output and blood pressure, but not if the labyrinth is damaged.[19,93]

The final major area in which the autonomic influences of the canals and otolith organs have been delineated is that of oculomotor control. Rotation about an earth-vertical axis that stimulates the horizontal canals excites the parasympathetic near-reflex triad (pupillary constriction, increased accommodation, and ocular convergence), whereas utricular stimulation has been shown to dilate the eyes and reduce accommodation (see Reference 85). The utricular influence over oculomotor control is believed to be carried out by means of parasympathetic inhibition rather than sympathetic excitation.[61]

In line with these differences between the utricular and other vestibular end-organs in their autonomic influences, recent physiological evidence suggests that direct connections exist between the medial and inferior vestibular nuclei — which receive heavy inputs from the horizontal canals and saccules, respectively — and the solitary tract nucleus and dorsal motor nucleus of the vagus nerve.[1,45,116] These latter regions, which are themselves highly interconnected, are important for the emetic response and parasympathetic control over cardiac output.[1,9] (Some medial and inferior vestibular neurons also have influences on sympathetic outflow; these neurons are believed to receive otolithic inputs; see Reference

117 and Chapter 6). Additional circuits that could mediate canal-autonomic interactions are ones linking the medial vestibular nucleus with the nodulus and uvula of the cerebellum, which could be involved in modulating vasodepressor and emetic responses.[85] By contrast, otolithic interconnections with different brainstem (e.g., the locus coeruleus, caudal raphe nucleus, and rostral ventrolateral medulla) and perhaps cerebellar regions appear to be responsible for most vestibulo-sympatho-excitatory effects.[63,83,85,114,115] Otolithic effects on oculomotor autonomic control may be routed through inhibitory circuits that ultimately impact the parasympathetic Edinger-Westphal nucleus.[61]

A modification of vestibularly based theories of motion sickness such as the one just described is the "sensory conflict" theory detailed by Reason and Brand.[87] According to this account, motion sickness is not the result of excessive stimulation of any one vestibular system or systems but rather of sensory conflict involving the vestibular organ, whether it involve intralabyrinthine conflict, visual-vestibular conflict, or perhaps even vestibular-proprioceptive conflict. The sensory conflict theory of motion sickness is attractive in many respects and can explain a large number of TMS findings, although it has been somewhat difficult to operationalize[112] and fails in some situations.[68] As will be shown in the following section, the predictions of the "sensory conflict" theory do not generally match the motion sickness patterns found in altered gravitational states, although the specific conflicts generated by the altered otolithic inputs in these states may be of greater relevance.[76]

2 AUTONOMIC FINDINGS IN UNUSUAL GRAVITATIONAL ENVIRONMENTS

This section will highlight the altered gravitational environments in which autonomic changes have been most widely studied. I will focus on the changes in autonomic functioning associated with *sustained* gravitational alterations (0-G and hyper-G), which most research has addressed. As noted earlier, there are several other factors beside vestibular ones—principally bodily fluid redistribution and musculoskeletal adaptations—that contribute to autonomic effects under altered gravitational conditions, but the vestibularly mediated changes will be the focus of this review.

2.1 Sustained 0-G (Weightlessness)

A wide-ranging set of bodily changes occur following long-term (i.e., several days or more) exposure to a weightless environment, the most important of which involve the autonomic, musculoskeletal, and vestibular systems and the corresponding set of perceptual and oculomotor adaptations that are linked to disturbances of the latter two systems.

The autonomic deficiencies during spaceflight that occur after the initial few days are consistent with a generalized sympathetic deficit,[91] although some sympathetic responses may be transiently elevated because of a generalized stress response upon entering the weightless environment. Clinical sympathetic deficiency is known to be associated with several cardiovascular changes observed during spaceflight, including (1) reduced catecholamine levels,[25,54] (2) decreased blood volume,[15,27,55,91,101] a partial consequence of the reduced sympathetic influence on renal sodium-retention, and (3) decreased orthostatic tolerance,[15,27,91] reflecting the reduced cardiac output and vasoconstriction associated with sympathetic decline. Decreased red blood cell production and an impaired ability to dissipate heat through sweating are also characteristic of both sympathetic deficiency and spaceflight,[15,27,91] as are the declines in glucose levels, insulin production, and other indices of sympathetically influenced energy metabolism following several weeks of weightlessness.[54] However, substantial changes in heart rate and blood pressure are *not* observed during spaceflight,[27,91] and actual *increases* in these measures occur immediately following the return to earth because of the presumed supersensitivity of sympathetic receptors that develops as overall sympathetic activity declines in 0-G.[91]

Despite the equivocal changes in heart rate and blood pressure during 0-G, it is likely that the catecholaminergic deficiencies contribute greatly to the hypovolemia given that other fluid retention hormones (e.g., antidiuretic hormone [ADH] and aldosterone) are not always decreased in the first few weeks.[49,54,55] One hypothesis[49] as to why these other hormones are not decreased in weightlessness is that they are part of the general stress response to the launch and entry into space and the SMS that later results. Altered sympathetic activity may be a primary means of reducing the excessive cephalic fluid pressures that would otherwise contribute to excessive ventricular filling; in turn, the ensuing hypovolemia could account for much of the orthostatic intolerance upon return to a 1-G environment, which is somewhat reduced by increased fluid intake prior to the return to earth.[71] Musculoskeletal decline, sympathetically mediated baroreceptor reflex impairment, and perhaps even changes in vestibulo-sympathetic reflexes could additionally contribute to the reduced orthostatic tolerance experienced immediately postflight.[15,27,91]

As mentioned earlier, the bedrest and water-immersion models are believed to provide reasonably good simulations of many of the cardiac and musculoskeletal changes that occur during 0-G spaceflight, but they definitely cannot account for the SMS autonomic syndrome, which afflicts as many as 50-75% of all astronauts. Space sickness is a phenomenon that has both similarities as well as differences with motion sickness on earth.[8] A linkage between SMS and TMS is supported by the following evidence: (1) SMS can be somewhat predicted from a multiple regression analysis of several terrestrial measures,[89] although no single measure has good predictive ability; (2) SMS is composed of many of the same symptoms accompanying TMS (depressed appetite, malaise, lethargy, gastrointestinal discomfort, and vomiting), although prodromal nausea and other autonomic signs (e.g., pallor, sweating) may be reduced or absent;[102] (3) adaptation to SMS confers a dramatic immunity to subsequent TMS;[76,102] and (4) many drugs that protect against motion sickness (e.g., scopolamine and promethazine) are also beneficial in alleviating SMS.[34,76] On the other hand, some of the salient differences between TMS and SMS are as follows: (1) the time-course of SMS is usually much longer than that of TMS;[8] (2) passive motions (e.g., Coriolis cross-coupled stimulation) that are normally quite provocative on earth are not disturbing in the 0-G environment,[18] whereas active head movements (especially pitch ones) that ordinarily do not lead to TMS are powerful antecedents to SMS;[52,76] (3) nausea and other autonomic activity that precede emesis during TMS are frequently absent during SMS;[8,102] and (4) the predictive ability of any single TMS test for susceptibility to SMS is weak.[89]

The above differences between SMS and other forms of motion sickness do not necessarily justify dissociating the two phenomena etiologically and clinically. For example, the reduced skin pallor and sweating during SMS relative to TMS have been attributed to the rostral fluid shifts and the drier space-cabin environment, respectively (Reference 76; but see Reference 102), and TMS can also develop over extended periods of time. Moreover, the relatively poor predictive relationship between TMS and SMS can be explained by several factors,[76] not the least of which is the fact that many astronaut/pilots presumably underwent adaptation to more traditional forms of motion sickness during their pilot training. For example, the relationship between traditional motion sickness and that experienced due to the fluctuating weightlessness of parabolic flight is much closer for student pilots and passengers than for experienced pilots.[47,59] It should also be noted that certain types of TMS cannot readily be predicted from other more traditional forms either; for example, experienced pilots who presumably have a high tolerance for cross-coupled and other types of nauseogenic motion stimulation are the ones more likely to experience symptoms following flight simulator training.[66]

Rather, the major physiological difference between SMS and other forms of motion sickness may be that *only a parasympathetic activation occurs in the former whereas both parasympathetic and sympathetic activation occur in the latter.* Such a distinction, which is supported by neuroendocrine data showing lowered noradrenergic activity in those subjects who experience emesis without prodromal nausea during SMS,[48] could explain why traditional sympathetic correlates of motion sickness (pallor, sweating, etc.) are rarely observed in SMS, whereas the emetic response and several other cholinergic (parasympathetic) signs (decreased appetite, lethargy, drowsiness, headache) are not only present but possibly even exaggerated in SMS.[102] (See Chapter 9 for additional discussion of the concept that the stress response of motion sickness involves increased sympathetic activity while the stomach emptying phenomena involve increased parasympathetic activity.) Another important difference between the two motion sickness syndromes lies in the release of the potent nauseogenic hormone ADH (vasopressin), which is dramatically increased during TMS but much less consistently so (if at all) during SMS[49,54,55] due presumably to its role in the overall hypovolemic response to 0-G. As for the ability of active head movements to quickly lead to emesis in SMS but not TMS, this may be explained on the basis of the contribution of the otolith organs to pitch and roll head movements in a 1-G environment[44] as opposed to a 0-G environment.

The putative role of altered graviceptor inputs in the occurrence of both SMS and at least part of the general sympathetic decline during spaceflight is consistent with the prevailing view among researchers that one of the most profound consequences of prolonged exposure to weightlessness is a reduced sensitivity of the otoliths to gravity inputs[79,88,107] in conjunction with a corresponding tendency for otolithic shearing to be interpreted as due to linear acceleration rather than changes in the head relative to gravity.[79,88] This conclusion is based on the following supporting evidence, including (1) the greater dependence on visual inputs to maintain spatial orientation and postural control postflight;[78,79,88,118] (2) the much poorer postflight ability to use vestibular inputs (primarily otolithic) in controlling posture;[78,79,88,118] (3) the presence of inflight and postflight perceptual illusions such as bodily inversion, bodily translation while rolling, and illusory vertical motion of the support surface[50,90,118] as well as elimination of others (the "hilltop" illusion[65]), all of which point to diminished graviceptor input and/or a reinterpretation of otolithic inputs as reflecting translation rather than head tilt relative to gravity; (4) reduced torsional (roll) eye movement gains postflight[14,16,90,107] and a tendency for ocular reflexes to compensate more for linear acceleration than for head tilt;[90] (5) changes in the vertical vestibulo-ocular reflex (VOR),[16,88] vestibulo-ocular velocity storage,[18] and optokinetic reflexes[16] in a manner consistent with reduced graviceptor inputs; and (6) synaptic changes in utricular hair cells.[92] (Disturbances in otolithically mediated velocity-storage mechanisms[44,64] may, in turn, underlie the reduced postrotatory time-constants [known as "dumping"] in 0-G and the reduced susceptibility to Coriolis cross-coupling stimulation.[18]) Finally, the role of the otoliths in SMS is further supported by findings that individuals with the largest amount of behaviorally measured otolithic asymmetry on earth are most susceptible to SMS,[53,60] possibly because of an even greater discrepancy in 0-G between semicircular activity and otolithic activity due to the symmetrical (i.e., absent) graviception experienced in 0-G.

Not all of the above changes can be attributed to adaptation of the otolith organs given that profound changes in the position-sensing capabilities of the muscle proprioceptors (which are also unloaded in 0-G) also affect postural control, perceptual stability, and head-eye coordination.[50-52] Altered proprioception has been implicated in nauseogenic head movements in both 0-G and hyper-G,[51,52] in the enhanced susceptibility to Coriolis cross-coupled stimulation in the hyper-G phase of parabolic flight environment[18] (see Section 2.3), and in the clinical syndrome known as "cervical vertigo".[102] It is not clear as

TABLE 1

Some Opposite Physiological Effects of Prolonged 0-G and Hyper-G Exposure

	0-G	Hyper-G
Musculoskeletal mass (load-bearing)	Decreased	Increased
Plasma volume	Decreased	Increased
Red blood cell production	Decreased	Increased
Cardiovascular reflexes	Decreased	Increased
Energy metabolism	Decreased	Increased
Sympathetic activity	Decreased	Increased

Note: See Reference 16, text, for references.

to how altered proprioception actually affects the occurrence of SMS and other types of motion sickness, but preliminary evidence suggests that motion sickness is not experienced in transient 0-G conditions by *labyrinth-defective individuals;*[47] hence, a disturbance in the integration of proprioceptive inputs with vestibular ones probably underlies the alleged proprioceptive contribution to SMS.

Obviously, many important aspects of the autonomic reaction to prolonged 0-G spaceflight remain poorly understood. But, the relative dearth of motion sickness symptoms upon return to the 1-G terrestrial environment[102] — which is associated with increased otolithic loading but similar (albeit) reversed fluid shifts, cardiovascular re-adaptation, and a comparable amount of sensory conflict — clearly indicates that *unloading of the otolith organs plays a major and perhaps even critical role in the autonomic imbalances associated with SMS.*

2.2 Sustained Hyper-G (Centrifugation)

A large literature has arisen concerning the anatomical and physiological effects of long-term hyper-G, part of which has been reviewed elsewhere.[12,95] In the past few years, several studies have been performed to assess the *aftermath* of chronic hyper-G as a model of SMS in humans,[5,6,43,62,74] which will be the focus of this section.

In general, direct and indirect measures of autonomic functioning suggest that chronic hyper-G effects produced by prolonged exposure to centrifugal forces — usually 2-3 $+G_z$ or G_x for a period of several hours or longer — are, in most instances, the reverse of those produced by prolonged exposure to weightlessness (see Table 1). For example, hyper-G leads to increased skeletal and muscle extensor mass (especially in the lower limbs), increased energy metabolism, increased blood volume, increased red blood cell production, decreased temperature, and activation of various stress hormones.[12,95] Chronic hyper-G exposure also produces effects on cardiovascular reflexes,[20] presumably due to the increased ventricular filling and, to some extent, the heightened sympathetic arousal. Many of the metabolic changes associated with prolonged hyper-G exposure are due to increased sympathetic action, given that they are abolished by lesions of the adrenal medulla,[57,77] which supplies norepinephrine and epinephrine peripherally. Most metabolic and other physiological effects of hyper-G exposure are, as in the case of 0-G exposure, apparently not mediated by the vestibular system.[111] The mirror-like physiological effects of prolonged exposure to 0-G vs. hyper-G environments have led researchers to view hyper- and hypo-gravitational effects as lying on opposite ends of a continuum relative to a 1-G setpoint.[12,95]

Recent reports[5,6,62,74] have noted the apparent similarity between the effects of returning from a prolonged hyper-G environment to the normal 1-G environment and the effects of moving from the 1-G terrestrial environment to prolonged 0-G weightlessness. Those individuals who have experienced both environments have noted the remarkable similarity between the perceptual disturbances and visceral discomfort produced by prolonged 0-G and posthyper-G exposure.[5,6,74] Presumably, this similarity exists because of the well-documented adaptation to prolonged hyper-G,[12] which then proves maladaptive for the subsequent return to 1-G. Even more intriguing is the fact that the percentage of those individuals susceptible to motion sickness in each environment is about the same (50%), with a similar rank ordering of susceptibilities.[6] These SMS-like symptoms can persist for as long as 6 hours or more following the cessation of the chronic acceleration;[5,6,74] by contrast, cardiovascular, respiratory, and metabolic changes—some of which (e.g., lowered blood pressure)[5,100] may reflect a transient sympathetic inhibition — return to normal within 20 min after even high (>5-G), sustained hyper-G exposure.[12] Most studies have accordingly focused on the perceptual and vestibular-autonomic aspects of chronic hyper-G SMS. For example, nausea and even emesis frequently occur upon return from hyper-G, and the predisposing factors (e.g., a rapid head movement in pitch) are similar to those that lead to SMS.[5,13,62,74]

The basis for the post hyper-G effects is presumably an attenuated otolithic output due to adaptation during the hyper-G phase, although not all tests support this conclusion.[5] The putative otolithic "deficit" is reflected in (1) poor overall postural control with the eyes closed, ataxia, and disturbances in locomotion;[6,26,43,62,74] (2) a greater reliance on visual cues for postural control;[6,62,74] (3) greater sensitivity to tilted environments and other spatial orientation illusions such as walking up stairs that remain perceptually "level";[5,6,74] (4) displacement of the subjective vertical from the true gravitational vertical,[6] which is known to follow vestibular cortical damage;[35] and (5) disruption of "righting reflexes" in animals.[26] Although there are reports of humans perceiving themselves to be as much as 90° from their true orientation,[6] no inversion illusions equivalent to those encountered during spaceflight have been noted. However, the downward reversals of animals trying to surface from underwater[26] and the curious repeated tumbling movements of chickens[96] following a return from hyper-G may be manifestations of orientation inversion illusions.

The remarkable similarity between SMS and post hyper-G sickness suggests that it is not otolithic deficits per se that underlie SMS but rather the specific loss of *graviceptor function* (since the otoliths can still provide information concerning linear acceleration of the head in space in 0-G). A further indication of the specificity of graviceptor adaptation in the induction of posthyper-G sickness is that prolonged exposure to G_y results in nauseogenic postexposure head movements in the expected *roll* plane whereas G_x stimulation leads to greater nauseogenic movements in *pitch*.[6] However, changes in utricular synapses following prolonged hyper-G are not identical to those occurring following weightlessness: whereas utricular synapses are reduced following hyper-G,[17] as would be expected in adapting to chronic gravitoinertial over-stimulation, they are increased in 0-G.[92]

In partial support of the proposed linkage between utricular denervation and parasympathetic disinhibition, a recent study showed that chronic hyper-G exposure produces large increases in c-fos activity in the solitary tract nucleus and also in the inferior olive,[46] which may connect with sympatho-inhibitory cerebellar regions (e.g., the nodulus and uvula).[85] However, the c-fos results also point to large increases in neuronal activity levels in the locus coeruleus, which could reflect the sympathetic reaction to hyper-G as opposed to the presumed parasympathetic rebound following the return to the 1-G environment. The exact contribution of hyper-G vs. posthyper-G effects on the neurochemical changes could not be estimated from the c-fos study, however, given that brain extraction occurred between 10 and 60 min following the G-exposure.

In summary, the return to 1-G following exposure to chronic hyper-G centrifugation shows promise as a model for many of the autonomic changes that occur during weightlessness. It may especially complement the bedrest and immersion models, which can better simulate the general autonomic adaptation to the 0-G environment but cannot mimic the graviceptor changes that are specifically hypothesized to underlie SMS. However, many questions remain concerning the effects of hyper-G on direct measures of autonomic functioning, and more careful documentation of the duration and magnitude of the posthyper-G autonomic impairment is required before this model's general applicability can be ascertained.

2.3 Transient 0-G and Hyper-G

Major autonomic changes also occur during transient exposure to unusual gravitational environments, most (but not all) of which are consistent with those observed following prolonged exposure to those same environments. This section will focus only on autonomic effects resulting from reduced and increased gravitational exposure.

The predominant source of knowledge concerning transient hyper- and hypo-gravity effects is parabolic flight studies, which date back to the 1950s.[28,59] The parabolic flight procedure repeatedly exposes pilots to 0-G and ~2-G environments using alternating pushover and pull-up flight maneuvers (see Figure 2). The cardiovascular effects of such acceleratory changes are only modest, although transient increases of as little as 2–3 G_z do produce short-term increases in some cardiac measures (see later discussion). A much more salient aspect of the parabolic flight encounter is the motion sickness symptoms produced in about 50% of all subjects.[47,59] It was discovered early on that both the hyper-G and weightless portions of the parabolic environment both contribute to the motion sickness, especially if subjects are not restrained.[47] It has more recently been demonstrated that pitch and, to a lesser extent, roll movements are nauseogenic in both the hypo- and hyper-gravity phases of parabolic flight, whereas yaw movements are the least likely to produce motion sickness.[52]

Parabolic-flight findings[52] regarding the nauseogenic potential of head movements in different planes of motion suggest that, as in the case of prolonged weightlessness, some interaction between the otoliths and the canals presumably underlies the development of parabolic flight sickness. Altered neck proprioceptive loading could also be a factor, given that simulated G-loading by means of excess head-weighting at 1-G has also been shown to be nauseogenic;[51] however, the increased G-loading must be related in some way to the neck proprioceptive influence over the vestibular system given that labyrinth-defective individuals apparently do not experience motion sickness during parabolic flight.[47] (One possibility is that greater cancellation of otolith outputs occurs due to the increased neck activity during hyper-G head movements, given that such inhibition is normally required during 1-G head movements to prevent vestibulospinal reflexes from inappropriately altering body posture during situations when head movement occurs and the body is stationary, (i. e., during combined neck proprioceptor and vestibular stimulation).[82] Another similarity with weightlessness is that Coriolis (cross-coupled) angular stimulation is much less nauseogenic in the 0-G phase of parabolic flight than in the 2-G phase; this may be related to the reduced velocity storage (see Chapter 2 by Schor and Tomko) that occurs in weightlessness.[18]

A second means of transiently altering the gravitoinertial loading on the subject is by means of centrifugation. In this case, the cardiovascular effects are quite pronounced, even at relatively low G-levels. For $+G_x$ stimulation and $+G_z$ exposures, for example, both heart rate and blood pressure transiently increase by about 25% over normal levels, and this increase is significantly less in both magnitude and duration in labyrinthectomized subjects (Figure 3). The increased heart rate and blood pressure are accompanied by a surge

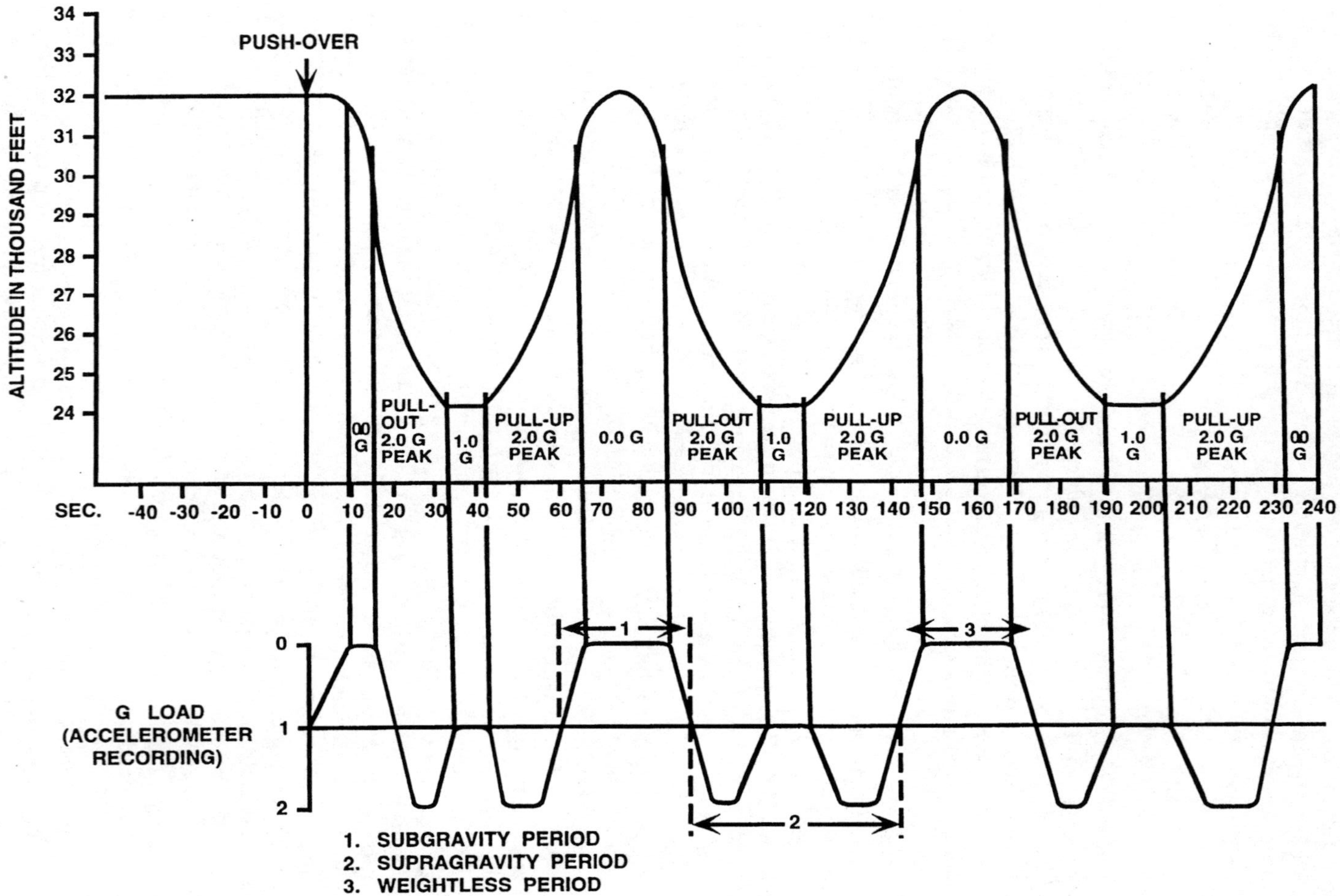

FIGURE 2
An illustration of how parabolic flight can generate fluctuating periods of 0-G and hyper-G (2-G) exposures in a KC-135 aircraft. (From Lackner, J. R. and Graybiel, A., *Aviat. Space Environ. Med.*, 58, A212, 1987, Figure 1. With permission.)

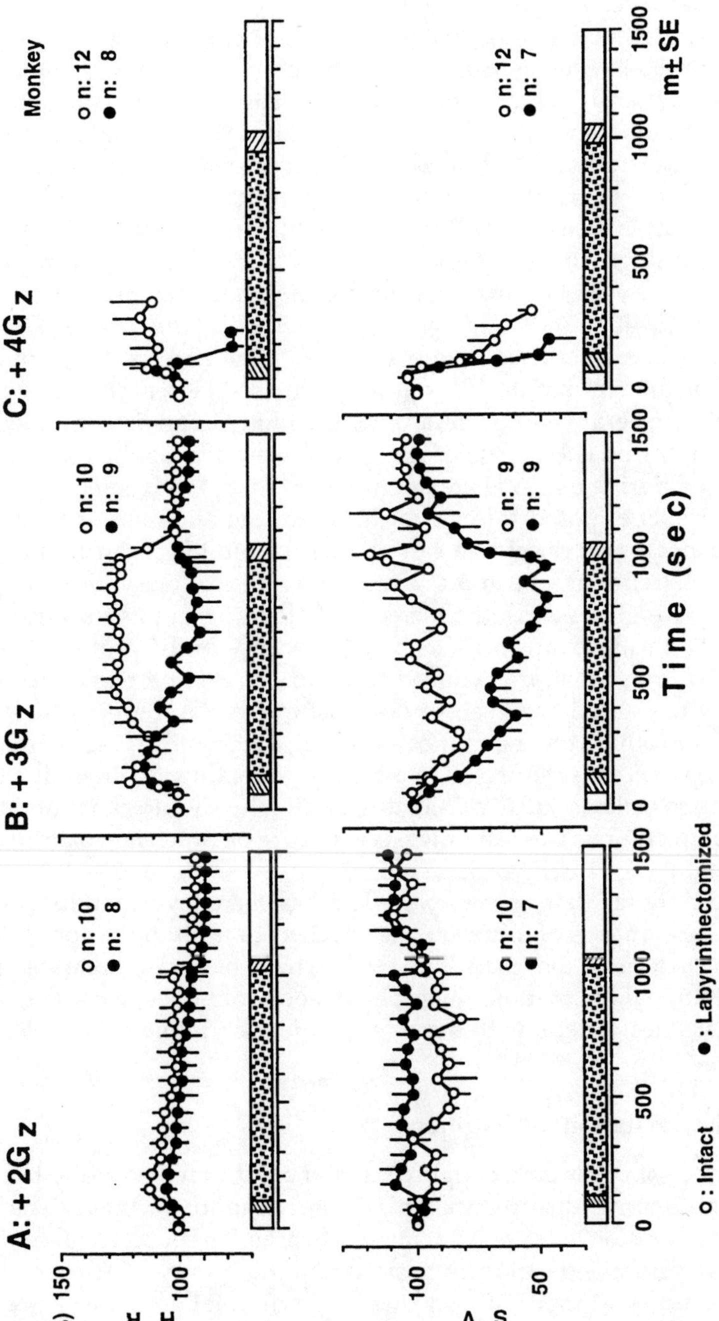

FIGURE 3
The role of vestibularly generated cardiovascular responses to sustained $+G_z$ acceleration in centrifuged monkeys. Note the reduced heart rate (HR) and stroke volume (SV) caused by G-exposure in labyrinthectomized (solid circles) as compared to control (open circles) monkeys during the accelerator stimulus (combined hatched and stippled interval). (From Satake, H. et al., *Acta Otolaryngol.*, Suppl. 481, 543, 1991, Figure 1. With permission.)

in sympathetic activation during G-loading, resulting in large increases in norepinephrine, epinephrine, and other stress hormones.[67] This activation somewhat compensates for the physical pressures on the hydrostatic column at high $+G_z$ levels, but overall cardiac output nonetheless drops and fails to maintain adequate cerebral perfusion at approximately 5 $+G_z$ in subjects without special G-protective measures.[12]

The decline in sympathetic activity following high-G exposure is, as noted earlier, fairly rapid. In comparison with the acceleratory portion of a transient high-G profile, subjects more frequently tend to experience nausea, vertigo, a feeling of lightness, and various tumbling sensations and illusions of body orientation during the deceleratory phase of a transient high-G exposure.[36,37] The unloading of orientation-sensing (graviceptor) inhibitory influences on the semicircular canals is hypothesized to result in the orientation illusions (and presumably nausea as well) that occur during the linear deceleration.[37]

In summary, the findings from transient acceleration paradigms are mostly consistent with the chronic 0-G and hyper-G results and the vestibular-autonomic model described in my recent paper.[85] In all three cases, *graviceptor unloading increases susceptibility* to nauseogenic head movements, especially in planes (i.e., pitch and roll) requiring integration of otolithic and canalicular inputs. By contrast, increased G-loading following spaceflight and during the accelerative stage of a transient high-G run *do not typically lead to motion sickness*, although this is less true of the 2-G phase of parabolic flight. This latter discrepancy may partly derive from the greater number of head movements performed in parabolic flight studies as opposed to the typical centrifuge situation, in which passive motions prevail and neck proprioceptive cancellation of otolith organ output is consequently much reduced. (Although head movements do not produce nausea upon returning to earth following weightlessness, the increased G-loading in this instance cannot be considered identical to a transient hyper-G exposure because of the profound sensory rearrangement and adaptation that are believed to occur during the prior spaceflight.) In contrast to the theory presented here, sensory conflict theory can perhaps better account for the similar nausea produced by head movements during both 0-G and 2-G parabolic flight, but it cannot explain the larger question of why G-unloading is generally much more likely to produce motion sickness than is G-loading: e.g., why should canal-otolith conflict during the *acceleratory* phase of a transient high-G exposure be much less nauseogenic than canal-otolith conflict during centrifuge *deceleration?*

Finally, it should be noted that a recent paradigm has been developed to move rapidly between $-G_z$ and $+G_z$ by means of a linear sled attached to a rotating room.[2] Short-term autonomic adaptations (termed the "push-pull effect") have been demonstrated using this device,[2] but its gravitational fluctuations are probably too rapid to be of use in demonstrating specific motion sickness effects with hypo-and hyper-G stimulation.

2.4 Other Altered Gravitational Conditions

There are a number of other unusual gravitational conditions that have been studied in connection with autonomic disturbances, particularly motion sickness. These include continuous "off-axis" rotation relative to gravity (such as barbecue rotation around an earth-horizontal axis), prolonged exposure to slow-rotating rooms (SRR), and continuous counter-rotation. The unusual aspect of sustained off-axis rotation is that once sufficient time has passed for the canals to cease their output, head rotation relative to gravity stimulates only the otolith organs presumably; however, no actual change in the gravity vector itself actually occurs in this highly nauseogenic situation. Prolonged SRR exposure also involves combined otolithic-canalicular stimulation that produces considerable motion sickness and especially a chronic drowsiness (the "sopite syndrome") that extends long after adaptation has reduced other motion sickness symptoms;[32] but, the SRR's rotation rates produce gravitoinertial vectors that are typically less than 1.1-G in magnitude, which only marginally qualifies the SRR as a truly altered gravitational state.

Continuous counter-rotation may be more justifiably considered an example of altered gravity in that the rotating gravitoinertial vector produced by centrifugal forces creates shearing in the otolith organs and a corresponding perception of continuous turning and tilting relative to gravity in a "cone-like" trajectory while at the same time avoiding stimulation of the semicircular canals (because the subject continuously yaws opposite to the platform's planetary yaw rotation).[31] Such a condition in which otolithic stimulation is presumably unaccompanied by canal activation has obvious implications for my previously discussed vestibular-autonomic theory. The results of the lone study[31] that formally investigated motion sickness in this environment revealed that varying degrees of nausea — but relatively little vomiting — are experienced by subjects during prolonged (up to 30 min) counter-rotation in the absence of vision.

The above counter-rotation result, along with similar ones from vertical (x-axis and z-axis) translation[42] and off-axis rotation[87] paradigms, demonstrate that predominantly otolithic vestibular stimulation can be nauseogenic. However, these findings do not necessarily contradict the notion that the *utricles* have a predominantly sympathetic action in humans, because (1) motion sickness presumably involves a combination of both sympathetic and parasympathetic activation, as noted earlier in this chapter and in Chapter 9; (2) the saccules, which contribute to the motion sickness resulting from z-axis stimulation,[42] are also stimulated by the continually cycling gravitoinertial vector during counter-rotation; and (3) allegedly nauseogenic interactions between the otolith organs and visceral graviceptors[108] may also occur due to the continually cycling G-vector.

3 A THEORY OF VESTIBULAR-AUTONOMIC INTERACTIONS AND ITS IMPLICATIONS

In this section, the general theory of vestibular-autonomic interactions presented elsewhere[85] and discussed briefly throughout this chapter will be further elaborated upon. The general basis of vestibular-autonomic interactions as well as the implications of this theory for living both in prolonged altered gravity as well as with reduced graviception during aging will be discussed.

3.1 The Basis for Vestibular-Autonomic Interactions

A significant vestibular role in autonomic control has repeatedly been demonstrated throughout this chapter and book, as well as in recent reviews.[85,113] It is likely that vestibular inputs affect both sympathetic and parasympathetic function, in both excitatory and inhibitory ways. The general theory that I put forth in a previous paper posited widely divergent roles of different vestibular systems in autonomic control, and the preceding review of the autonomic effects of altered gravity is largely consistent with this proposed dichotomy in that hyper-G exposure (associated with utricular excitation) generally stimulates the sympathetic system whereas hypo-G exposure (as well as the aftermath of prolonged hyper-G stimulation) apparently impairs both utricular graviception and sympathetic arousal.

As noted in previous reviews,[85,113] the vestibular system's involvement in motion sickness and specific cardiovascular compensations (e.g., the increased orthostatic demands of head-up tilt in quadrupeds) presumably underlie much of its widespread interaction with the autonomic nervous system. However, the graviceptor-autonomic link could additionally derive from the more fundamental problem of humans in maintaining orthostasis while upright in a 1-G environment. It is well known that humans are marginally adapted to the cardiovascular demands of the upright posture,[97] with venous return barely sufficient under normal conditions to maintain adequate cerebral perfusion while standing (as opposed to moving, which reduces venous compliance and promotes venous return). Indeed, slight perturbations such as prolonged standing on a hot day (which lowers venous return by means of vasodilation) can easily result in inadequate cerebral perfusion and syncope. A major increase in sympathetic output occurs in response to the orthostatic challenge of moving from the supine to the standing position.[91,97] Although mechanical pressure exerted on the antigravity muscles is known to enhance cardiovascular output,[97] the otolith organs also play an important role both in stimulating the antigravity extensor muscles to maintain stance and in sensing the postural changes in moving to an upright posture; hence, they are ideally suited to help boost cardiac output to compensate for the reduced venous return.

According to the above hypothesis, then, the same neurovestibular/neurochemical systems that are involved in antigravity extension should stimulate sympathetically mediated cardiac output as well. General evidence for such a linkage is provided by the important otolithic projections to monoaminergic (noradrenergic and serotonergic) brainstem regions[83,114,115,117] that help to maintain sympathetic arousal as well as antigravity extension;[23] indeed, markedly reduced neuronal activity in these regions[41] may account for the parasympathetic dominance[3] and collapse of extensor rigidity[81] during sleep. Yates and Miller[117] demonstrated that most vestibulo-sympathetic outputs originate not from the lateral vestibular nucleus — which is important in activating postural reflexes — but from otolithic recipient neurons in the inferior and medial nuclei. However, Yates and Miller[117] studied a quadrupedal species (the cat) in which orthostasis may be challenged specifically by *pitch movements*. In bipeds, different vestibular receptors may be important for signaling movements that may jeopardize venous return to the heart. Moreover, activation of different pathways in vestibulo-extensor and vestibulo-sympathetic reflexes does not necessarily preclude a common underlying otolithically based mechanism.

3.2 Implications for Long-Duration Spaceflight

Although many of the physiological changes during spaceflight begin to level off or even cease after a few days, many others (e.g., red blood cell production, oxygen uptake, and musculoskeletal changes) do not.[10,15] Various types of countermeasures have been employed to overcome the physiological changes caused by prolonged weightlessness, including lower-body negative pressure[27] — which reverses the cephalic fluid shifts — and various exercise regimens that counteract the lower-body musculoskeletal changes, reduce venous compliance, and promote venous return.[15] None of the current countermeasures involves stimulation of the vestibular system via artificial gravity, nor are any of these measures totally effective in preventing the physiological deconditioning that occurs in the weightless environment. Hence, it is unlikely that an optimal battery of spaceflight countermeasures can be designed without including some form of vestibular stimulation.

The use of artificial gravity in overcoming the effects of spaceflight was first proposed in the last century (see Reference 28) and continues to be debated. Various proposals include passive centrifugation[11] and centrifugation powered by lower-body movement (which would combine the beneficial effects of both lower-body exercise and artificial gravity). From the standpoint of the vestibular-autonomic theory proposed here, it is critical that centrifugal and other countermeasures involving vestibular stimulation minimize otolithic adaptation in order to avoid the creation of a sympathetic deficiency. Any centrifugation or other form of otolithic stimulation at low-to-moderate intensities (<2-G) for 30 min or less should avoid severe adaptation of the otoliths.[43,62]

Graviceptor stimulation during prolonged spaceflight may prove even more beneficial for maintaining neurochemical stasis in the central nervous system (CNS). Vestibular mechanisms may be only one of many bodily systems that mediate autonomic changes during spaceflight, but they may be a much more significant component of the *neurochemical regulation of autonomic activity by the CNS*. It has been shown, for example, that 50% of all neurons in the locus coeruleus respond to whole-body tilt in the same manner as do neurons in the lateral vestibular nucleus,[83] although the coerulear neurons may also respond to nongraviceptor inputs. Because of the important otolithic projections to monoaminergic brainstem regions such as the locus coeruleus and raphe nucleus,[83,115] it is likely that deficient graviceptor activity could produce severe psychological consequences, as suggested by the depression and anxiety that are frequently found in vestibular disorders.[21,39,84] Although little data are available concerning the psychological consequences of spaceflight, short-duration (<2-wk) missions do not appear to lead to psychological consequences. However, many psychological decrements have been observed in long-term bedrest studies[110] and the psychological risks of long-term spaceflight are undoubtedly even greater. Moreover, increases in rapid-eye-movement (REM) sleep have been reported during both short-term spaceflight[86] and long-term bedrest,[110] which possibly reflect a decrement in brainstem monoaminergic activity given that excess REM sleep also occurs in depressive disorders and other conditions associated with decreased CNS noradrenergic and serotonergic levels.[29,41] Any spaceflight countermeasure that includes a lower-body exercise regimen that leads to increased monoaminergic levels in the CNS[22] would presumably produce psychological benefits above and beyond its well-documented ability to partially alleviate the cardiovascular and musculoskeletal deconditioning of simulated and actual spaceflight.[15,40,56]

3.3 Implications for Aging

The physiological effects of normal and pathological aging resemble those of prolonged exposure to weightlessness in many respects,[70] including musculoskeletal atrophy and degeneration, calcium and other mineral loss, various postural deficits, and a general sympathetic decline that is manifested in a reduced orthostatic tolerance.[98] It is unclear as to how many of these autonomic and other physiological changes can at least partly be ascribed to the vestibular hair cell losses that reach substantial proportions in the elderly,[69,94] but the postural deficits in particular may be a partial consequence of the vestibular losses.[73,94] Brainstem and cerebellar structures that are intimately connected with the vestibular system and play an important role in autonomic control also degenerate during both normal and pathological aging, particularly the locus coeruleus[106] and cerebellar vermis.[94] It is conceivable that damage to these vestibular projection regions contributes to the autonomic deficits found in the elderly, given the cerebellar and coerulear degeneration in CNS syndromes that are associated with both otoneurological and autonomic disturbances (e.g., Shy-Drager syndrome).[75,119]

Virtually no research has attempted to correlate vestibular deficiencies with the autonomic impairments found in the elderly. It should be stressed that because *both* canal and otolith functioning are compromised during aging[69,94] — unlike in altered gravity environments — it is not easy to predict how vestibularly mediated autonomic deficits would be manifested in the elderly. Nor is it clear to what extent vestibular deficits contribute to neurochemical deficits associated with psychological dysfunction in the elderly, although the frequent psychological problems (notably depression and anxiety) manifested in other populations with otoneurological deficits[21,39,84] make it probable that some psychological consequences may ensue from the age-related decline in vestibular output.

4 CONCLUSIONS

The foregoing theoretical review has shown that the vestibular system contributes in at least two ways to the autonomic syndromes associated with exposure to unusual gravitational conditions. The most significant consequence of altered labyrinthine (otolithic) output involves the autonomic symptoms of space and motion sickness, which cannot be explained by any other simulated model of prolonged 0-G exposure. It is likely that the otolith organs also play a role (albeit a lesser one) in cardiovascular adaptation to altered gravity, although relatively few studies have addressed this issue. The autonomic changes produced by altered gravitational states may be generally viewed as reflecting heightened sympathetic activity during hyper-G exposure and diminished sympathetic activity during prolonged 0-G exposure or in the aftermath of prolonged hyper-G exposure. This relationship is consistent with the "continuity principle" of gravitational physiology[12,95] as well as my recent vestibulo-autonomic model[85] that hypothesizes that utricular inputs (which are more greatly stimulated in hyper-G and diminished in hypo-G conditions) exert an important sympatho-excitatory influence in humans.

Perhaps an even more critical aspect of the vestibular involvement in the autonomic changes associated with altered gravity are the CNS neurochemical imbalances that presumably serve as the substrate for the vestibularly induced autonomic imbalances. Although the consequences of such imbalances for overall neuropsychological functioning are still largely unknown, they will hopefully receive much greater attention in future vestibular research.

ACKNOWLEDGMENTS

I wish to thank Drs. Russell Burton, Victor Convertino, James Lackner, David Ludwig, Paul Werchan, and Bill Yates for their various contributions to this manuscript. The views expressed in this chapter are those of the author and do not necessarily reflect the views of the United States Air Force or the Department of Defense.

REFERENCES

1. Balaban, C. D. and Beryozkin, G., Vestibular nucleus projections to nucleus tractus solitarius and the dorsal motor nucleus of the vagus nerve: potential substrates for vestibulo-autonomic interactions, *Exp. Brain Res.*, 98, 200, 1994.
2. Banks, R. D., Grissett, J. D., Turnipseed, G. T., Saunders, P. L., and Rupert, A. H., The "push-pull effect," *Aviat. Space Environ. Med.*, 65, 699, 1994.
3. Baust, W. and Bohnert, B., The regulation of heart-rate during sleep, *Exp. Brain Res.*, 7, 169, 1969.
4. Berry, P., The stomach — both a target and potential contributor in space sickness, in *Nausea and Vomiting: Recent Research and Clinical Advances*, Kucharczyk, J., Stewart, D. J., and Miller, A. D., Eds., CRC Press, Boca Raton, FL, 1991, 43.
5. Bles, W., Bos, J. E., Furrer, R., et al., *Space Adaptation Syndrome Induced by a Long Duration +3G_x Centrifuge Run* (IZF 1989-25), TNO Institute for Perception, Soesterberg, The Netherlands, 1989.
6. Bles, W. and de Graaf, B., Postural consequences of long duration centrifugation, *J. Vest. Res.*, 3, 87, 1993.
7. Bonde-Petersen, F., Suzuki, Y., and Sadamoto, T., Cardiovascular effects of simulated zero-gravity in humans, *Acta Astronautica*, 10, 657, 1983.
8. Borison, H. L., Is space sickness a form of motion sickness? *Physiologist*, 28, S-25, 1985.
9. Brizzee, K. R., The central nervous connections involved in motion induced emesis, in *Motion and Space Sickness*, Crampton, G. H., Ed., CRC Press, Boca Raton, FL, 1990, 9.

10. Buchanan. P. and Convertino, V. A., A study of the effects of prolonged simulated microgravity on the musculature of the lower extremities in man: an introduction, *Aviat. Space Environ. Med.*, 60, 649, 1989.
11. Burton, R. R. and Meeker, L. J., Physiologic validation of a short-arm centrifuge for space application, *Aviat. Space Environ. Med.*, 63, 476, 1992.
12. Burton, R. R. and Smith, A. H., Adaptation to acceleration environments, in *Handbook of Physiology: Adaptation to the Environment*, Fregly, M. J. and Blatteis, C. M., Eds., Oxford University Press, Oxford, 1995, chap. 40.
13. Clark, C. C., Observations of a human experiencing 2 G for 24 hours, *Aerospace Med.*, 31, 300, 1960.
14. Clarke, A. H., Teiwes, W., and Scherer, H., Evaluation of the torsional VOR in weightlessness, *J. Vest. Res.*, 3, 207, 1993.
15. Convertino, V. A., Exercise and adaptation to microgravity environments, in *Handbook of Physiology: Adaptation to the Environment*, Fregly, M. J. and Blatteis, C. M., Eds., Oxford University Press, Oxford, 1995, chap. 39.
16. Dai, M., McGarvie, L., Kozlovskaya, I., Raphan, T., and Cohen, B., Effects of spaceflight on ocular counterrolling and the spatial orientation of the vestibular system, *Exp. Brain Res.*, 102, 45, 1994.
17. Daunton, N. G., Ross, M. D., Fox, R. A., Corcoran, M. L., Cutler, L. D., and Wu, L. C., Effects of chronic hyper-gravity on the righting reflex and vestibular endorgan in the rat, *Soc. Neurosci. Abstr.*, 17, 316, 1991.
18. DiZio, P., Lackner, J. R., and Evanoff, J. N., The influence of gravitoinertial force level on oculomotor and perceptual responses to Coriolis, cross-coupling stimulation, *Aviat. Space Environ. Med.*, 58, A218, 1987.
19. Doba, N. and Reis, D. J., Role of the cerebellum and vestibular apparatus in regulation of orthostatic reflexes in the cat, *Circ. Res.*, 34, 9, 1974.
20. Duling, B. R., Effects of chronic centrifugation at 3 *g*'s on cardiovascular reflexes of the rat, *Am. J. Physiol.*, 213, 466, 1967.
21. Eagger, S., Luxon, L. M., Davies, R. A., Coelho, A., and Ron, M. A., Psychiatric morbidity in patients with peripheral vestibular disorder: a clinical and neuro-otological study, *J. Neurol. Neurosurg. Psychiatr.*, 55, 383, 1992.
22. Elam, M., Svensson, T. H., and Thoren, P., Brain monoamine metabolism is altered in rats following spontaneous, long-distance running, *Acta Physiol. Scand.*, 130, 313, 1987.
23. Engberg, I., Lundberg, A., and Ryall, R. W., Is the tonic decerebrate inhibition of reflex paths mediated by monoaminergic pathways? *Acta Physiol. Scand.*, 72, 123, 1968.
24. Fareh, J., Cottet-Emard, J.-M., Fagette, S., Pequignot, J.-M., and Gharib, C., Central and peripheral norepinephrine turnover after hindlimb suspension in the rat, *Aviat. Space Environ. Med.*, 64, 386, 1993.
25. Fareh, J., Cottet-Emard, J.-M., Pequignot, J.-M., et al., Norepinephrine content in discrete brain areas and neurohypophysial vasopressin in rats after a 9-d spaceflight, *Aviat. Space Environ. Med.*, 64, 507, 1993.
26. Fox, R. A., Daunton, N., Corcoran, M., and Wu, L., Time-course of behavioral adaptation to hypergravity, in *Posture and Gait: Control Mechanisms*, Vol. 1, Woollacott, M. and Horak, F., Eds., University of Oregon Books, Eugene, 1992, 178.
27. Frey, M. A. B., Charles, J. B., and Houston, D. E., Weightlessness and response to orthostatic stress, in *Circulatory Response to the Upright Posture*, Smith, J. J., Ed., CRC Press, Boca Raton, FL, 1990, 65.
28. Gerathewohl, S. J., Personal experiences during short periods of weightlessness reported by sixteen subjects, *Acta Astronautica*, 2, 203, 1956.
29. Gillin, J. C., Sleep studies in affective illness: Diagnostic, therapeutic, and pathophysiological implications, *Psychiatr. Ann.*, 13, 367, 1983.
30. Goodall, M., McCally, M., and Graveline, D. E., Urinary adrenaline and noradrenaline response to simulated weightless state, *Am. J. Physiol.*, 206, 431, 1964.
31. Graybiel, A. and Johnson, W. H., A comparison of the symptomatology experienced by healthy persons and subjects with loss of labyrinthine function when exposed to unusual patterns of centripetal force in a counter-rotating room, *Ann. Otol. Rhin. Laryngol.*, 72, 357, 1963.
32. Graybiel, A. and Knepton, J., Sopite syndrome: a sometimes sole manifestation of motion sickness, *Aviat. Space Environ. Med.*, 47, 873, 1976.
33. Graybiel, A. and Lackner, J. R., Rotation at 30 RPM about the Z axis after 6 hours in the 10° head-down position: effect on susceptibility to motion sickness, *Aviat. Space Environ. Med.*, 50, 390, 1979.
34. Graybiel, A. and Lackner, J. R., Treatment of severe motion sickness with antimotion sickness injections, *Aviat. Space Environ. Med.*, 58, 773, 1987.
35. Gresty, M. A., Bronstein, A. M., Brandt, T., and Dieterich, M., Neurology of otolith function: peripheral and central disorders, *Brain*, 115, 647, 1992.

36. Grissett, J. D., *Mathematical Model for Interaction of Canals and Otoliths in Perception of Orientation, Translation, and Rotation* (NAMRL Spec. Rep. 93-5), Naval Aerospace Medical Research Laboratory, Pensacola, FL, 1993.
37. Guedry, F. E., Rupert, A. H., McGrath, B. J., and Oman, C. M., The dynamics of spatial orientation during complex and changing linear and angular acceleration, *J. Vest. Res.*, 2, 259, 1992.
38. Guyton, A. C., *Textbook of Medical Physiology*, 6th ed., W.B. Saunders, Philadelphia, 1981, chap. 20.
39. Hallam, R. S. and Stephens, S. D. G., Vestibular disorder and emotional distress, *J. Psychosom. Res.*, 29, 407, 1985.
40. Hargens, A. R., Recent bed rest results and countermeasure development at NASA, *Acta Physiol. Scand.*, 1994, 150, 103.
41. Hobson, J. A., Lydic, R., and Baghdoyan, H. A., Evolving concepts of sleep cycle generation: from brain centers to neuronal populations, *Behav. Brain Sci.*, 9, 371, 1986.
42. Igarashi, M., Himi, T., Kulecz, W. B., and Kobayashi, K., Role of otolith endorgans in the genesis of vestibular-visual conflict sickness (pitch) in the squirrel monkey (first report), *Aviat. Space Environ. Med.*, 58, A207, 1987.
43. Igarashi, M., Nakazato, T., Maeda, T., et al., Effect of artificial G-load on postural stability: a human centrifuge study, *Jpn. J. Aerospace Environ. Med.*, 31,73, 1994.
44. Igarashi, M., Takahashi, M., Kubo, T., Levy, J. K., and Homick, J. L., Effect of macular ablation on vertical optokinetic nystagmus in the squirrel monkey, *ORL*, 40, 312, 1978.
45. Ito, J. and Honjo, I., Central fiber connections of the vestibulo-autonomic reflex arc in cats, *Acta Otolaryngol.*, 110, 379, 1990.
46. Kaufman, G. D., Anderson, J. H., and Beitz, A. J., Fos-defined activity in rat brainstem following centripetal acceleration, *J. Neuroscience*, 12, 4489, 1992.
47. Kellogg, R. S., Kennedy, R. S., and Graybiel, A., Motion sickness symptomatology of labyrinthine defective and normal subjects during zero gravity maneuvers, *Aerospace Med.*, 36, 315, 1965.
48. Kohl, R. L., Hormonal responses of metoclopramide-treated subjects experiencing nausea or emesis during parabolic flight, *Aviat. Space Environ. Med.*, 58, A266, 1987.
49. Kohl, R. L., Endocrinology of space/motion sickness, in *Motion and Space Sickness*, Crampton, G. H., Ed., CRC Press, Boca Raton, FL, 1990, 65.
50. Lackner, J. R., Orientation and movement in unusual force environments, *Psychol. Sci.*, 4, 134, 1993.
51. Lackner, J. R. and Dizio, P., Altered sensory-motor control of the head as an etiological factor in space-motion sickness, *Percept. Mot. Skills*, 68, 784, 1989.
52. Lackner, J. R. and Graybiel, A., Head movements in low and high gravitoinertial environments elicit motion sickness: implications for space motion sickness, *Aviat. Space Environ. Med.*, 58, A212, 1987.
53. Lackner, J. R., Graybiel, A., Johnson, W. H., and Money, K. E., Asymmetric otolith function and increased susceptibility to motion sickness during exposure to variations in gravitoinertial acceleration level, *Aviat. Space Environ. Med.*, 58, 652, 1987.
54. Leach, C. S., Fluid control mechanisms in weightlessness, *Aviat. Space Environ. Med.*, 58, A74, 1987.
55. Leach, C. S., Altchuler, S. I., and Cintron-Trevino, N. M., The endocrine and metabolic responses to space flight, *Med. Sci. Sports Exerc.*, 15, 432, 1983.
56. Leach, C. S., Hulley, S. B., Rambaut, P. C., and Dietlein, L. F., The effect of bedrest on adrenal function, *Space Life Sci.*, 4, 415, 1973.
57. Leon, H. A. and Chackerian, M. J., Nutritional and hormonal factors concerned with the regulation of liver protein synthesis during acute centrifugation stress, *Endocrinology*, 82, 429, 1968.
58. Lindsay, J. R., Oppenheimer, M. J., Wycis, H. T., and Spiegel, E. A., Receptor apparatus of the vestibulovasomotor reaction, *Arch. Otolaryngol.*, 42, 257, 1945.
59. Loftus, J. P., Motion sickness in the C-131B, in *Symposium on Motion Sickness with Special Reference to Weightlessness* (AMRL-TDR-63-25), Aerospace Medical Research Laboratories, Wright-Patterson Air Force Base, OH, 1963, 3.
60. Markham, C. H. and Diamond, S. G., A predictive test for space motion sickness, *J. Vest. Res.*, 3, 289, 1993.
61. Markham, C. H., Estes, M. S., and Blanks, R. H. I., Vestibular influences on ocular accommodation in cats, *Equilibrium Res.*, 3, 102, 1973.
62. Martin, E. J. and Albery, W. B., *Emulation of Space Motion Sickness (SMS) on the Dynamic Environment Simulator Centrifuge* (AL/CF-TR-1993- 0007), Air Force Materiel Command, Wright-Patterson Air Force Base, OH, 1993.
63. Martner, J., Cerebellar influences on autonomic mechanisms. An experimental study in the cat with special reference to the fastigial nucleus, *Acta Physiol. Scand.*, Suppl. 425, 1, 1975.
64. Matsuo, V. and Cohen, B., Vertical optokinetic nystagmus and vestibular nystagmus in the monkey: up-down asymmetry and effects of gravity, *Exp. Brain Res.*, 53, 197, 1984.
65. Merfeld, D. M., Christie, J. R. I., and Young, L. R., Perceptual and eye movement responses elicited by linear acceleration following spaceflight, *Aviat. Space Environ. Med.*, 65, 1015, 1994.

66. Miller, J. W. and Goodson, J. E., Motion sickness in a helicopter simulator, *Aerospace Med.*, 31, 204, 1960.
67. Mills, F. J., The endocrinology of stress, *Aviat. Space Environ. Med.*, 56, 642, 1985.
68. Money, K. E., Motion sickness, *Physiol. Rev.*, 50, 1, 1970.
69. Nadol, J. B., Jr. and Schuknecht, H. F., Pathology of peripheral vestibular disorders in the elderly, *Am. J. Otolaryngol.*, 11, 213, 1990.
70. Nicogossian, A., The human space enterprise in the 21st century, *Aviat. Space. Environ. Med.*, 65, 1149, 1994.
71. Nicogossian, A. E., Charles, J. B., Bungo, M. W., and Leach-Huntoon, C. S., Cardiovascular function in space flight, *Acta Astronautica*, 24, 323, 1991.
72. Norre, M. E. and Degroote, M., Influence of caloric and rotation testing upon bloodpressure and pulse rate, in *Vertigo, Nausea, Tinnitus and Hypoacusia in Metabolic Disorders*, Claussen, C.-F., Kirtane, M. V., and Schlitter, K., Eds., Excerpta Medica, Amsterdam, 1988, 139.
73. Norre, M. E., Forrez, G., and Beckers, A., Vestibular dysfunction causing instability in aged patients, *Acta Otolaryngol.*, 104, 50, 1987.
74. Ockels, W. J., Furrer, R., and Messerschmid, E., Simulation of space adaptation syndrome on earth, *Exp. Brain Res.*, 79, 661, 1990.
75. Ohashi, N., Nakagawa, H., Kanda, K., and Mizukoshi, K., Otoneurological manifestations of the Shy-Drager syndrome, *Eur. Arch. Otorhinolaryngol.*, 248, 150, 1991.
76. Oman, C. M., Lichtenberg, B. K., Money, K. E., and McCoy, R. K., M.I.T./Canadian vestibular experiments on the Spacelab-1 mission. 4. Space motion sickness: Symptoms, stimuli, and predictability, *Exp. Brain Res.*, 64, 316, 1986.
77. Oyama, J. and Platt, W. T., Metabolic alterations in rats exposed to acute acceleration stress, *Endocrinology*, 76, 203, 1965.
78. Paloski, W. H., Black, F. O., Reschke, M. F., Calkins, D. S., and Shupert, C., Vestibular ataxia following shuttle flights: effects of microgravity on otolith-mediated sensorimotor control of posture, *Am. J. Otol.*, 14, 9, 1993.
79. Parker, D. E., Human vestibular function and weightlessness, *J. Clin. Pharmacol.*, 31, 904, 1991.
80. Patil, N. P., Schneider, D., Claussen, C.-F., and Popivanova, C., Cardiac reactions in neurootological patients during vestibular stimulation, in *Vertigo, Nausea, Tinnitus and Hypoacusia in Metabolic Disorders*, Claussen, C.-F., Kirtane, M. V., and Schlitter, K., Eds., Excerpta Medica, Amsterdam, 1988, 149.
81. Pompeiano, O., The neurophysiological mechanisms of the postural and motor events during desynchronized sleep, *Proc. Assoc. Res. Nerv. Ment. Dis.*, 45, 351, 1967.
82. Pompeiano, O., Experimental central nervous system lesions and posture, in *Vestibular and Visual Control on Posture and Locomotor Equilibrium*, Igarashi, M. and Black, F. O., Eds., Karger, Basel, 1985, 218.
83. Pompeiano, O., Manzoni, D., Barnes, C. D., Stampacchia, G., and D'Ascanio, P., Responses of locus coeruleus and subcoeruleus neurons to sinusoidal stimulation of labyrinth receptors, *Neuroscience*, 35, 227, 1990.
84. Pratt, R. T. C. and McKenzie, W., Anxiety states following vestibular disorders, *Lancet*, 2, 347, 1958.
85. Previc, F. H., Do the organs of the labyrinth differentially influence the sympathetic and parasympathetic systems? *Neurosci. Biobehav. Rev.*, 17, 397, 1993.
86. Quadens, O. and Green, H., Eye movements during sleep in weightlessness, *Science*, 225, 221, 1984.
87. Reason, J. T. and Brand, J. J., *Motion Sickness*, Academic Press, London, 1975, chaps. 4 and 5.
88. Reschke, M. F., Bloomberg, J. J., Harm, D. L., and Paloski, W. H., Space flight and neurovestibular adaptation, *J. Clin. Pharmacol.*, 34, 609, 1994.
89. Reschke, M. F., Homick, J. L., Ryan, P., and Moseley, E. C., Prediction of the space adaptation syndrome, in *Motion Sickness: Mechanisms, Prediction, Prevention and Treatment* (AGARD CP-372), AGARD, Neuilly-Sur-Seine, France, 1984, 26-1.
90. Reschke, M. F. and Parker, D. E., Effects of prolonged weightlessness on self-motion perception and eye movements evoked by roll and pitch, *Aviat. Space Environ. Med.*, 58, A153, 1987.
91. Robertson, D., Convertino, V. A., and Vernikos, J., The sympathetic nervous system and the physiological consequences of spaceflight: a hypothesis, *Am. J. Med. Sci.*, 308, 126, 1994.
92. Ross, M. D., Morphological changes in rat vestibular system following weightlessness, *J. Vest. Res.*, 3, 241, 1993.
93. Satake, H., Matsunami, K., and Miyata, H., The vestibulo-autonomic function viewed from cardiac responses in centrifuged monkeys, *Acta Otolaryngol.*, Suppl. 481, 543, 1991.
94. Sloane, P. D., Baloh, R. W., and Honrubia, V., The vestibular system in the elderly: clinical implications, *Am. J. Otolaryngol.*, 10, 422, 1989.
95. Smith, A. H., Principles of gravitational biology, in *Foundations of Space Biology and Medicine*, Calvin, M. and Gazenko, O. G., Eds., National Aeronautics and Space Administration, Washington, D.C., 1975, 129.

96. Smith, A. H. and Burton, R. R., Chronic acceleration of animals, in *Gravity and the Organism*, Gordon, S. A. and Cohen, M. J., Eds., University of Chicago Press, Chicago, 1971, 371.
97. Smith, J. J. and Ebert, T. J., General response to orthostatic stress, in *Circulatory Response to the Upright Posture*, Smith, J. J., Ed., CRC Press, Boca Raton, FL, 1990, 1.
98. Smith, J. J., and Porth, C. J. M., Age and the response to orthostatic stress, in *Circulatory Response to the Upright Posture*, Smith, J. J., Ed., CRC Press, Boca Raton, FL, 1990, 121.
99. Spiegel, E. A., Effect of labyrinthine reflexes on the vegetative nervous system, *Arch. Otolaryngol.*, 44, 61, 1946.
100. Sudoh, M., Shioda, K., Kohno, M., Ikawa, S., Kawakami, K., and Saiki, H., Cardiopulmonary response of rats to centrifugal accelerations, *Physiologist*, 30, S-135, 1987.
101. Thornton, W. E., Moore, T. P., and Pool, S. L., Fluid shifts in weightlessness, *Aviat. Space. Environ. Med.*, 58, A86, 1987.
102. Thornton, W. E., Moore, T. P., Pool, S. P., and Vanderploeg, J., Clinical characterization and etiology of space motion sickness, *Aviat. Space Environ. Med.*, 58, A1, 1987.
103. Uchino, Y., Kudo, N., Tsuda, K., and Iwamura, Y., Vestibular inhibition of sympathetic nerve activities, *Brain Res.*, 22, 195, 1970.
104. Vanderploeg, J. M., Near vision acuity measurements of space shuttle crewmembers, *Aviat. Space Environ. Med.*, 56, 492, 1985.
105. Vernikos, J., Metabolic and endocrine changes, in *Inactivity: Physiological Effects*, Sandler, H. and Vernikos, J., Eds., Academic Press, Orlando, FL, 1986, 99.
106. Vijayashankar, N. and Brody, H., A quantitative study of the pigmented neurons in the nuclei locus coeruleus and subcoeruleus in man related to aging, *J. Neuropath. Exp. Neurol.*, 38, 490, 1979.
107. von Baumgarten, R. J., European vestibular experiments on the Spacelab-1 mission. 1. Overview, *Exp. Brain Res.*, 64, 239, 1986.
108. von Gierke, H. E. and Parker, D. E., Differences in otolith and abdominal viscera graviceptor dynamics: implications for motion sickness and perceived body position, *Aviat. Space Environ. Med.*, 65, 747, 1994.
109. Wang, S. C., Chinn, H. I., and Renzi, A. A., Experimental motion sickness in dogs: role of abdominal visceral afferents, *Am. J. Physiol.*, 190, 578, 1957.
110. Winget, C. M. and DeRoshia, C. W., Psychosocial and chronophysiological effects of inactivity and immobilization, in *Inactivity: Physiological Effects*, Sandler, H. and Vernikos, J., Eds., Academic Press, Orlando, 1986, 123.
111. Wunder, C. C., Milojevic, B., and Eberly, L., Growth and food consumption of labyrinthectomized hamsters during chronic centrifugation at 5*g* and 6*g*, *Nature*, 210, 177, 1966.
112. Yardley, L., Motion sickness and perception: a reappraisal of the sensory conflict approach, *Br. J. Psychol.*, 83, 449, 1992.
113. Yates, B. J., Vestibular influences on the sympathetic nervous system, *Brain Res. Rev.*, 17, 51, 1992.
114. Yates, B. J., Goto, T., and Bolton, P. S., Responses of neurons in the rostral ventrolateral medulla of the rat to natural vestibular stimulation, *Brain Res.*, 601, 255, 1993.
115. Yates, B. J., Goto, T., Kerman. I., and Bolton, P. S., Responses of caudal medullary raphe neurons to natural vestibular stimulation, *J. Neurophysiol.*, 70, 938, 1993.
116. Yates, B. J., Grélot, L., Kerman, I. A., Balaban, C. D., Jakuç, J., and Miller, A. D., Organization of vestibular inputs to nucleus tractus solitarius and adjacent structures in cat brain stem, *Am. J. Physiol.*, 267, R974, 1994.
117. Yates, B. J. and Miller, A. D., Properties of sympathetic reflexes elicited by natural vestibular stimulation: Implications for cardiovascular control, *J. Neurophysiol.*, 71, 2087, 1994.
118. Young, L. R., Oman, C. M., Merfeld, D., et al., Spatial orientation and posture during and following weightlessness: human experiments on Spacelab Life Sciences 1, *J. Vest. Res.*, 3, 231, 1993.
119. Ziegler, M. G., Lake, C. R., and Kopin, I. J., The sympathetic-nervous-system defect in primary orthostatic hypotension, *New Engl. J. Med.*, 296, 293, 1977.

11

Panic, Phobia, and Vestibular Dysfunction

Rolf G. Jacob, Joseph M. Furman, and James M. Perel

CONTENTS

0-8493-7668-1/96/$0.00+$.50

LIST OF ABBREVIATIONS

ACTH	Adrenocorticotrophic hormone
Ag	Group of patients with panic disorder with moderate to severe agoraphobia[69]
Ag-I	Scale in part I of the SitQ reflecting non-SMD agoraphobic avoidance
β-CCE	β-Carboline ethyl ester
BZD	Benzodiazepine
Ca^{2+}	Calcium ion
cAMP	Cyclic adenosine monophosphate
CCK	Cholecystokinin
CO_2	Carbon dioxide
ENT	Ear, nose, and throat patient sample,[75] consisting of patients with vestibular dysfunction or patients evaluated for hearing loss
GABA	γ- aminobutyric acid
IV	Intravenous
LVN	Lateral vestibular nucleus
MCPP	*m*-Chlorphenylpiperazine
N	Normal control group[69]
NPY	Neuropeptide Y
P	Group of patients with uncomplicated panic disorder[69]
sd	Standard deviation
SitQ	Situational Characteristics Questionnaire[70]
SMD	Space and motion discomfort
SMD-I	Scale in part I of the SitQ measuring space and motion discomfort
SMD-II	Scale in part II of the SitQ measuring space and motion discomfort
SymQ	Symptom questionnaire[70]
SSRI	Selective serotonin reuptake inhibiting antidepressant drug
t	Student's *t*-test
5-HT	5-Hydroxytryptamine, serotonin

1 INTRODUCTION

Panic disorder is a psychiatric disorder characterized by recurrent panic attacks (a sudden onset of symptoms of fear). Characteristics of panic attacks include a high salience of autonomic symptoms, including heart palpitations, sweating, trembling, hot or cold flashes, gastrointestinal distress, shortness of breath, feeling like choking, chest discomfort, numbness or tingling sensations (paresthesias), and dizziness or faintness. Some of these autonomic symptoms, including shortness of breath, chest discomfort, paresthesias, and feeling dizzy or faint, are generally attributed to hyperventilation, a behavior in which these patients are known to engage excessively. Accompanying the autonomic symptoms are cognitive symptoms of "feelings of unreality" (derealization, depersonalization) and fear of "going crazy" or even of death. Typically there is a strong urge to flee or escape from the situation in which the panic attack takes place. After one or more attacks, patients begin to worry about future episodes, and this worry is a major component of panic disorder.

Formally, a *panic attack* is defined as an episode of fear or discomfort, in which four or more of the ten autonomic symptoms and three cognitive symptoms listed above develop abruptly and reach a peak within about 10 minutes.[5] *Panic disorder* is characterized by recurrent experiences of panic attacks, with at least one attack accompanied by one of the following: (1) worry about future attacks, (2) worry about the implications of an attack (e.g., having a heart attack or brain tumor), and (3) major changes in behavior related to the attacks.[5]

An important complication of panic disorder is the development of agoraphobic avoidance behaviors, which develop in order to reduce the chance of a panic attack. If such avoidance behaviors become chronic, a variant of panic disorder called *panic disorder with agoraphobia* has developed. Not all patients develop agoraphobia, and those without agoraphobic avoidance are diagnosed as having panic disorder without agoraphobia or "uncomplicated panic disorder."

Patients with panic disorder and agoraphobia typically avoid situations such as traveling or driving, supermarkets, shopping malls, movie theaters, bridges, tunnels and/or stadiums. Formally, *agoraphobia* is defined as the fear of places from which escape would be difficult or embarrassing or in which help might not be readily available. An individual with agoraphobia either avoids these situations or endures them with extreme distress.[5]

Panic disorder has a lifetime prevalence of 2%.[41] Panic disorder, with or without agoraphobia, is one of a number of anxiety disorders in the psychiatric diagnostic system.[5] The other anxiety disorders include certain other phobic conditions, generalized anxiety disorder, post-traumatic stress disorder, obsessive compulsive disorder, and anxiety related to medical conditions (e.g., hyperthyroidism). *Generalized anxiety disorder* is characterized by chronic anxiety and worrying about minor matters. The other phobic conditions include *specific phobia, social phobia,* and *agoraphobia without panic.* A major difference between panic disorder with agoraphobia and other phobic conditions relates to the source of the fear: in panic disorder with agoraphobia the source of the fear is perceived as being inside the patient (i.e., the possibility of a panic attack). In specific phobia, on the other hand, the perceived source of the fear is in an external object — for example, a dog in the case of dog phobia. Fear of flying is an interesting condition that could either be due to panic disorder with agoraphobia or represent a specific phobia. In the former case, the patient would report fear of a panic attack during the flight, while in the latter case, the patient would report a preoccupation with the possibility of crashing.[96] Agoraphobia without panic is a condition in which agoraphobic avoidance occurs for reasons other than fear of panic. For example, some patients with vestibular disorders have no clear panic attacks but nevertheless avoid stores, shopping malls, stadiums, or heights because the spatial characteristics of these situations induce vestibular symptoms (space and motion discomfort, to be discussed later). Such patients might receive a diagnosis of either agoraphobia without panic, or anxiety related to a medical condition. For further discussion of such differential diagnostic issues see Jacob, et al.[67]

Panic disorder is of interest in otoneurology because a high prevalence of vestibular abnormalities have been reported in several studies.[62,68,72,125,143,144] Most of these studies included several tests of vestibular and/or balance function. A statistic that could be determined from a majority of these studies was the proportion of subjects with abnormal results on one or more of these tests. The results of studies from which this statistic could be determined are listed in Table 1. With one exception,[134] the prevalence of panic patients with vestibular or balance abnormalities was remarkably high (71–88%).

Recently, we completed a study in which we compared the prevalence of vestibular dysfunction in the following conditions: (1) panic disorder without or with only limited agoraphobic avoidance (P); (2) panic disorder with agoraphobia (Ag); (3) non-panic anxiety disorders (Npa), that is, generalized anxiety disorder, or social phobia or depression with symptoms of anxiety; (4) depression without anxiety (D); and (5) normal controls (N). The normal controls had been screened psychiatrically but had not undergone any screening for vestibular abnormalities prior to the study. The subjects received the same vestibular test battery that was used in the laboratory routinely for clinical diagnosis of vestibular disorders, including dynamic posturography, rotational testing, ocular motor screening, positional testing, and caloric testing. We found that (1) laboratory vestibular abnormalities were common in all groups; (e.g., about 40% of members in the normal control group [group N] had abnormalities on at least one of the tests); and (2) more than 90% of the agoraphobics had vestibular abnormalities, a rate significantly higher than any of the other groups.[69]

These results raise several questions including:

1. What is the role of vestibular dysfunction in panic disorder in general and panic disorder with agoraphobia in particular?
2. What is the significance of vestibular abnormalities in the control groups, particularly the normal controls?

In order to appreciate the interrelationship between vestibular dysfunction and panic disorder, specifically agoraphobia, we will first bring the reader up to date on current thinking concerning the etiology of panic disorder. We will then discuss possible psychological and neurological linkages to the vestibular system. Finally, we will see how such linkages fit with data examining clinical correlates of laboratory vestibular dysfunction both in the patients with panic disorder and normal controls.

2 CURRENT THEORIES OF PANIC DISORDER

Most pathophysiological theories of panic disorder have been stimulated by observations that administration of certain pharmacologically active substances trigger panic attacks in patients with panic disorder. For a panicogenic agent to deserve serious interest, the following criteria are considered: (1) the panic symptoms triggered by the agent are similar to those of natural attacks; (2) the agent should trigger panic attacks specifically in patients with panic disorder as opposed to other psychiatric disorders; (3) drugs effective in the treatment of panic disorder should block attacks; and (4) the panicogenic effect should be replicable.[140] None of the agents that have been studied completely fulfills each

TABLE 1

Prevalence of Vestibular Abnormalities in Panic Disorder

Source	Diagnosis	Selected for dizziness	N	Vestibular tests[a] included for: VOR				VSR	Prevalence any abnormality (%)	Other results
Jacob et al., 1993[68] 1985[72,b]	Panic disorder	Yes	21	O	P	C	R	S	86	
Jacob et al., 1988[71]	Panic disorder	Yes	8	O	P	C	R	S	83	
Swinson et al., 1993[134]	Panic disorder	Yes	15	O	P	C	hfr	—	0	hfr: Increased gain
	Normal controls		10						0	
Sklare et al., 1990[125]	Panic disorder	No	17	O	—	C	hs	—	71	
Yardley et al., 1994[144,c]	Panic disorder	No	17	O	—	C	R	D	88	
	Normal controls		17						42	

Note: The only studies included are those from which the statistic "at least one vestibular abnormality" could be determined.

[a] Codes for tests included for vestibular ocular reflex (VOR): O, ocular motor screening including spontaneous and gaze nystagmus; P, positional testing; C, caloric testing; R, rotational testing (low frequency); hs, head-shaking nystagmus; hfr, rotational testing including high-frequency stimuli (0.32–3.25 Hz). Codes for tests of vestibular spinal reflex: S, static platform posturography; D, dynamic platform posturography.

[b] At least one abnormality result published in Jacob et al.[68]

[c] As calculated from Table 6, including spont. nystagmus, caloric test (RVR or DP >29%), sinusoidal rotation, and posturography.

of these criteria. In the next sections we will discuss the neuropharmacology of panic disorder. In addition, we will discuss brainstem abnormalities in regulation of respiration and cardiovascular function. Next, we will review cognitive theories of panic. Finally, we will review the factors that contribute to the development of agoraphobia.

2.1 The Neuropharmacology of Panic Disorder

Neuropharmacological theories of panic disorder have focused on the transmitter systems norepinephrine, serotonin, GABA, and cholecystokinin. It is unlikely that any one of the systems targeted by the challenge agent constitutes the only substrate of anxiety.[39,116] More likely is an interaction of intercellular signals integrated by second messengers such as cyclic adenosine monophosphate (cAMP) and the calcium ion (Ca^{2+}). Such interactions produce various changes in different regions of the brain and body that, taken together, constitute the clinical state of panic disorder.[56]

2.1.1 Noradrenergic Transmission

The noradrenergic theory of panic disorder postulates that panic disorder involves dysregulation in central norepinephrine systems. The original stimulus for this theory came from observations that electrical stimulation of the locus coeruleus induced behaviors highly similar to fear in monkeys.[115] The notion that simple locus coeruleus overactivity is responsible for panic attacks is not generally accepted today, however. For example, in studies on humans with surgically implanted permanent electrodes in the locus coeruleus, no manifestations of anxiety were described after stimulation.[77,86] The patients had received this treatment for relief of severe spasticity or intractable epilepsy. Studies of coeruleus-evoked potentials in the ambulatory rat[118] suggest that this nucleus signals the presence of novel stimuli. The signal from locus coeruleus is then projected to the forebrain where it affects arousal, attention, and memory formation.

Patients with panic disorder, as well as patients with affective disorders such as depression or other anxiety disorders, exhibit altered functioning of adrenergic and noradrenergic systems. Both central α and central and peripheral β receptors have been implicated. The presynaptic α_2 antagonist yohimbine elicits panic attacks in patients with panic disorder, and some of the neuroendocrine effects of yohimbine can be blocked by clonidine, an α_2 adrenoceptor agonist.[8]

The somatic symptoms of anxiety and panic are suggestive of activation of peripheral, sympathetically mediated physiological responses; however, increased cardiosympathetic or adrenomedullary activity at rest or following experimental stressors in panic disorder is not consistently observed. In fact, there is evidence of *diminished* functioning of the peripheral β-adrenergic receptor system in panic disorder. It has been found that lymphocyte β adrenoreceptor density is significantly *reduced* in patients with panic disorder.[91] The pathophysiological significance of these observations is unclear. In animal studies, downregulation of β adrenoreceptors appears to be adaptive, since it is associated with lesser degrees of stress-induced organic pathology.

Neuropeptide Y (NPY) may be another factor in the catecholamine dynamics in panic states. NPY is a modulatory transmitter[55,65] that remains in circulation for a longer time than norepinephrine or epinephrine. NPY acts at presynaptic sites to inhibit norepinephrine release. As a result, NPY attenuates stress-associated increases in norepinephrine. Costored with various transmitters, NPY is present in many parts of the central nervous system. The anxiolytic and sedative effects of NPY involve Y receptors in the central nucleus of the amygdala; conversely, anxiety is associated with inhibition of central Y-receptor expression.

2.1.2 Serotonergic Transmission

The serotonin theory of panic disorder[76,107] is supported by (inconsistent) findings that panic patients tend to experience panic attacks during challenges with some serotonin agonists such as fenfluramine or *m*-chlorphenylpiperazine (MCPP). Furthermore, agents with selective serotonin (5-HT) reuptake inhibiting effects (SSRI's), including clomipramine, fluvoxamine, and fluoxetine, reduce panic attacks. The dorsal raphe is a serotonergic nucleus that has been implicated in panic disorder. For a review of the distribution of serotonergic nerve terminals in the central nervous system, see Steinbusch.[133]

It is unclear whether serotonin has a specific action in the sequence of events that lead to anxiety or panic.[28] Different 5-HT receptors and different 5-HT pathways sometimes perform opposing functions.[42] The majority of cortical and subcortical 5-HT innervation originates in the raphe nucleus. The 5-HT_{1A} autoreceptors in the raphe nucleus play a pivotal role in *reducing* the amount of 5-HT increase seen outside the raphe. Using (±)fluoxetine and (±)citalopram (the most selective binding agent for the 5-HT transporter), *acute* administration of SSRI has been shown to decrease raphe nucleus 5-HT neuron firing.[15] This decrease in firing has been shown to be blocked by a 5-HT_{1A} antagonist.[60] Additionally, the systemic administration of 5-HT_{1A} agonist will shut down raphe 5-HT neuron firing.

Chronic administration of SSRIs has different effects. After chronic administration of (±)citalopram, the firing rate slowly returned to near the control range after 14 days.[38] This has been interpreted as a progressive down-regulation of the 5-HT_{1A} autoreceptor in the raphe, resulting from an adaptive response to the chronically increased levels of 5-HT. Chronic SSRI treatment would result in adaptive changes to 5-HT neurotransmission including both down-regulation of the somatodendritic autoreceptor, which restores the acutely suppressed firing rate, and down-regulation of the terminal autoreceptor, which likely allows increased 5-HT release and overall neurotransmission.

Most *non-SSRI, tricyclic antidepressants* produce significant 5-HT reuptake site blockade at therapeutic doses, in addition to varying degrees of noradrenergic uptake blockade.[126] This may explain the antipanic and antiphobic utility of lower dosing and steady-plasma level range of 100 to 140-ng/ml regimens of imipramine.[94] One classic tricyclic antidepressant with little 5-HT reuptake blockade is desipramine (DMI), and the panic-inhibiting effect of DMI has not been clear-cut.[50,90,93]

To date, the most convincing evidence of the utility of 5-HT receptor ligands in the treatment of anxiety disorders concerns 5-HT_{1A} receptor *agonists* and 5-HT_{2A}, and 5-HT_{2C} receptor *antagonists*. For example, drugs are currently under trial that combine an agonist activity at 5-HT_{1A} receptors and an antagonist activity at 5-HT_{2C} receptors, and that have anxiolytic effects that are especially potent in rats and pigeons.

2.1.3 GABAergic Transmission and the Benzodiazepine Receptor

The benzodiazepine drugs reduce symptoms of anxiety and panic. Inverse benzodiazepine agonists such as β-carboline ethyl ester (β-CCE) have the effect of increasing anxiety. These observations have stimulated interest in this benzodiazepine (BZD) receptor. At least three types of BZD receptors (or possibly three different conformations of the same receptor) have been identified.[107] The w_1 receptors are found in the limbic system, brainstem arousal centers, cerebellum, and hippocampus; w_2 receptors are present in the cortex and spinal cord; and w_3 receptors are found in the periphery. Medications that interact only with w_1 receptors, such as zolpidem, might have a more specific anxiolytic and/or hypnotic action, depending upon dosage, with fewer incidental effects.

The BZD receptors are closely associated with receptors for γ-aminobutyric acid (GABA). There are two receptors for GABA. The $GABA_A$ receptors are coupled to chloride (Cl) channels, and the $GABA_B$ receptors are coupled to Ca^{2+} and possibly cAMP. The $GABA_B$ receptors may help to regulate release of other transmitters. The $GABA_A$ receptor interacts with $GABA_B$, which increases the frequency of openings of the Cl channel to which it is linked. This augments the net influx of negatively charged Cl ions, increasing the net negative charge within the neuron (i.e., hyperpolarizing it), and making it more difficult for any excitatory influence to cause depolarization.

The benzodiazepine receptor is linked specifically to the $GABA_A$ receptor. Upon interaction with an agonist, the BZD receptor increases the affinity of the GABA receptor for its transmitter, thereby augmenting the ability of GABA to hyperpolarize the neuron. In brain systems involved in anxiety, BZD receptor agonists decrease the likelihood of depolarization, thus decreasing the release of transmitters such as norepinephrine, dopamine, and serotonin.[63]

2.1.4 Cholecystokinin (CCK)

A panicogenic substance that most recently attracted attention is *cholecystokinin* and related peptides.[16,107] Originally, cholecystokinin attracted interest because the excitatory effects of cholecystokinin were antagonized by benzodiazepine receptor agonists, suggesting a role of cholecystokinin in anxiety. In animal models, CCK agonists may provoke behaviors reminiscent of anxiety, whereas CCK receptor antagonists are reported to possess anxiolytic effects.[57] In humans, administration of cholecystokinin produces panic attacks, particularly in patients with panic disorder. Conversely, a cholecystokinin antagonist, L-365,260, has been found to have antipanic properties.[17]

Several CCK fragments have been identified that may differ in their affinity for the various CCK binding sites in the brain and periphery. CCK_{8S}, the most abundant CCK fragment, has equal affinity for sites designated at CCK_A and CCK_B receptors, while the tetrapeptide CCK_4 has a 100-fold-higher affinity for CCK_B receptors. Neuroanatomical studies have revealed that CCK_B receptors are most abundant in the hippocampus. Moreover, there is a close functional relationship between CCK- and GABA-containing neurons that can be completely abolished by benzodiazepines. It has been suggested that peripherally administered CCK_4 acts centrally, after crossing the blood-brain barrier, through excitation of deep cortical and hippocampal neurons, of the amygdala and possibly of other central areas related to anxiety.

The hypothalamus-pituitary-adrenal axis is readily stimulated by CCK agonists. Recently, Abelson et al.[1] have reported that IV administration of pentagastrin, a synthetic analogue of cholecystokinin tetrapeptide (CCK_4), not only produces more panic attacks in patients (12/22) than in controls (1/22), but also leads to significant rises in ACTH and cortisol levels in patients and controls.[142] These findings suggest a greater sensitivity for CCK receptor agonists in patients suffering from panic disorder than in healthy controls.

2.2 Brainstem Autonomic Regulation, Including Respiration

Inhalation of carbon dioxide elicits panic attacks in a greater proportion of panic patients than in normal or non-panic psychiatric controls.[20,52,54,106,108] First-degree relatives (parents, siblings and children) of patients with panic disorder were observed to react more than controls but less than the patients themselves.[109] The panic attacks are associated with increased minute ventilation and inspiratory drive.[52] Respiratory changes also comprise the most reliable physiological indicator of panic elicited by challenge agents other than carbon dioxide, such as lactate.[105] Patients with panic disorder frequently hyperventilate,[106] and naturally occurring panic attacks tend to be associated with an increase in tidal volume.[105] Patients with panic disorder also have been observed to show more subtle changes in respiratory behavior such as breathing more irregularly[20] or engaging in chest rather than abdominal breathing.[105] Caffeine, another panicogenic agent in panic patients, increases respiratory sensitivity to hypercapnia, as well as respiratory sensitivity to hypoxia and production of CO_2 during exercise.[40]

The increased sensitivity to CO_2 and the tendency to hyperventilate may be caused by a hypersensitive autonomic control mechanism in the brainstem.[106] Klein[83] has proposed that a hypersensitive "suffocation alarm" system represents the core abnormality in panic disorder. However, panic patients also have (less consistently) been shown to have some cardiovascular differences from normal controls, suggesting a more generalized dysregulation of brainstem autonomic circuits. For example, a challenge dose of clonidine resulted in greater decreases in sitting and standing diastolic blood pressure in panic patients than in controls.[26] Conversely, a yohimbine challenge was associated with greater increases in systolic blood pressure in patients compared to controls,[27] although in a replication study the blood pressure changes with yohimbine or clonidine were not as significant.[30] Patients with generalized anxiety disorder or depression did not show similar changes in blood pressure.[29,58] Yohimbine was also associated with increases in standing heart rate variability in the spectral mid-frequency band, consistent with greater Meyer waves.[150] Some investigators have found differences between panic patients and controls in cardiovascular reactivity[61,97] (i.e., cardiovascular responsivity to psychological or behavioral stimuli). However, others have found no differences in reactivity.[131] One study examining respiratory sinus arrhythmia in patients with panic disorder found no differences from controls in this measure of "vagal tone."[7]

Integrating the neuropharmacological and autonomic findings in panic disorder, Gorman et al.,[53] in 1989, hypothesized that the source of panic is related to "irritable foci" in the brainstem, including the locus coeruleus, dorsal raphe nucleus, and/or medullary chemoreceptors such as the nucleus reticularis gigantocellularis. *Brainstem autonomic dysregulation* is an alternative term frequently used to describe the underlying problem in panic disorder. Such dysregulation might involve the cardiovascular system and/or the respiratory system. To date, the most robust findings in the literature appear to be those indicating altered control of respiration.

2.3 Cognitive Factors in Panic Disorder

A problem with panic challenge research has been a relative failure to consider cognitive factors that might contribute to the emergence of panic attacks. For example, Margraf et al.,[94] in 1985, criticized the published literature on the efficacy of lactate infusions in affecting panic attacks. The arguments were based on the following grounds: (1) confound with differences in baseline anxiety levels between patients and controls; (2) inadequate operational criteria for panic attacks and inadequate control strategies, including non-blinded observers, different infusion rates for lactate and the control solution and different pre-infusion information, and a tendency of the panic patients to respond to placebo infusions with panic (36% of patients vs. 0% of controls responded to placebo, compared to 56% of patients vs. 9% of controls panicking with lactate).

Such concerns are not limited to lactate infusions. For example, Albus et al.,[2,3] addressing the methodological concerns raised above, administered yohimbine in the context of instructions aimed at minimizing panic attacks. While the patients experienced increased anxiety and "panicky" feelings compared to controls, none of the patients actually reported a panic attack. Sanderson, et al.[117] administered 5% CO_2 as a panic challenge agent using an apparatus that deceptively signalled the possibility of patient control over the concentration of CO_2. In the absence of illusion of control, eight out of ten panic patients panicked, while in the presence of illusion of control, only two out of ten patients panicked.

In a recent review, D. M. Clark[33] argues that panic attacks are cognitively mediated because of evidence for the following: (1) panic patients tend to catastrophically misinterpret certain bodily sensations; (2) thoughts based on such misinterpretations accompany panic attacks; and (3) experimental manipulations of cognitive set influence the probability of panic.

An argument against a cognitive theory of panic is that physical symptoms typically precede the catastrophizing cognitions in panic attacks[79,99,151] and that panic attacks can occur during sleep. On the other hand, it is clear that the cognitive factors can modify the escalating sequence of events leading to a full-blown panic attack.

2.4. The Transition to Agoraphobia

The development of agoraphobia represents a significant complication of panic disorder, one that is associated with a high degree of social disability. Some patients with panic disorder develop very little avoidance, while others develop very extensive avoidance to the point of being home bound. The factors contributing to the development of agoraphobia are poorly understood. Female sex is a significant risk factor,[6,36,49,93,141] a finding usually attributed to a woman's traditional social roles providing increased opportunities to engage in avoidance behaviors.

Agoraphobics may have an increased prevalence of history of childhood anxiety disorders.[6,37,84] Some studies found that agoraphobics had a greater number of panic symptoms or more severe symptoms[36,130,136] but this is not a universal finding.[6,80,139] Finally some studies found differences in panic symptom profiles, but the findings have been inconsistent across studies. Agoraphobia has been associated with symptoms of derealization[22,119] dizziness and faintness,[36,119,136] palpitations, feelings of unreality, sweating and shaking but less dizziness,[80] or chest pain and dyspnea.[78]

A set of relatively robust findings is that of cognitive correlates with agoraphobia. Katerndahl[78] found "anticipatory anxiety" to be the most immediate independent correlate of agoraphobia. Perceived lack of controllability of panic[136] or labeling of panic sensations as life-threatening[18] was found to be important. Agoraphobic avoidance is also associated with anticipation of negative evaluation by others; agoraphobics often score higher than uncomplicated panic patients in measures of interpersonal sensitivity or social anxiety or report greater concerns about social ridicule.[6,21,36,44,111,136] Furthermore, a cognitive mediation of agoraphobic avoidance is suggested by the fact that one theme of agoraphobic avoidance is the perception of lack of availability of help in case of a panic attack. This theme of agoraphobic behavior can be conceptualized as an "approach" rather than "avoidance" behavior, the approach gradient being toward situations where help is likely to be available.[73]

There are few biological differences between agoraphobics and patients with uncomplicated panic disorder. Panic patients with and without agoraphobia differed with respect to vulnerability for panic with sodium bicarbonate,[51] a suggestion that has been viewed as consistent with agoraphobia being a more severe variant of panic disorder.[83] Patients with uncomplicated panic disorder had decreased (more abnormal) response of lymphocyte cAMP to isoproterenol than agoraphobics, suggesting less down-regulation of the β-receptor.[91,92] The significance of the latter finding is unclear.

According to Gorman et al.,[53] the prefrontal cortex is essential for developing agoraphobic avoidance. Alternatively, agoraphobic avoidance can be interpreted as a conditioned fear of situations in which panic attacks have occurred. The central nucleus of the amygdala is essential for the formation of conditioned fear responses.[85] In addition, the hippocampus contributes to assessing the context in which a fear-inducing stimulus occurs. The cortex is not essential for fear conditioning but serves the purpose of interpreting more intricate stimuli. We are emphasizing the role of the amygdala here because, as we will see in the next section, the amygdala has major connections with the parabrachial nucleus, a major pontine center of somatic-visceral integration which also receives significant vestibular input.[11]

3 THE RELATIONSHIP BETWEEN PANIC DISORDER AND VESTIBULAR DYSFUNCTION

In the preceding section we saw that there were both psychological and biological theories of panic disorder. Consequently, we can examine the relationship between panic disorder and vestibular dysfunction from both a psychological and a biological perspective. We will discuss the psychological models first.

3.1 Psychological Linkages: Somatopsychic and Psychosomatic Models

Both the somatopsychic and psychosomatic models assume a *psychological* or a *behavioral mediation* of the relationship between panic disorder and vestibular dysfunction. The two models differ with respect to which of the two represents the cause and which the effect.[66] Somatopsychic models assign a primary causal role to the vestibular dysfunction, while psychosomatic models imply that psychological or behavioral effects on the vestibular system are primary. Both models can be applied (1) to the relationship between vestibular dysfunction and panic attacks; and (2) within panic disorder, to the relationship between vestibular dysfunction and agoraphobia.

3.1.1 Panic Attacks, Anxiety, and Hyperventilation

According to the *somatopsychic model*, panic patients tend to catastrophically reinterpret unfamiliar somatic sensations as implying danger.[33] In panic patients with vestibular dysfunction, the initial panic attacks may have developed in response to catastrophic misinterpretation of vestibular sensations. After the experience of one or several panic attacks, patients become more vigilant for signs of future attacks. Such vigilance may increase the risk of panic attacks even in the absence of further vestibular sensations.

The *psychosomatic models* are based on findings of increases in vestibular responses with increased arousal.[35,48] In addition, hyperventilation is known to produce altered responses on the rotational[137] and positional (but not caloric) tests.[101] Three different psychosomatic models seem possible.

First, vestibular test abnormalities in panic disorder may represent false positive findings, i.e., they constitute measurement errors that occurred because the subjects were anxious or hyperventilated. According to this "trivial" psychosomatic model, laboratory vestibular abnormalities, if observed, would be clinically irrelevant.

Second, vestibular function may truly be affected by anxiety or panic. According to this "clinical" psychosomatic model, patients would suffer from (reversible) vestibular dysfunction whenever they are anxious (i.e., during a large portion of their normal daily activities).

A third psychosomatic model suggests that behavioral factors reactivate a dormant vestibular lesion. According to this model, vestibular lesions become asymptomatic as a result of central compensation. Anxiety or hyperventilation might alter vestibular processing by resetting central compensatory mechanisms, resulting in a re-emergence of vestibular symptoms. This third psychosomatic model thus presupposes psychosomatic influences on existing vestibular lesions. Therefore, it can be conceptualized as a "hybrid" somatopsychic/psychosomatic model.

3.1.2 Space and Motion Discomfort and the Transition to Agoraphobia

A *psychosomatic* explanation for why agoraphobics in our recent study (i.e., group Ag) had more vestibular dysfunction than patients with uncomplicated panic disorder (group P) would be that agoraphobia is the more severe disorder of the two. Consequently, psychosomatic effects such as hyperventilation may have played a greater role in group Ag than in group P. To test this hypothesis, one would have to measure respiration or other correlates of anxiety in the two groups. Comparing anxiety levels measured by the Spielberger State Anxiety Inventory[129] at each vestibular test occasion, however, we found that members of group P and Ag did *not* differ substantially with respect to self-reported anxiety symptoms.[69]

A *somatopsychic* explanation would be that agoraphobic patients begin to avoid certain situations because of situation-specific symptomatology related to vestibular dysfunction. The mechanism for such situation specificity could be as follows: the vestibular system subserves the function of spatial orientation and maintenance of upright balance. The main inputs to the vestibular system are stimuli in the visual, proprioceptive, and vestibular sensory modalities. In patients with vestibular dysfunction, information from the vestibular channel is likely to be erroneous, causing adaptive changes that result in increased reliance on visual and proprioceptive spatial information. For example, when exposed to illusory movements in the visual field, patients with vestibular dysfunction responded with greater increases in body sway compared to normals, suggesting increased sensitivity to visual spatial stimuli.[114] As a consequence of this altered sensory integration of spatial information, patients may develop an unusual sensitivity to situations in which visual or proprioceptive cues are inadequate or misleading. Examples of such situations include: (1) long visual distances (e.g., shopping malls, stadiums, heights); (2) confusing or complex visual stimuli (e.g., Navaho rugs, striking geometric patterns on wallpapers or floors); (3) combinations of visual, proprioceptive, and vestibular cues that are unusual or complex (e.g., walking down supermarket aisles, reading while riding in a car); (4) excessive vestibular stimulation (e.g., dancing, vigorous head movements); and (5) movements involving neck extension and reorientation with respect to gravity (e.g., looking up at tall buildings, closing eyes in shower).[71,74,75]

We refer to the situational specificity of vestibular symptoms as *space and motion discomfort*. We have found that space and motion discomfort is common in patients with vestibular disorders compared to patients evaluated for hearing loss.[75] Space and motion discomfort also constitutes a theme in agoraphobic avoidance. Agoraphobics might avoid heights, open spaces, long corridors, tunnels, or shopping malls simply because the visual spatial cues in these situations elicit, or at some earlier time elicited, vestibular symptoms.

3.2 Neurophysiological and Neuroanatomical Linkages

When the initial clinical observations of vestibular abnormalities in panic disorder were published in 1985,[72] relatively little was known about the neurobiology of panic disorder. Even less was known about the neuroanatomical or neurophysiological basis for a relationship between panic disorder and vestibular dysfunction. Progress was possible thanks to increased understanding of the neurobiology of panic disorder in psychiatry (as exemplified in previous sections of this chapter) and increased awareness of vestibular-autonomic connections in neuro-otology (as exemplified in many of the chapters in this book). In addition, some cross-fertilization between the two fields has occurred. For example, the work by Schuerger and Balaban,[121] leading to the identification of a neural pathway from the locus coeruleus to the lateral vestibular nucleus, was in part inspired by the clinical findings of vestibular dysfunction in panic disorder. The fact that this projection was primarily to the lateral vestibular nucleus contributed to a focus on the vestibulo-spinal reflex in a subsequent study examining the effect of optic flow stimuli on body sway.[74]

Connections between the vestibular system and structures of interest for panic disorder include (1) reciprocal connections between the locus coeruleus and the lateral vestibular nucleus; (2) serotonergic pathways; (3) pathways serving vestibular-autonomic reflexes involving respiration and the circulatory system; and (4) integration of vestibular information with other visceral sensory modalities at higher levels, such as the nucleus parabrachialis. The current state of knowledge concerning these connections was recently reviewed by Balaban.[10,11]

3.2.1 Networks Involving the Locus Coeruleus

Mediated by the coeruleospinal, coeruleo-reticular, coeruleo-cerebellar fibers, and recently discovered coeruleoreticular fibers,[121] the locus coeruleus has a role in the control of posture. It exerts a moderating influence on spinal reflexes by increasing the excitability of spinal motoneurons.[46,112]

Vestibular Influences on Locus Coeruleus — It has been debated whether the locus coeruleus receives input from the vestibular system. Aston-Jones et al.[9] reported that input to the locus coeruleus was limited and came from only two major sources, the nucleus gigantocellularis and nucleus prepositus hypoglossi. On the other hand, Cederbaum and Aghajanian[23] found that the locus coeruleus of the rat receives afferent input from a number of areas in the forebrain, spinal cord, and brainstem, including the contralateral vestibular and fastigial nuclei. Similarly, Fung et al.[47] confirmed the existence of vestibular connections with the locus coeruleus but found that a large portion of the vestibulo-coerulear connections originated in the ipsilateral lateral vestibular nucleus (LVN). Functionally, cells in the locus coeruleus were found to respond to stimulation both of the labyrinth[113] and to neck rotation.[13]

Projections from the Locus Coeruleus to the Vestibular System — The presence of connections from the locus coeruleus to the vestibular nuclei has not been generally appreciated,[31] in part because early fluorescence histochemical studies had indicated a lack of noradrenergic terminals in the vestibular nuclei. Gorman et al.[53] hypothesized that the locus coeruleus would affect vestibular processing via its projections to the cerebellum.[103] However, the cells in the lateral vestibular nucleus respond with excitation to iontophoretically applied norepinephrine and *d*-amphetamine; conversely, cells in the medial vestibular nucleus were inhibited.[80] In a more recent study, Licata et al.[88] found 85% of vestibular neurons tested to be responsive to noradrenaline. Most of the responses were inhibitory (86%), and the majority of the responsive neurons were located in the lateral and superior vestibular nucleus. Clonidine mimicked the noradrenaline responses in a dose-dependent manner, and yohimbine was effective in blocking noradrenaline or clonidine evoked responses, suggesting a primary involvement of the α_2 receptor.

Recently, Schuerger and Balaban[121] showed, in the rat, that the locus coeruleus and subcoeruleus had dense projections to the dorsal lateral vestibular nucleus, and less dense connections with the superior vestibular nucleus, while there were relatively few connections with the medial and inferior vestibular nuclei. Similar observations were made in rabbits and monkeys.[120,122] The projections to the lateral vestibular nucleus are of interest because this region is known to receive terminals from the utricular nerve and anterior and posterior ampullary nerves but not the horizontal ampullary nerve,[123,124] suggesting that these projections are responsive mainly to vertical canal and otolith inputs.[10]

3.2.2 Serotonergic Transmission

Recently, connections from the vestibular nuclei to the nucleus raphe magnus have been observed,[11] suggesting some effects of vestibular stimulation on the serotonergic system. The neurons in the raphe magnus and pallidus receive powerful vestibular input, typically pitch related.[146,147] Conversely, the vestibular nuclei receive serotonergic input.

Steinbusch[132] indicated that serotoninergic innervation of the vestibular nuclei was more prominent than noradrenergic innervation and located predominantly in the caudal part of the medial vestibular nucleus. Licata et al.[89] found that application of 5-HT to cells in the superior vestibular nucleus altered spontaneous firing rates, although differently in different specific cells (excitation in 48%, inhibition in 29%, and biphasic response in 10%). Serotonin applied to the lateral vestibular nucleus resulted in excitatory or biphasic responses.[87] Kishimoto et al.[82] found that conditioning stimuli from the dorsal raphe nucleus resulted in LVN inhibition only. Furthermore, they found these effects only in those LVN neurons that had ascending projections, suggesting that the effect of 5-HT is to inhibit the vestibulo-ocular reflex. Perhaps this hypothesis could explain a finding in the rotational and caloric test data of our recent study—that patients with panic disorder and agoraphobia (group Ag) consistently had the lowest gain on the rotational test and the lowest total velocity on the caloric test.

3.2.3 Projections from Vestibular Nuclei to Brainstem Autonomic Regions

Earlier, we described the panic disorder theory hypothesizing dysregulation of autonomic brainstem circuits. The vestibular system can influence such circuits, resulting in vestibular-autonomic reflexes. The original observations that vestibular stimulation could affect respiration and circulatory variables were made more than 70 years ago.[19,128,127] Projections from the medial and inferior vestibular nuclei to the nucleus tractus solitarius and the ventral respiratory group constitute an anatomic basis for such effects.[12,98,148] Changes in the activity of respiratory nerves have been observed as a result of electrical stimulation of the vestibular nerve; these responses were abolished by lesions in the medial and inferior vestibular nucleus.[149] Similarly, the cells in the rostral ventrolateral medulla involved in cardiovascular control respond to natural vestibular stimulation, the main input coming from the otoliths.[146]

3.2.4 The Parabrachial Nucleus Network and Visceral Sensory Integration

Balaban[11] recently found extensive projections from the superior, medial, and inferior vestibular nuclei to the parabrachial nucleus. The parabrachial nucleus is a major way station for visceral afferent input, including taste, respiration, and circulation.[14,64,102] Included among its multiple inputs are cholecystokininergic pathways from the area postrema.[59] It is also a major recipient of noxious somatosensory information, suggesting that the parabrachial nucleus is an area where somato-autonomic integration takes place.[14,144] Efferent pathways from the parabrachial nucleus to the lower brainstem participate in the control of circulation and respiration.[24,25,45,135] The parabrachial nucleus also has reciprocal connections with the central nucleus of the amygdala,[4,45,100] which is implicated in anxiety and fear learning. This direct vestibular projection to ascending visceral pathways, melded with descending inputs from the prefrontal, infralimbic, and insular cortex, the central nucleus of the amygdala, and several nuclei in the hypothalamus,[100,145] may provide the major substrate for psychiatric manifestations of vestibular dysfunction, including panic disorder and phobia.

4 LINKAGE BETWEEN PANIC, PHOBIA, AND VESTIBULAR DYSFUNCTION: SOME EMPIRICAL DATA

To illustrate the theoretical notions of the preceding sections, we will present data on the relationship between laboratory vestibular dysfunction and symptoms. The data come from two different populations: (1) the patients with panic disorder without agoraphobia and normal controls (groups P and N) from the study described earlier (Jacob et al.[69]); (2) outpatients at an ear, nose, and throat (ENT) clinic evaluated for dizziness or mild hearing loss examined by D. B. Clark et al.[32] and Jacob et al.[75] Patients in the ENT sample either had vestibular disorders verified by vestibular tests or physical examination, or came to the clinic for evaluation of mild hearing loss with presumed normal vestibular function. In each of these groups (P, N, and ENT), there was a prevalence of laboratory vestibular dysfunction of 40–50%, offering an opportunity to compare the symptoms of subgroups characterized by normal and abnormal laboratory vestibular function in each group. All these subjects had provided data on a questionnaire measuring degree of space and motion discomfort (see Section 3.1.3), and the subjects in groups N and P filled out an additional survey, the Symptom Questionnaire (SymQ). For the actual questions in the two questionnaires, see Jacob and Lilienfeld.[70]

The *Situational Characteristics Questionnaire* (SitQ) consists of two parts. *Part I* elicits information on the degree of discomfort in certain aspects of situations such as riding in a car, bus, escalator, or elevator, or going through a tunnel. For example, for the situation of "riding in a car," one item in the questionnaire elicits the difference between the degree of discomfort experienced when sitting in the back seat compared to sitting in the front seat. It was expected that patients with vestibular dysfunction, because of their relative dislike of sitting in the back seat, would show a particularly large back seat–front seat difference. Part I yields scores on two scales, the SMD-I and Ag-I, thought to reflect discomfort related to space and motion, on the one hand, and "other" agoraphobic avoidance, on the other (e.g., being far vs. being close to the exit while in a supermarket). Our expectation was that patients with vestibular dysfunction would have high scores on the SMD-I but not on the Ag-I, a prediction that was confirmed in group ENT (see below).

The items of *Part II* of the SitQ are in the traditional Likert format; items are rated on a 4-point scale. Items in Part II include "looking up at tall buildings," "closing eyes in shower," and "aerobic exercise." Part II provides scores for a second space and motion discomfort scale, the SMD-II.

The *Symptom Questionnaire (SympQ)* comes in two versions, a "panic or anxiety" version for members of group P and a "stress" version for members of group N. The stress version was used for individual members of group N because none of them had symptoms of anxiety or panic. Either version elicits the prevalence of specific symptoms "during" panic or anxiety/stress or "between" panic or anxiety/stress, using a 4-point Likert-type scale (1-4) anchored by the descriptors "never or hardly ever," "sometimes," "often," and "very often or almost constantly." For example, the "during attack" part of the panic version instructs subjects to "please indicate how commonly you experience the symptoms below *when you are very anxious* (i.e., having a panic attack)"; the "between attacks" version instructs subjects to "please indicate how commonly you experience the symptoms below *when you are not anxious* (i.e., *between* panic attacks)."

Certain items of the SympQ had been *a priori* identified as belonging to one of two scales: the "vestibular symptom scale" and the "non-dizziness scale." The *non-dizziness scale* included items such as "heart palpitations" that represent autonomic sensations that are not usually thought to be a consequence of vestibular dysfunction. The *vestibular symptom scale* included the items of "room spinning," "spinning in head," "veering," "tendency to fall," "feeling like falling," and "nausea." The symptoms of dizziness and lightheadedness were not included in the vestibular symptom scale because these symptoms could clearly have been a consequence of hyperventilation.

Thus, each of the questionnaires included scales that had been identified *a priori* and that were thought to be related to vestibular dysfunction, i.e., the SMD-I and the SMD-II of the Situational Characteristics Questionnaire and the "between panic/stress" and "during panic/stress" Vestibular Symptom scales of the Symptom questionnaire. We first conducted *hypothesis testing analyses*, presented elsewhere[69,75] but reviewed here, to examine our hypothesis that individuals with vestibular dysfunction within a certain group would have higher scores on scales measuring vestibular symptoms and space and motion discomfort.

Following the prospective analyses described above, we conducted further *exploratory item analyses*, examining which items (irrespective of the scale to which they originally belonged) appeared to differ in the subgroups characterized by the presence or absence of vestibular dysfunction. These exploratory analyses involved first calculating the effect sizes (difference divided by the standard deviation of the difference[34]) for the vestibular/non-vestibular subgroup difference for each item. An effect size of greater or equal to .4 was taken as a cutoff for a difference being noteworthy for that item. According to Cohen,[34] an effect size of 0.8 is considered "large," an effect size of 0.5 is considered "moderate," and an effect size of 0.3 is considered "small."

Often we found that items associated with relatively large effect sizes (i.e., those that showed large differences between vestibular and non-vestibular subjects) in one group (for example, group P) did not necessarily show large effect sizes for another group (for example, group N). In concluding the exploratory analyses, we attempted to identify items that would consistently be endorsed by vestibular subjects *across* groups. We will be presenting these items and suggest that they might be used in alternative scales that could identify vestibular subjects in ways that would not be as affected by concomitant psychiatric diagnoses as the original scales.

4.1 Symptoms and Laboratory Vestibular Dysfunction

Within group P, the between-attacks vestibular symptom scale scores proved to be significantly higher in patients with abnormal vestibular function (Jacob et al.[69]). On the other hand, there were no significant differences in vestibular symptoms *during* panic attacks between patients with and without vestibular dysfunction.

With respect to group N, for periods of stress, there were no significant differences between subjects with and without vestibular dysfunction on any of the symptom scales. However, surprisingly, for periods "between" times of stress, *normal controls with vestibular dysfunction had significantly higher scores on the "non-dizzy" scale* (mean ± SD = 1.17 ± 0.20 and 1.04 ± 0.08, respectively, $t(43) = 2.79$; $p = .008$ two-tailed). Figure 1 is a box plot of this finding. The non-dizzy scale contains 11 symptoms that in clinical practice could hardly be considered as "vestibular": trembling, heart palpitations, hot or cold flashes, dry mouth, pressure in chest, blurred vision, numbness, knot in stomach, sweating, feeling disconnected or unreal, and lump in throat.

Table 2 lists the actual "between" panic/stress items that were relatively highly related to vestibular dysfunction (effect size of difference between the subgroups = 0.4 or more). The items were rank-ordered by effect size.

In *group P*, four of the six items on the vestibular symptom scale differentiated the vestibular subgroups above the criterion level, including "room spinning," " nausea," " veering" and "tendency to fall." In addition, the related symptoms of "dizziness," "lightheadedness," and "wobbly legs" were associated with vestibular dysfunction, as was the symptom of "numbness." Surprisingly, the between-attacks symptom of "ringing in the ears" was indicative of normal vestibular function in this group. Another item, "heart palpitations," was also a marker for normal vestibular function, but the effect size did not reach our criterion level of an effect size of ≥0.4.

For *group N*, the only item from the vestibular symptom scale that appeared was "room spinning." The remaining symptoms included "sweating," "dry mouth," "knot in stomach," "lump in throat," and "hot or cold flashes." There was no item that was associated

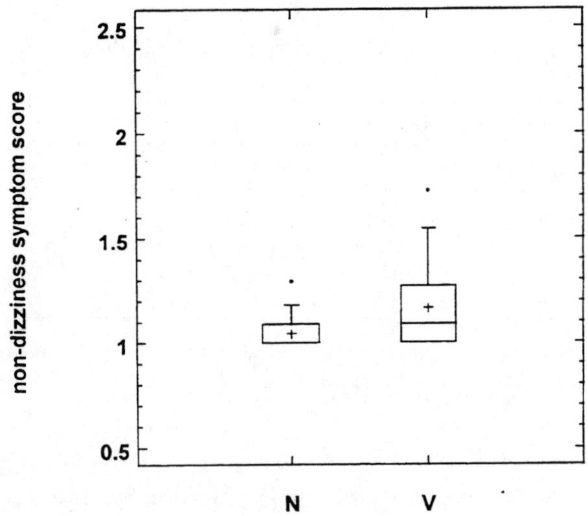

FIGURE 1
Box plots of non-dizziness symptom scores for healthy normal control subjects with abnormal (V) and normal (N) laboratory vestibular function. Upper and lower limit of box represents the 75th and 25th percentile, respectively; the line in the middle of a box represents the median; the "+" represents the mean; the "whiskers" connect the highest points within 1.5 × interquartile range; the dots represent outlying data points.

TABLE 2
Between Attacks or "Stress" SympQ Items That Differentiated Among Individuals with and without Vestibular Dysfunction in Group P and N (effect size ≥0.4)

Panic disorder		Normal controls	
Abnormal[b]	Normal[b]	Abnormal	Normal
Room spinning[a]	Ringing	Sweating	None
Dizziness		Dry mouth	
Wobbly legs		Knot in stomach	
Numbness		Lump in throat	
Nausea[a]		Hot or cold flashes	
Veering[a]		*Room spinning*	
Lightheadedness			
Tendency to fall[a]			

[a] *Italicized* items are included in the vestibular symptom scale.

[b] Normal and Abnormal refer to subgroups with and without vestibular dysfunction.

with normal vestibular function at criterion level (effect size of 0.4), but one item, "heart palpitations" (not shown in the table), was associated with normal vestibular function just below criterion level (effect size of 0.3).

4.1.1 *Optimized Vestibular Dysfunction Symptom Scale*

To identify items that would consistently discriminate between subgroups with and without vestibular dysfunction in both groups, we first averaged the effect sizes for groups P and N. Those items that had a low average effect size or were associated with normal vestibular function in one of the two groups were excluded. By these selection criteria we

TABLE 3

Symptoms between Panic/Stress Included in the Optimized Vestibular Symptom Scale

Positive weight (+1)			Negative weight (–1)
Room spinning	Dizziness	Disorientation	Heart palpitations
Wobbly legs	Nausea	Feeling like fainting	
Dry mouth	Feeling like falling	Spinning in head	
Veering	Hot or cold flashes	Blurred vision	
Sweating	Lump in throat	Lightheadedness	
Numbness	Blacking out	Tendency to fall	

Note: Items are listed in order of average effect sizes of differences between vestibular and non-vestibular subgroups in groups P and N.

TABLE 4

Scores on the Optimized Vestibular Dysfunction Symptom Scale in Group P and Group N

	P		N	
Vestibular function	**Abnormal**	**Normal**	**Abnormal**	**Normal**
Mean	1.34	1.10	0.98	0.91
s.d.	0.39	0.13	0.13	0.05
Number of individuals	16	13	18	27
Difference				
Eff Size	0.80		0.76	
t	2.14		2.50	
p (one-tailed)	0.02		0.009	

excluded the following symptoms: "trembling," "pressure in chest," "knot in stomach," "feeling disconnected," "ear fullness," and "ringing." The lack of usefulness of the ear-related symptoms ("fullness" and "ringing") seems noteworthy. Furthermore, one item consistently was associated with normal vestibular function in both groups, albeit below criterion level: "heart palpitations." This item was therefore included in the optimized vestibular dysfunction symptom scale but given a weight of –1. The new "optimized vestibular symptom scale," able to discriminate between individuals with and without vestibular dysfunction in both groups P and N, is presented in Table 3. It includes many symptoms not normally considered "vestibular," such as "hot or cold flashes" and "lump in throat."

The means and standard deviations for groups P and N on this optimized symptom scale are presented in Table 4. The effect size of the vestibular subgroup differences on the optimized vestibular symptom scale are close to 0.8 ("large") in both groups.

4.1.2 Vestibular Dysfunction, Autonomic Symptoms, and Panic Attacks

The above results permit the following two conclusions: first, vestibular dysfunction is associated not only with vestibular symptoms, but also with autonomic symptoms. This conclusion is based primarily on the results with group N. The finding of autonomic symptoms such as "hot or cold flashes" being a marker of vestibular dysfunction — at least in individuals who do not have autonomic symptoms for other reasons such as anxiety — appears consistent with the vestibular-autonomic connections described earlier (Section 3).

Second, vestibular dysfunction may play a role in determining the manifestations of panic disorder. A simplified model of the relationship between vestibular dysfunction and panic attacks that would account for our finding is shown in Figure 2. The autonomic

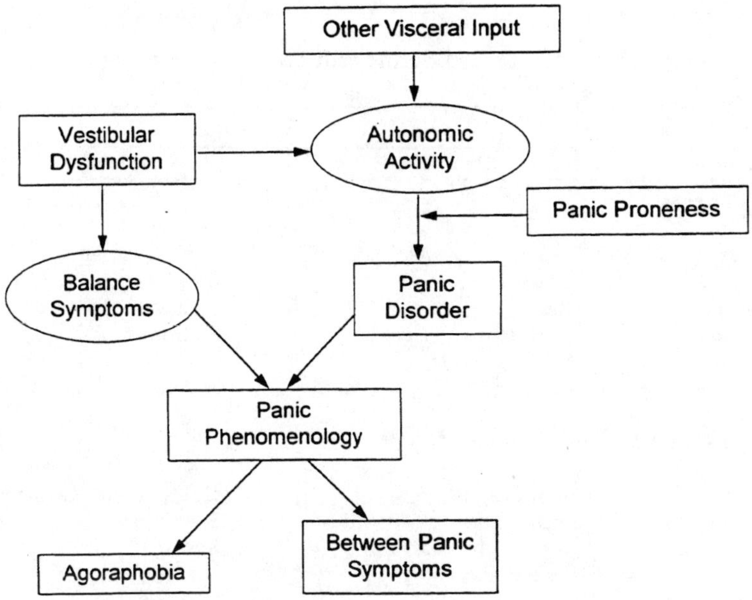

FIGURE 2
The role of vestibular dysfunction in panic disorder.

symptoms related to vestibular dysfunction, when combined with other visceral input, might, in some cases, trigger panic attacks. However, not all individuals with vestibular dysfunction develop panic disorder. In fact none of the individuals in group N had a history of panic attacks. A predisposition for panic (i.e., panic proneness) appears to be required for panic attacks to develop. Panic proneness might be related to a genetic predisposition for anxiety. Thus, according to this model, vestibular dysfunction is not a sufficient or even necessary condition for panic disorder but may constitute a risk factor for its manifestation.

Furthermore, once the panic disorder is established, vestibular dysfunction may play a role in the phenomenology of the disorder. Patients in group P with vestibular dysfunction exhibited more vestibular symptoms *between attacks* than those without vestibular dysfunction. Furthermore, as we will discuss further in the next section, vestibular dysfunction may contribute to the development of agoraphobia. On the other hand, the symptoms *during* the panic attacks themselves are largely unaffected by the presence of vestibular dysfunction. The panic symptoms per se seem to be related to other factors and may be so intense that they mask the symptomatic manifestation of vestibular dysfunction.

4.2 Space and Motion Discomfort

The SitQ was administered not only to the psychiatric patients discussed in the previous section, but also to a sample of ENT patients evaluated for dizziness or hearing loss. Both of the two space and motion discomfort scales (SMD-I and SMD-II) were associated with significant differences when comparing patients with and without vestibular dysfunction in the ENT sample[75] as well as group P.[69] There were no significant differences between vestibular and non-vestibular subjects in group N on the SMD-I or SMD-II. However, as we will see below, the lack of difference with respect to the SMD-I was related to the results of one particular item.

Exploratory item analyses were done for both the SMD-I and the SMD-II. The latter proved relatively uninteresting and will not be presented here. Table 5 shows which of the

TABLE 5

SMD-I Items Differentiating Among Individuals with Normal and Abnormal Vestibular Dysfunction in Groups ENT, P, and N

	Characteristic	
Situation	**High SMD**	**Low SMD**
ENT Patients		
Tunnels	Looking at lights flashing by	Looking at end
Riding	Back seat	Front seat
Elevators	Glass	Standard
Elevators	Moving	Stationary
Riding	Winding roads	Straight roads
Buses	Standing still	Moving
Normal Control Subjects (Group N)		
Elevators	Stopping	Going at steady speed
Elevators	Stopping	Starting
Buses	Standing on platform	Sitting
Riding	Narrow roads	Wide roads
Riding	Bumpy roads	Smooth roads
[*Riding in a car*	*Downhill*	*Uphill*]
Panic Disorder (Group P)		
Supermarkets	Looking at items on shelf	End of aisle
Tunnels	Curved	Straight
Movies	Sitting in front	Sitting in back
Buses	Standing on platform	Sitting
Riding in car	Crooked roads	Straight roads
Riding	Downhill	Uphill
Riding	Back seat	Front seat
Escalators	Up	Down
[*Fields*	*Middle*	*Edge*]

Note: Items in *italics* predicted in the wrong direction for this group.

SMD-I items were associated with noteworthy differences between individuals with and without vestibular dysfunction within each of the study groups. Items that are italicized and bracketed predicted in the "wrong" direction, that is, contrary to our prediction, they showed a higher score for individuals with normal vestibular function. In the ENT sample, six of the items discriminated between those with normal vs. those with abnormal vestibular dysfunction in the predicted direction. For group P, eight items showed differences between those with normal and those with abnormal vestibular function in the predicted direction, while one item (field: middle vs. edge) was associated with a difference in the unpredicted direction. That is, contrary to our original prediction, patients with vestibular dysfunction reported more discomfort standing near the "edge" as opposed to the "middle" of an open field.

As already mentioned, for Group N, the SMD-I total score showed no significant differences between vestibular and non-vestibular subjects. However, in this group, seven items showed an association with vestibular dysfunction in the predicted direction, whereas one item, "riding downhill vs. uphill," exhibited a difference in the unpredicted direction (i.e., subjects in group N with vestibular dysfunction reported more discomfort riding uphill than downhill).

4.2.1 *Optimized SMD-I Scale*

We were able to derive an optimized SMD-I scale that would be associated with significant subgroup differences in all three groups (P, N, ENT). Three of the original 18 items of the SMD-I were eliminated because they were associated with erroneous predictions at or below criterion level in some groups ("fields: standing in the middle vs. edge," "fields: open vs. enclosed," and "riding: downhill vs. uphill"). Furthermore, we included one item from the Ag-I ("riding: limited access roads vs. unlimited access roads"), because this item showed a substantial association with vestibular dysfunction in group P. Including this item can also be justified on theoretical grounds, because limited-access roads often show the wide and featureless quality that has been described, for example, for the "motorists' vestibular disorientation syndrome." The items of the "optimized" 16-item SMD-I are listed in Table 6.

A statistical comparison of the scores according to the optimized version of the SMD-I for each vestibular subgroup within groups P, N, and ENT appears in Table 7. As can be seen, there are significant differences in the expected direction between the vestibular subgroups in each of the major groups; the effect sizes of the subgroup difference ranged between 0.6 and 0.8. It can also be seen that the means for group N in both subgroups are lower than the ones for groups P and ENT. Thus, any prediction of vestibular dysfunction based on the scale would still need to be made in the context of the individual's concomitant diagnosis. The optimized version of the SMD-I represents an improvement, a more "robust" version of the SMD-I. It required only four changes among the 18 items of the original scale. Nevertheless, the validity of this revised scale obviously would need to be established on a validation sample of new subjects.

4.2.2 *Vestibular Dysfunction, Space and Motion Discomfort, and Phobic Conditions*

The results of the preceding section indicated that space and motion discomfort, as measured by the optimized SMD-I, was a marker of vestibular dysfunction in ENT patients, in patients with panic disorder (group P), and even in group N. Thus, the individuals in group N with vestibular dysfunction had an increased prevalence both of autonomic symptoms (as discussed in the earlier section) and of space and motion discomfort. It appears that vestibular dysfunction associated with space and motion discomfort is a common phenomenon that, when combined with other predisposing factors, could lead to certain anxiety disorders. In the absence of predisposing factors, vestibular dysfunction may merely lead to the subtle space and motion symptoms observed in group N.

Figure 3 illustrates the anxiety conditions in which space and motion discomfort might play a role. First, as already discussed, space and motion discomfort might be a mechanism by which patients with panic disorder develop agoraphobia. Such a mechanism would account for the large difference in the prevalence of vestibular dysfunction between patients with panic disorder with and without agoraphobia found in our recently completed study.[69] Space and motion discomfort might also explain the association between agoraphobic avoidance and posturographic abnormalities found by Yardley et al.[143] If these results prove to be robust, vestibular dyfunction would contribute to the solution of the enigma, discussed earlier, of why some panic patients develop agoraphobia and others do not.

In Figure 3, we also raise the possibility that vestibular dysfunction associated with space and motion discomfort may play a role in other anxiety disorders besides panic disorder, namely, agoraphobia without panic or height phobia (see Section 1). A vestibular

TABLE 6
Items in Optimized SMD-I Scale

	Characteristic	
Situation	**SMD Aspect**	**Non-SMD Aspect**
Riding in a Car		
1.	Bumpy roads	Smooth
2.	Winding roads	Straight
3.	Narrow roads	Wide
4.	Limited access roads	Unlimited access roads
5.	Back seat	Front seat
6.	Reading	Looking out window
Riding in a bus		
7.	Standing on platform	Sitting
8.	Moving	Standing still
Supermarkets		
9.	Looking at items on shelf	Looking at end of aisle
Tunnels		
10.	Curved	Straight
11.	Looking at lights on side	Looking at light at end
Movies		
12.	Front	Back
Elevators		
13.	Moving	Stationary
14.	Glass	Standard
15.	Stopping	Riding at steady speed
16.	Stopping	Starting

involvement in height phobia was already suggested 40 years ago by Pogany.[110] We have evaluated a number of patients with agoraphobia without panic as well as patients with height phobia and found vestibular abnormalities (unpublished).

To test the theory that space and motion discomfort observed in patients with anxiety disorders might be associated with an increased sensitivity to visual balance cues, we selected patients with anxiety disorders and clinically significant space and motion discomfort (including fear of heights) for a study in which we assessed the effect of optic flow stimuli on body sway.[74] Subjects stood on a force platform while monocularly viewing a screen on which moving visual environments were projected. Some of the visual scenes would move in a repetitive pattern at a frequency of 0.3 Hz. Anterior-posterior body sway was recorded and decomposed into frequency bands by spectral analysis. Sway in the 0.3-Hz frequency band was used as an indicator for sway elicited by the movements of the visual scene. We found significant differences between patients and controls with respect to (1) overall levels of sway and (2) movement-induced sway, and (3) a trend toward greater persistence of movement-induced sway after cessation of the moving optic flow stimuli.

TABLE 7

Revised (Optimized) SMD-I Scale: Scores for Normals and Abnormals within Three Study Groups

	P[a]		N[b]		V/H[c]	
Vestibular Function	**Abn[d]**	**Norm[e]**	**Abn[d]**	**Norm[e]**	**Abn[d]**	**Norm[e]**
Mean	3.7	1.5	1.6	0.8	3.4	1.4
s.d.	3.3	3.0	1.3	1.3	2.9	2.0
Number of individuals	16	14	26	17	32	29
Difference						
Effect size[f]	0.68		0.62		0.81	
t	1.91		1.98		3.2	
p (one-tailed)[g]	<.05		<.05		<.01	

[a] P, Panic disorder with no or mild agoraphobia.

[b] N, Normal subjects without anxiety disorders.

[c] V/H, Balance patients with vestibular dysfunction (V) vs. patients with hearing loss (H, excluding two probable agoraphobics).

[d] Abn, Vestibular function abnormal (at least one abnormality).

[e] Norm, Vestibular function normal.

[f] Effect size, difference divided by standard deviation of the difference.

[g] p, One-tailed probability.

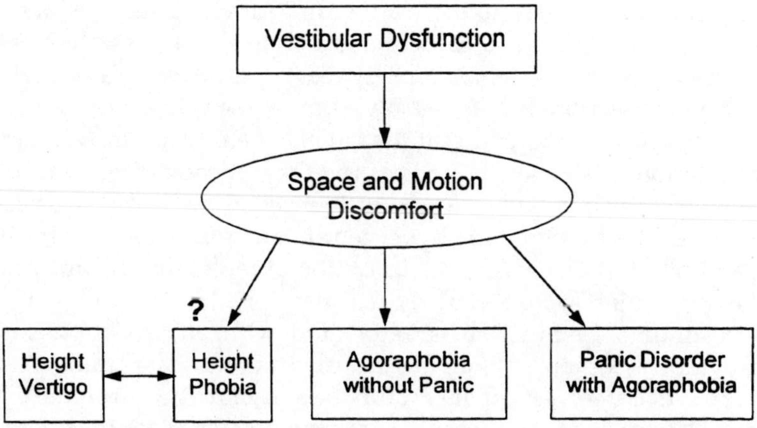

FIGURE 3
Space and motion discomfort and anxiety disorders.

This experiment thus appears to confirm that space and motion discomfort in anxiety disorders is associated with increased sensitivity to visual balance stimuli. However, because normal controls were used in the experiment, the effect of space and motion discomfort was confounded with the presence or absence of anxiety disorders. A further step toward testing the unique role of space and motion discomfort would be to compare optic flow sensitivity in patients with anxiety disorders with space and motion discomfort to patients without space and motion discomfort, and to compare healthy normals with space and motion discomfort to healthy normals without space and motion discomfort.

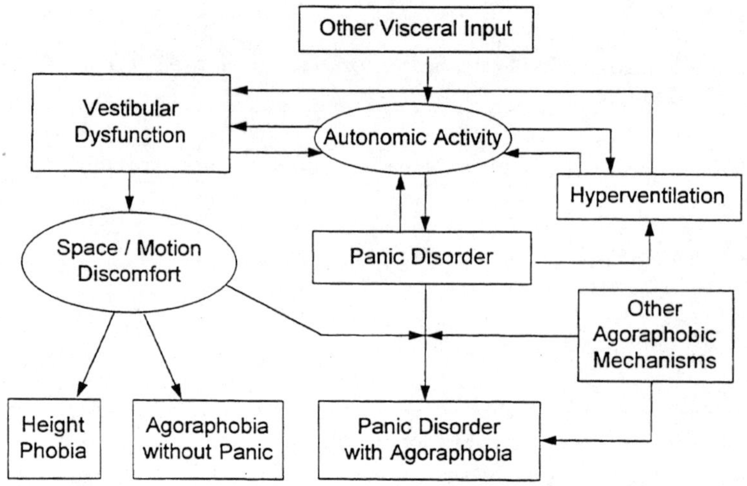

FIGURE 4
Overview of relationships between vestibular dysfunction and certain anxiety disorders.

5 CONCLUSION

The relationships among global symptoms, situational specificity of symptoms, and vestibular dysfunction observed in our studies provide clues to the role of the vestibular system in certain anxiety disorders. Vestibular dysfunction, by virtue of vestibular-autonomic connections, or due to convergence with other visceral afferents, may trigger panic attacks. Also, mediated by space and motion discomfort, vestibular dysfunction facilitates certain phobic avoidance patterns including agoraphobia and possibly height phobia.

The role of the vestibular system in anxiety disorders nevertheless has been oversimplified in the previous section. Figure 4 shows a more complete overview. In particular, in the previous section we assumed that the causal chain of events always begins with vestibular dysfunction. However, in light of the theoretical background discussed earlier, a more likely scenario is one in which influences in the reverse direction also take place. For example, the behavior of hyperventilation may affect vestibular function. Furthermore, activity in the locus coeruleus or dorsal raphe nucleus may be propagated to parts of the vestibular system. Finally, as mentioned earlier, this model does not explain why vestibular dysfunction is associated with an anxiety disorder in some individuals and not in others. Other factors, such as anxious predisposition, are involved. However, even if vestibular dysfunction is not the ultimate cause of panic disorder, height phobia, or agoraphobia, it nevertheless might contribute to the phenomenology of some of them.

6 SUMMARY

This chapter reviews possible connections between vestibular dysfunction and anxiety disorders. Psychologically mediated linkages include somatopsychic effects of vestibular symptoms as well as psychosomatic effects on vestibular function (e.g., the effect of hyperventilation). Situationally specific vestibular symptoms may result in space and motion discomfort, which may contribute to the development of agoraphobia. Neurologically mediated linkages include (1) reciprocal connections between the locus coeruleus and the vestibular nuclei, particularly the lateral vestibular nucleus; (2) serotonergic pathways involving the dorsal raphe nucleus; (3) pathways from the vestibular nuclei to brainstem circuitry mediating vestibular-autonomic and vestibular-respiratory reflexes; and (4) convergence of vestibular afferents and other visceral afferents in the parabrachial nucleus, which also receives input from the limbic system and amygdala.

We present illustrative data relating normal vs. abnormal laboratory vestibular dysfunction to symptomatology in normal controls, otoneurological patients, and patients with panic disorder. We concluded that (1) vestibular dysfunction may be a risk factor for panic attacks; (2) vestibular dysfunction in patients with panic disorder contributes to the patients' symptom profiles, in particular vestibular symptoms between panic attacks; (3) vestibular dysfunction may be a risk factor for the development of agoraphobia as a complication of panic disorder; (4) vestibular dysfunction associated with space and motion discomfort may contribute to other anxiety disorders, in particular height phobia and agoraphobia without panic.

ACKNOWLEDGMENTS

We would like to thank Carey Balaban, Ph.D., for his input concerning neuroanatomy, Sue Russell, M.F.A., for her editorial assistance, and David Williams, M.S., for help in preparing the figures.

REFERENCES

1. Abelson, J. L., Glitz, D., Cameron, O. G., Lee, M. A., Bronzo, M., and Curtis, G. C., Blunted growth hormone response to clonidine in patients with generalized anxiety disorder, *Arch. Gen. Psychiatr.*, 48, 157, 1991.
2. Albus, M., Zahn, T. P., and Breier, A., Anxiogenic properties of yohimbine. I. Behavioral, physiological and biochemical measures, *Eur. Arch. Psychiatr. Clin. Neuroscience*, 241, 337, 1992.
3. Albus, M., Zahn, T. P., and Breier, A., Anxiogenic properties of yohimbine. II. Influence of experimental set and setting, *Eur. Arch. Psychiatry Clin. Neurosci.*, 241, 345, 1992.
4. Alden, M., Besson, J., and Bernard, J., Organization of the efferent projections from the pontine parabrachial area to the bed nucleus of the stria terminalis and neighboring regions: APHA I study in the rat, *J. Comp. Neurol.*, 341, 289, 1994.
5. American Psychiatric Association. *Diagnostic and Statistical Manual of Mental Disorders*, 4th ed., American Psychiatric Association, Washington, D.C., 1994.
6. Aronson, T. A. and Logue, C. M., On the longitudinal course of panic disorder: developmental history and predictors of phobic complications, *Compr. Psychiatr.*, 28, 344, 1987.
7. Asmundson, G. J. G. and Stein, M. B., Vagal attenuation in panic disorder: an assessment of parasympathetic nervous system function and subjective reactivity to respiratory manipulations, *Psychosom. Med.*, 56, 187, 1994.
8. Asnis, G. M. and van Praag, H. M., The norepinephrine system in panic disorder, in *Panic Disorder: Clinical, Biological, and Treatment Aspects*, Asnis, G. M. and van Praag, H. M., Eds., John Wiley & Sons, New York, 1995, 119.
9. Aston-Jones, G., Ennis, M., Pieribone, V. A., Nickell, W. T., and Shipley, M. T., The brain nucleus locus coeruleus: restricted afferent control of a broad efferent network, *Science*, 234, 734, 1986.
10. Balaban, C. D., Neuroanatomical and physiological substrate relating panic disorder and vestibular dysfunction, paper presented at the *Anxiety Disorders Assoc. Am. 15th Annu. Conf.*, Pittsburgh, April 7, 1995.
11. Balaban, C. D., Vestibular nucleus projections to the parabrachial nucleus: implications for vestibular influences on the autonomic nervous system, submitted for publication.
12. Balaban, C. D. and Beryozkin, G., Vestibular nucleus projections to nucleus tractus solitarius and the dorsal motor nucleus of the vagus nerve: potential substrates for vestibulo-autonomic interactions, *Exp. Brain Res.*, 98, 200, 1994.
13. Barnes, C. D., Manzoni, D., Pompeiano, O., Stampacchia, G., and D'Ascanio, P., Responses of locus coeruleus and subcoeruleus neurons to sinusoidal neck rotation in decerebrate cat, *Neuroscience*, 31, 371, 1989.

14. Bernard, J. F., Huang, G. F., and Besson, J. M., The parabrachial area: electrophysiological evidence for an involvement in visceral nociceptive processes, *J. Neurophysiol.*, 71, 1646, 1994.
15. Blier, P., Chaput, Y., and de Montigny, C., Long-term 5-HT reuptake blockade, but not monoamine oxidase inhibition, decreases the function of terminal 5-HT autoreceptors: an electrophysiological study in the rat brain, *Naunyn-Schmiedebergs Arch. Pharmacol.*, 337, 246, 1988
16. Bradwejn, J. and Koszycki, D., Cholecystokinin and panic disorder, in *Panic Disorder: Clinical, Biological, and Treatment Aspects*, Asnis, G. M. and van Praag, H. M., Eds., John Wiley & Sons, New York, 1995, 233.
17. Bradwejn, J., Koszycki, D., Couetoux du Tertre, A., van Megen, H., den Boer, J., Westenberg, H., and Annable, L., The panicogenic effects of cholecystokinin-tetrapeptide are antagonized by L-365, 260, a central cholecystokinin receptor antagonist, in patients with panic disorder, *Arch. Gen. Psychiatr.*, 51, 486, 1994.
18. Breier, A., Charney, D. S., and Heninger, G. R., Agoraphobia with panic attacks: development, diagnostic stability, and course of illness, *Arch. Gen. Psychiatr.*, 43, 1029, 1986.
19. Brown, T. G., Die Atembewegungen des Frosches und ihre Beeinflussung durch die nervosen Zentren und durch das Labyrinth, *Pfluger's Archiv.*, 13, 192, 1909.
20. Bystritsky, A. and Shapiro, D., Continuous physiological changes and subjective reports in panic patients: a preliminary methodological report, *Biol. Psychiatr.*, 32, 766, 1992.
21. Cameron, O. G. and Hudson, C. J., Influence of exercise on anxiety level in patients with anxiety disorders, *Psychosomatics*, 27, 720, 1986.
22. Cassano, G. B., Petracca, A., Perugi, G., Toni, C., Tundo, A., and Roth M., Derealization and panic attacks: a clinical evaluation on 150 patients with panic disorder/agoraphobia, *Compr. Psychiatr.*, 30, 5, 1989.
23. Cedarbaum, J. M. and Aghajanian, G. K., Afferent projections to the rat locus coeruleus as determined by a retrograde tracing technique, *J. Comp. Neurol.*, 178, 1, 1978.
24. Chamberlin, N. L. and Saper, C. B., Topographic organization of cardiovascular responses to electrical and glutamate microstimulation of the parabrachial nucleus in the rat, *J. Comp. Neurol.*, 326, 245, 1992.
25. Chamberlin, N. L. and Saper, C. B., Topographic organization of respiratory responses to glutamate microstimulation of the parabrachial nucleus in the rat. *J. Neuroscience*, 14, 6500, 1994.
26. Charney, D. S. and Heninger, G. R., Abnormal regulation of noradrenergic function in panic disorders: effects of clonidine in healthy subjects and patients with agoraphobia and panic disorder, *Arch. Gen. Psychiatr.*, 43, 1042, 1986.
27. Charney, D. S., Heninger, G. R., and Breier, A., Noradrenergic function in panic anxiety: effects of yohimbine in healthy subjects and patients with agoraphobia and panic disorder, *Arch. Gen. Psychiatr.*, 41, 751, 1984.
28. Charney, D. S., Krystal, J. J., Southwick, S. M., Nagy, L. M., Woods, S. W., and Heninger, G. R., Serotonin function in panic and generalized anxiety disorders, *Psychiatr. Ann.*, 20, 593, 1990.
29. Charney, D. S., Woods, S. W., and Heninger, G. R., Noradrenergic function in generalized anxiety disorder. Effects of yohimbine in healthy subjects and patients with generalized anxiety disorder, *Psychiatr. Res.*, 27, 173, 1989.
30. Charney, D. S, Woods, S. W., Krystal, J. H., Nagy L. M., and Heninger, G. R., Noradrenergic neuronal dysregulation in panic disorder: the effects of intravenous yohimbine and clonidine in panic disorder patients, *Acta Psychiatr. Scand.*, 86, 273, 1992.
31. Chikamori, Y., Sasa, M., Fujimoto, S., Takaori, S., and Matsuoka, I., Locus coeruleus-induced inhibition of dorsal cochlear nucleus neurons in comparison with lateral vestibular nucleus neurons, *Brain Res.*, 194, 53, 1980.
32. Clark D. B., Hirsch, B. E., Smith, M. G., Furman, J. M., and Jacob, R. G., Panic in otolaryngology patients presenting with dizziness or hearing loss, *Am. J. Psychiatr.*, 151, 1223, 1994.
33. Clark, D. M., Cognitive mediation of panic attacks induced by biological challenge tests, *Adv. Behav. Res. Ther.*, 15, 75, 1993.
34. Cohen, J., *Statistical Power Analysis for the Behavioral Sciences*, Lawrence Erlbaum, New York, 1988.
35. Collins, W. E., Habituation of vestibular responses with and without visual stimulation, in *Vestibular System, Part 2, Psychophysics, Applied Aspects and General Interpretations*, Kornhuber, H. H., Ed., Springer-Verlag, Berlin, 1974, 369.
36. DeJong, G. M. and Bouman T. K., Panic disorder: a baseline period: predictability of agoraphobic avoidance behavior, *J. Anx. Dis.*, 9, 185, 1995.
37. Delito, J. A., Perugi, G., Maremmani, I., Mignani, V., and Cassano, G. B., The importance of separation anxiety in the differentiation of panic disorder from agoraphobia, *Psychiatr. Dev.*, 3, 227, 1986.
38. de Montigny, C., Chaput, Y., and Blier, P., Modification of serotonergic neuron properties by long-term treatment with serotonin reuptake blockers, *J. Clin. Psychiatr.*, 51 (Suppl. B), 4, 1990.

39. Dubovsky, S. L., Generalized anxiety disorder: new concepts and psychopharmacologic therapies, *J. Clin. Psychiatr.*, 51, 3, 1990.
40. D'Urzo, A. D., Jhirad, R., Jenne, H., Avendano A., Rubenstein, I., D'Costa, M., and Goldstein, R. S., Effect of caffeine on ventilatory responses to hypercapnia, hypoxia, and exercise in humans, *J. Appl. Physiol.*, 68, 322, 1990.
41. Eaton, W. W. and Keyl, P. M., The epidemiology of panic, in *Panic Disorder: Clinical, Biological, and Treatment Aspects*, Asnis, G. M. and van Praag H. M., Eds., John Wiley & Sons, New York, 1995, 50.
42. Eison, M. S., Serotonin: a common neurobiologic substrate in anxiety and depression, *J. Clin. Psychopharmacol.*, 10 (Suppl.), 26S, 1990.
43. Feil, K. and Herbert, H., Topographic organization of spinal and trigeminal somatosensory pathways to the rat parabrachial and Kolliker-Fuse nuclei, *J. Comp. Neurol.*, 353, 506, 1995.
44. Fleming, B. and Faulk, A., Discriminating factors in panic disorder with and without agoraphobia, *J. Anx. Dis.*, 3, 209, 1989.
45. Fulwiler, C. E. and Saper, C. B., Subnuclear organization of the efferent connections of the parabrachial nucleus in the rat, *Brain Res. Rev.*, 7, 1984.
46. Fung S. J., Chan, J. Y. H., Manzoni, D., White, S. R., Lai, Y., Strahlendorf, H. K., Zhuo, H., Liu, R., Reddy, V. K., and Barnes, C. D., Cotransmitter-mediated locus coeruleus action on motoneurons, *Brain Res. Bull.*, 35, 423, 1994.
47. Fung, S. J., Reddy, V. K., Bowker, R. M., and Barnes, C. D., Differential labeling of the vestibular complex following unilateral injections of horseradish peroxidase into the cat and rat locus coeruleus, *Brain Res.*, 401, 347, 1987.
48. Furman, J. M. R., O'Leary, D. P., and Wolf, J. W., Changes in the horizontal vestibulo-ocular reflex of the Rhesus monkey with behavioral pharmacological alerting, *Brain Res.*, 206, 490, 1981.
49. Garvey, M. J. and Tuason, V. B., The relationship of panic disorder to agoraphobia, *Compr. Psychiatr.*, 25, 529, 1984.
50. Gentil, V., Lotufo-Neto, F., Andrade, L., Cordas, T., Bernik, M., Ramos, R., Maciel, L., Migakawa, E., and Gorenstein, C., Clomipramine, a better reference drug for panic/agoraphobia. I. Effectiveness comparison with imipramine, *J. Psychopharmacol.*, 7, 316, 1993.
51. Gorman, J. M., Battista, D., Goetz, R. R., Dillon, D. J., Liebowitz, M. R., Fyer, A. J., Kahn, J. P., Sandberg, D., and Klein, D. F., A comparison of sodium bicarbonate and sodium lactate infusion in the induction of panic attacks, *Arch. Gen. Psychiatr.*, 46, 145, 1989.
52. Gorman, J. M., Fyer, M. R., Goetz, R., Askanazi, J., Liebowitz, M. R., Fyer, A. J., Kinney, J., and Klein, D. F., Ventilatory physiology of patients with panic disorder, *Arch. Gen. Psychiatr.*, 45, 31, 1988.
53. Gorman, J. M., Liebowitz, M. R., Fyer, A. J., and Stein, J., A neuroanatomical hypothesis for panic disorder, *Am. J. Psychiatr.*, 146, 148, 1989.
54. Gorman, J. M., Papp, L. A., Coplan, J. D., Martinez, J. M., Lennon, S., Goetz, R. R., Ross, D., and Klein, D. F., Anxiogenic effects of CO_2 and hyperventilation in patients with panic disorder. *Am. J. Psychiatr.*, 151, 547, 1994.
55. Grundemar, L. and Hakanson, R., Neuropeptide Y effector systems: perspectives for drug development, *Trends Pharmacol. Sci.*, 151, 153, 1994.
56. Hamon, M., Neuropharmacology of anxiety: perspectives and prospects, *Trends Pharmacol. Sci.*, 15, 36, 1994.
57. Harro, J., Vassar, E., and Bradwejn, J., CCK in animal and human research on anxiety, *Trends Pharmacol. Sci.*, 14, 244, 1993.
58. Heninger, G. B. R., Charney, D. S., and Price, L. H., Alpha2 adrenergic receptor sensitivity in depression: the plasma MHPG, behavioral, and cardiovascular responses to yohimbine, *Arch. Gen. Psychiatr.*, 45, 718, 1988.
59. Herbert, H. and Saper, C. B., Cholecystokinin-, galanin-, and corticotropin-releasing factor-like immunoreactive projections from the nucleus of the solitary tract to the parabrachial nucleus in the rat, *J. Comp. Neurol.*, 293, 591, 1990.
60. Hjorth, S., Serotonin 5-HT_{1A} autoreceptor blockade potentiates the ability of the 5-HT reuptake inhibitor citalopram to increase nerve terminal output of 5-HT *in vivo*, a microdialysis study, *J. Neurochem.*, 60, 776, 1993.
61. Hoehn-Saric, R., McLeod, D. R., and Zimmerli, W. D., Psychophysiological response patterns in panic disorder, *Acta Psychiatr. Scand.*, 83, 4, 1991.
62. Hoffman, D. L., O'Leary, D. P., and Munjack, D. J., Autorotation test abnormalities of the horizontal and vertical vestibulo-ocular reflexes in panic disorder, *Otolaryngol. Head Neck Surg.*, 110, 259, 1994.
63. Hommer, D. W., Skolnick, P., and Paul, S. M., The benzodiazepine/GABA receptor complex and anxiety, in *Psychopharmacology: The Third Generation of Progress*, Meltzer, H. Y., Ed., Raven Press, New York, 1987, 977.

64. Horst, H., Moga, M. M., and Saper, C. B., Connections of the parabrachial nucleus with the nucleus of the solitary tract and the medullary reticular formation in the rat, *J. Comp. Neurol.*, 293, 540, 1990.
65. Irwin, M., Brown, M., Patterson, T., Hauger, R., Mascovich, A., and Grant, I., Neuropetide T and natural killer cell activity: findings in depression and Alzheimer caregiver stress, *FASEB J.*, 5, 3106, 1991.
66. Jacob, R. G., Panic disorder and the vestibular system, *Psychiatr. Clin. North Am.*, 11, 361, 1988.
67. Jacob, R. G., Furman, J. M., and Balaban, C. D., Psychiatric aspects of vestibular disorders, in *Handbook of Neurotology/Vestibular System*, Baloh, R. W. and Halmagyi, G. M., Eds., Oxford University Press, New York, in press.
68. Jacob, R. G., Furman, J. M. R., Clark, D. B., Durrant, J. D., and Balaban C. B., Psychogenic dizziness, in *The Vestibulo-Ocular Reflex and Vertigo*, Sharpe, J. A. and Barber, H. O., Eds., Raven Press, New York, 1993, 305.
69. Jacob, R. G., Furman, J. M., Durrant, J. D., and Turner, S. M., Panic, agoraphobia and vestibular dysfunction: clinical test results, *Am. J. Psychiatr.*, submitted.
70. Jacob, R. G. and Lilienfeld, S. O., Panic disorder: diagnosis, medical and psychological assessment, in *Panic Disorder and Agoraphobia: A Guide for the Practitioner*, Walker, J. R. and Ross, C. A., Eds., Brooks/Cole, Pacific Groves, CA, 1991, 16.
71. Jacob, R. G., Lilienfeld, S. O., Furman, J. M. R., Turner, S. M., Space and motion phobia in panic disorder with vestibular dysfunction, *J. Anx. Dis.*, 3, 117, 1988.
72. Jacob, R. G., Moller, M. B., Turner, S. M., and Wall, C. W., III, Otoneurological examination of panic disorder and agoraphobia with panic attacks: a pilot study, *Am. J. Psychiatr.*, 142, 715, 1985.
73. Jacob R. G. and Rapport, M. D., Panic disorder, in *Behavioral Treatment of Anxiety Disorders*, Turner, S. M., Ed., Plenum Press, New York, 1984, 187.
74. Jacob, R. G., Redfern, M. S., and Furman, J. M., Optic flow-induced sway in anxiety disorders associated with space and motion discomfort, *J. Anx. Dis.*, 9, 411, 1995.
75. Jacob, R. G., Woody, S. R., Clark, D. B., Lilienfeld, S. O., Hirsch, B. E., Kucera, G. D., Furman, J. M., and Durrant, J. D., Discomfort with space and motion: a possible marker of vestibular dysfunction assessed by the Situational Characteristics Questionnaire, *J. Psychopathol. Behav. Assess.*, 15, 299, 1993.
76. Kahn, R. S., Westenberg, H. G. M., and Moore, C., Increased serotonin function and panic disorder, in *Panic Disorder: Clinical, Biological, and Treatment Aspects*, Asnis, G. M. and van Praag, H. M., Eds., John Wiley & Sons, New York, 1995, 151.
77. Kaitin, K. I., Bliwise, D. L., Gleason, C., Nino-Murcia, G., Dement, W. C., and Libet, B., Sleep disturbance produced by electrical stimulation of the locus coeruleus in a human subject, *Biol. Psychiatr.*, 21, 710, 1986.
78. Katerndahl, D. A., Factors in the panic-agoraphobia transition, *J. Am. B. Fam. Pract.*, 2, 10, 1989.
79. Katenrndahl, D. A., The sequence of panic symptoms, *J. Fam. Pract.*, 26, 49, 1988.
80. Katerndahl, D. A., Gabel, L. L., and Monk, J. S., Comparative symptomatology of phobic and nonphobic panic attacks, *Fam. Pract. Res. J.*, 6, 106, 1985.
81. Kirsten, E. B. and Sharma, J. N., Characteristics and response differences to iontophoretically applied norepinephrine, D-amphetamine and acetylcholine on neurons in the medial and lateral vestibular nuclei of the cat, *Brain Res.*, 112, 77, 1976.
82. Kishimoto, T., Sasa, M., and Takaori, S., Inhibition of lateral vestibular nucleus neurons by 5-hydroxytryptamine derived from the dorsal raphe nucleus, *Brain Res.*, 553, 229, 1991.
83. Klein, D. F., False suffocation alarms, spontaneous panics, and related conditions, *Arch. Gen. Psychiatr.*, 50, 306, 1993.
84. Laraia, M. T., Stuart, G. W., Frye, L. H., Lydiard, R. B., and Ballenger, J. C., Childhood environment of women having panic disorder with agoraphobia, *J. Anx. Dis.*, 8, 1, 1994.
85. LeDoux, J. E., Emotion, memory and the brain, *Sci. Am.*, 50, June 1994.
86. Libet, B. and Gleason C. A., The human locus coeruleus and anxiogenesis, *Brain Res.*, 634, 178, 1994.
87. Licata, F., LiVolsi, G., Maugeri, G., and Santangelo F., Effects of 5-hydroxytryptamine on the firing rates of neurons of the lateral vestibular nucleus in the rat, *Exp. Brain Res.*, 79, 293, 1990.
88. Licata, F., LiVolsi, G., Maugeri, G., Ciranna, L., and Santangelo, F., Effects of noradrenaline on the firing rate of vestibular neurons, *Neuroscience*, 53, 149, 1993.
89. Licata, F., LiVolsi, G., Maugeri, G., Ciranna, L., and Santangelo, F., Serotonin-evoked modifications of the neuronal firing rate in the superior vestibular nucleus: a microiontophoretic study in the rat, *Neuroscience*, 1993, 52, 941, 1993.
90. Lydiard, R. B., Preliminary results of an open, fixed-dose study of desipramine in panic disorder, *Psychopharmacol. Bull.*, 23, 139, 1987.
91. Maddock, R. J., Carter, C. S., Magliozzi, J. R., and Gietzen, D. W., Evidence that decreased function of lymphocyte beta adrenoreceptors reflects regulatory and adaptive processes in panic disorder with agoraphobia, *Am. J. Psychiatr.*, 150, 1219, 1993.

92. Maddock, R. J., Gietzen, D. W., and Goodman, T. A., Decreased lymphocyte beta- adrenoreceptor function correlates with less agoraphobia and better outcome in panic disorder, *J. Affect. Dis.*, 29, 27, 1993.
93. Maier, W. and Buller, R., One-year follow-up of panic disorder: outcome and prognostic factors, *Eur. Arch. Psychiatr. Neurol. Sci.*, 238, 106, 1988.
94. Margraf, J., Ehlers, A., and Roth, W. T., Sodium lactate infusions and panic attacks: a review and critique, *Psychosom. Med.*, 48, 23, 1985.
95. Mavissakalian, M. and Perel, J. M., Imipramine in panic disorder with agoraphobia: dose-ranging and plasma level response relationships, *Am. J. Psychiatr.*, 152, 673, 1995.
96. McNally, R. J. and Louro, C. E., Fear of flying in agoraphobia and simple phobia: distinguishing features, *J. Anx. Dis.*, 6, 319, 1992.
97. Middleton, H. C., Cardiovascular dystonia in recovered panic patients, *J. Affect. Dis.*, 19, 229, 1990.
98. Miller, A. D., Yamaguchi, T., Siniala, M. S., and Yates, B. J., Ventral respiratory group bulbospinal inspiratory neurons participate in vestibular-respiratory reflexes, *J. Neurphysiol.*, 73, 1303, 1995.
99. Mizobe, Y., Yamada, K., and Fujii, I., The sequence of panic symptoms, *Jpn. J. Psychiatr. Neurol.*, 46, 597, 1992.
100. Moga, M. M., Herbert, H., Hurley, K. M., Yasui, Y., Gray, T., and Saper, C. B., Organization of cortical, basal forebrain, and hypothalamic afferents to the parabrachial nucleus in the rate, *J. Comp. Neurol.*, 295, 624, 1990.
101. Monday, L. A. and Tetrault, L., Hyperventilation and vertigo, *Laryngoscope*, 109, 1003, 1980.
102. Norgren, R. and Leonard, C. M., Taste pathways in rat brainstems, *Science*, 173, 1136, 1971.
103. Olson, L. and Fuxe, K., On the projections from the locus coeruleus noradrenaline neurons: the cerebellar innervation, *Brain Res.*, 28, 165, 1971.
104. Page, N. G. R. and Gresty, M. A., Motorist's vestibular disorientation syndrome, *J. Neurol. Neurosurg. Psychiatr.*, 48, 729, 1985.
105. Papp, L. A. and Gorman, J. M., Respiratory neurobiology of panic, in *Panic Disorder: Clinical, Biological, and Treatment Aspects*, Asnis, G. M. and van Praag, H. M., John Wiley & Sons, New York, 1995, 255.
106. Papp, L. A., Klein, D. F., and Gorman, J. M., Carbon dioxide hypersensitivity, hyperventilation, and panic disorder. *Am. J. Psychiatr.*, 150, 1149, 1993.
107. Perel, J. M., Pharmacology of anxiety disorders, in *Anxiety Disorders, Psychological and Pharmacological Treatments*, Prien, R. and Mavissakalian, M., Eds., American Psychiatric Press, in press.
108. Perna, G., Barbini, B., Cocchi, S., Bertani, A., and Gasperini M., 35% CO_2 challenge in panic and mood disorders, *J. Affect. Dis.*, 33, 189, 1995.
109. Perna, G., Cocchi, S., Bertani, A., Arancio, C., and Bellodi L., Sensitivity to 35% CO_2 in healthy first-degree relatives of patients with panic disorder, *Am. J. Psychiatr.*, 152, 623, 1995.
110. Pogany, E., Uber den Hohenschwindel. *Monatsschrift der Ohrenheilkunde und Laryngorhinologie*, 92, 209, 1958.
111. Pollard, C. A. and Cox, G. L., Social-evaluative anxiety in panic disorder and agoraphobia, *Psychol. Rep.*, 62, 323, 1988.
112. Pompeiano, O., Relationship of noradrenergic locus coeruleus neurones to vestibulospinal reflexes, *Prog. Brain Res.*, 80, 329, 1989.
113. Pompeiano, O., Manzoni, D., Barnes, C. D., Stampacchia, G., and D'Ascanio, P., Responses of locus coeruleus and subcoeruleus neurons to sinusoidal stimulation of labyrinth receptors, *Neuroscience*, 35, 227, 1990.
114. Redfern, M. S. and Furman, J. M., Postural sway of patients with vestibular disorders during optic flow, *J. Vest. Res.*, 4, 221, 1994.
115. Redmond, D. E., Alterations in the function of the nucleus locus coeruleus: a possible model for studies of anxiety, in *Animal Models in Psychiatry and Neurology*, Hanin, I. and Usdin, H., Eds., Pergamon Press, Oxford, 1977, 293.
116. Rickels, K. and Schweizer, E. E., Current pharmacotherapy of anxiety and panic, in *Psychopharmacology: The Third Generation of Progress*, Meltzer, H. Y., Ed., Raven Press, New York, 1987, 1193.
117. Sanderson, W. C., Rapee, R. M., and Barlow, D. H., The influence of an illusion of control on panic attacks induced via inhalation of 5.5% carbon dioxide-enriched air, *Arch. Gen. Psychiatr.*, 46, 157–162, 1989.
118. Sara, S. J., Vankov, A., and Herve, A., Locus coeruleus-evoked responses in behaving rats: a clue to the role of noradrenaline in memory, *Brain Res. Bull.*, 35, 457–465, 1994.
119. Schneier, F. R., Fyer, A. J., Martin, L. Y., Ross, D., Mannuzza, S., Liebowitz, M. R., Gorman, J. M., and Klein, D. F., A comparison of phobic subtypes within panic disorder, *J. Anx. Dis.*, 5, 65, 1991.
120. Schuerger, R. J. and Balaban, C. D., The distribution of noradrenergic axons and cell bodies in the vestibular nuclei (VN) of rabbits, *FASEB J.*, 7(Abstr.), A396, 1993.

121. Schuerger, R. J. and Balaban, C. D., Immunohistochemical demonstration of regionally selective projections from locus coeruleus to the vestibular nuclei in rats, *Exp. Brain Res.*, 92, 351, 1993.
122. Schuerger R. J. and Balaban, C. D., Noradrenergic projections to the vestibular nuclei (VN) in monkey, *Soc. Neurosci. Abs.*, 19(Abstr.), 136, 1993.
123. Siegborn, J. and Grant, G., Brainstem projections of different branches of the vestibular nerve. An experimental study by transganglionic transport of horseradish peroxidase in the cat. I. The horizontal ampullar and utricular nerves, *Arch. Ital. Biol.*, 121, 237, 1983.
124. Siegborn, J., Yingcharoen, K., and Grant, G., Brainstem projections of different branches of the vestibular nerve: an experimental study by transganglionic transport of horseradish peroxidase in the cat. II. The anterior and posterior ampullar nerves, *Anat. Embryol.*, 184, 291, 1991.
125. Sklare, D. A., Stein, M. B., Pikus, A. M., and Uhde T. W., Dysequilibrium and audiovestibular function in panic disorder: symptom profiles and test findings, *Am. J. Otol.*, 11, 338, 1990.
126. Sleight, A. J., Smith, R. J., Marsden, D. A., and Palfreyman, M. G., The effects of chronic treatment with amitriptyline and MDL 72394 on the control of 5-HT release *in vivo*, *Neuropharmacology*, 28, 477, 1989.
127. Spiegel, E. A., Vestibular mechanisms, in *Medical Physics*, Glasser, O., Ed., Year Book, Chicago, 1944, 1638.
128. Spiegel, E. A. and Démétriades, Th. D., Beitrage zum studium des vegetativen nervensystems. III. Mitteilung. Der einfluss des vestibularapparates auf das gefassystem, *Pfluger's Archiv.*, 196, 185, 1922.
129. Spielberger, C. D., Gorsuch, R. L., and Lushene, R. E., *The State-Trait Anxiety Inventory: Test Manual for Form X*, Consulting Psychologists Press, 1970.
130. Starcevic, V., Kellner, R., Uhlenhuth, E. H., and Pathak, D., The phenomenology of panic attacks in panic disorder with and without agoraphobia, *Compr. Psychiatr.*, 34, 36, 1993.
131. Stein, M. B. and Asmundson, G. J. G., Autonomic function in panic disorder: cardiorespiratory and plasma catecholamine responsivity to multiople challenges of the autonomic nervous system, *Biol. Psychiatr.*, 36, 548, 1994.
132. Steinbusch, H. W. M., Distribution of histaminergic neurons and fibers in rat brain, *Acta Otolaryngol.*, Suppl. 479, 12, 1991.
133. Steinbusch, H. W. M., Distribution of serotonin-immunoreactivity in the central nervous system of the rat — cell bodies and terminals, *Neuroscience*, 6, 557, 1981.
134. Swinson, R. P., Cox, B. J., Rutka, J., Mai, M., Kerr, S., and Kuch K., Otoneurological functioning in panic disorder patients with prominent dizziness, *Compr. Psychiatr.*, 34, 127, 1993.
135. Takayama, K. and Miura, M., Respiratory responses to microinjection of excitatory amino acid agonists in ventrolateral regions of the lateral parabrachial nucleus in the cat, *Brain Res.*, 604, 217, 1993.
136. Telch, M. J., Brouillard, M., Telch, C. F., Agras, W. S., and Taylor, C. B., Role of cognitive appraisal in panic related avoidance, *Behav. Res. Ther.*, 27, 373, 1989.
137. Theunissen, E. J., Huygen, P. L., and Golgering, H. T., Vestibular hyperactivity and hyperventilation, *Clin. Otolaryngol.*, 11, 161, 1986.
138. Thyer, B. A., Himle, J., Curtis, G. C., Cameron, O. G., and Nesse R. M., A comparison of panic disorder and agoraphobia with panic attacks, *Compr. Psychiatr.*, 26, 208, 1985.
139. Turner, S. M., Williams, S. L., Beidel, D. C., and Mezzich J. E., Panic disorder and agoraphobia with panic attacks, covariation along the dimensions of panic and agoraphobic fear, *J. Abnorm. Psychol.*, 95, 384, 1986.
140. Uhde, T. W., Caffeine-induced anxiety: an ideal chemical model of panic disorder? in *Panic Disorder: Clinical, Biological, and Treatment Aspects*, Asnis, G. M. and van Praag, H. M., Eds., John Wiley & Sons, New York, 1995, 181.
141. Uhde, T. W., Boulenger, J-P., Roy-Byrne, P. P, Geraci, M. F., Vittone, B. J., and Post, R. M., Longitudinal course of panic disorder: clinical and biological considerations, *Prog. Neuro-Psychopharmacol. Biol. Psychiatr.*, 9, 39, 1985.
142. Van Megen, H. F. G. M., Den Boer, J. A., Westenberg, H. G. M., Haigh, J. R. M., and Traub, M., On the signficance of cholecystokinin receptors in panic disorder, *Prog. Neuro-Psychopharmacol. Biol. Psychiatr.*, 18, 1235, 1994.
143. Yardley, L., Britton, J., Lear, S., Bird, J., and Luxon, L. M., Relationship between balance system function and agoraphobic avoidance, *Behav. Res. Ther.*, 33, 435, 1995.
144. Yardley, L., Luxon, L., Lear, S., Britton, J., and Bird, J., Vestibular and posturographic test results in people with symptoms of panic and agoraphobia, *J. Audiol. Med.*, 3, 48, 1994.
145. Yasui, Y., Itoh, K., Takada, M., Mitani, A., Kaneko, T., and Mizuno N., Direct cortical projections to the parabrachial nucleus in the cat. *J. Comp. Neurol.*, 234, 77, 1985.
146. Yates, B. J., Goto, T., and Bolton, P. S., Responses of neurons in the rostral ventrolateral medulla of the cat to natural vestibular stimulation, *Brain Res.*, 601, 255, 1993.

147. Yates, B. J., Goto, T., Kerman, I., and Bolton, P. S., Responses of caudal medullary raphe neurons to natural vestibular stimulation, *J. Neurophysiol.*, 70, 938, 1993.
148. Yates, B. J., Grélot, L., Kerman, I. A., Balaban, C. D., Jakus, J., and Miller, A. D., Organization of vestibular inputs to nucleus tractus solitarius and adjacent structures in cat brain stem, *Am. J. Physiol.*, 36, R974, 1994.
149. Yates, B. J., Jakus, J., and Miller, A. D., Vestibular effects on respiratory outflow in the decerebrate cat, *Brain Res.*, 629, 209, 1993.
150. Yeragani, V. K., Berger, R., Pohl, R., Srinivasan, K., Balon, R., Ramesh. C., Weinberg, P., and Berchou, R., Effects of yohimbine on heart rate variability in panic disorder patients and normal controls: a study of power spectral analysis of heart rate, *J. Cardiovasc. Pharmacol.*, 20, 609, 1992.
151. Zucker, D., Taylor, C. B., Brouillard, M., Ehlers, A., Margraf, J., Telch, M., Roth, W. T., and Agras, W. S., Cognitive aspects of panic attacks: content, course and relationship to laboratory stressors, *Br. J. Psychiatr.*, 155:86, 1989.

12

Central Autonomic Syndromes

Susan M. Baser

CONTENTS

0-8493-7668-1/96/$0.00+$.50

LIST OF ABBREVIATIONS

ACh-E	Acetylcholinesterase
AVP	Arginine vasopressin
CNS	Central nervous system
CEN	Central nucleus of the amygdala
CGRP	Calcitonin gene-related peptide
CO_2	Carbon dioxide
DHE	Dihydroergotamine
DMV	Dorsal motor nucleus of the vagus
DNAB	Dorsal noradrenergic bundle
ILCC	Intermediolateral cell column
MAOI	Monoamine oxidase inhibition
MSA	Multiple system atropy, or Shy-Drager syndrome
NA	Nucleus ambiguus
NE	Norepinephrine
NPY	Neuropeptide-Y
NTS	Nucleus of the tractus solitarius
O_2	Oxygen
OPCA	Olivopontocerebellar atrophy
PAF	Pure autonomic failure (also known as idiopathic orthostatic hypotension)
PEV	Posturally evoked vomiting
RVL	Rostroventrolateral
SND	Striatonigral degeneration
VIP	Vasoactive intestinal polypeptide
VNAB	Ventral noradrenergic bundle

1 INTRODUCTION

Autonomic control of vital functions is accomplished through extensive innervation of body organs by the sympathetic and parasympathetic nervous systems. This chapter will first describe functional aspects of the anatomy and physiology of the autonomic nervous system, then discuss conditions and disease states that can alter autonomic function, and finally explore how these conditions may be treated. For a more comprehensive review of the anatomy and physiology of the autonomic nervous system, the reader is referred to Chapter 3 (Sved and Ruggiero) in this book.

Reflex adjustments after processing of information allow the body to compensate for numerous physiologic demands, such as those that occur during changes in posture and exercise. Circulatory, thermoregulatory, respiratory, gastrointestinal, urological, and sudomotor function is under the control and influence of the autonomic nervous system. Most viscera have dual innervation, except glands of the skin and peripheral blood vessels, which have only sympathetic innervation. The autonomic nervous system is composed of numerous reflex arcs that consist of afferent fibers, central integration pathways, and efferent neurons.[79] Receptors in the periphery convey information centrally via bipolar neurons, which have cell bodies located in dorsal root ganglia and specialized cranial nerve ganglia. Transmission of information from the central nervous system involves two neurons, preganglionic and postganglionic.

Many diseases of the autonomic nervous system are characterized by the presence of prominent orthostatic hypotension due either to problems in integration of neural signals that indicate a change in posture or to malfunction in the efferent limb of the baroreflex arc. Autonomic symptoms also occur secondary to malfunction of thermoregulatory, respiratory, gastrointestinal, urological, secretorimotor, and sudomotor systems. Poor temperature control, ataxic and apneustic breathing, sleep apnea, laryngeal dysfunction, and stridor may occur in many autonomic syndromes with central autonomic involvement. Prominent gastrointestinal symptoms such as pseudoobstruction, constipation, and diarrhea occur in many diseases characterized by an imbalance between the more peripheral components of sympathetic and parasympathetic nervous system innervation. Urological dysfunction with urinary urgency, incontinence, and impotence is a common finding in dysautonomias which involve either peripheral or central components of autonomic projections. Exploring the anatomy and physiology of the autonomic nervous system may explain the etiology of particular neurovegetative symptoms that result when one or more of the reflex components (afferent, integration, or efferent) do not function appropriately.

2 ANATOMY OF AUTONOMIC CONTROL

2.1 Afferent Connections

2.1.1 Blood Pressure Control

The autonomic nervous system buffers central blood pressure oscillations, with heart rate moving in a direction opposite to blood pressure. Normally, systolic and diastolic blood pressure move in the same direction over a 24 hour cycle, with little change in pulse pressure. A change in posture causes a complex series of adjustments mediated by baroreflexes to maintain cardiac output and arterial blood pressure.[113] Moving to an upright posture causes blood to shift into the abdomen and lower extremities, reducing venous return. This decrease in venous return could reduce cardiac output, that, if unopposed, would result in a fall in arterial pressure. The rapid reduction in central blood volume during a change to an upright posture also causes a fall in stroke volume caused by the decrease in cardiac filling pressure. In response to these postural changes, there is a reflex increase in heart rate, with withdrawal of vagal activity to the sinus node preceding the increased sympathetic activity. The changes in heart rate compensate (in part) for the reduction in stroke volume, and as a result, cardiac output declines only slightly. Further adjustments are the result of increased sympathetic outflow, which increases systemic vascular resistance.[7]

Systemic blood pressure is monitored peripherally by baroreceptors.[75] These mechanically sensitive afferents from the aorta (which course in the vagus nerve) and carotid sinus (which course in the glossopharyngeal nerve) are stimulated when stretched by increases in arterial pressure.[63] Other receptors located in large arteries include chemoreceptors that respond to changes in arterial oxygen, CO_2, and pH.[35,46] The pulmonary stretch receptors are responsible for the respiratory-cycled heart rate variation known as sinus arrhythmia.[112]

Information from the cardiovascular system is transmitted to the central nervous system (CNS) and integrated in the nucleus of the tractus solitarius (NTS). The NTS, which is located in the dorsomedial medulla adjacent to the dorsal motor nucleus of the vagus (DMV), receives a patterned input from afferents arising from cardiovascular, respiratory, and other visceral receptors.[7]

Information concerning body posture in space is transmitted to the CNS via vestibular afferents. A detailed description of vestibulo-sympathetic reflexes is provided by Yates in Chapter 6. Briefly, vestibular and autonomic pathways are linked by the presence of neural projections from the vestibular nucleus to brainstem areas that regulate circulation, such as the NTS and DMV. Additional neurophysiological studies show that vestibular inputs into the rostroventrolateral (RVL) medulla appear to come mostly from otolith receptors;[115] as described below, the RVL medulla plays a major role in controlling the excitability of sympathetic preganglionic neurons. The presence of otolith inputs to the RVL medulla is consistent with the hypothesis that this region mediates vestibulo-sympathetic reflexes involved in correcting posturally related changes in blood pressure. Electrical stimulation of the vestibular nerve alters activity recorded from sympathetic nerves in cats, first producing an increase, followed by an inhibition, of sympathetic discharges.[102] Experimentally, transection of the vestibular nerve impairs compensation for hypotension produced by tilt, suggesting that vestibular inputs influence orthostatic reflexes.[29]

Afferent information regarding changes in cardiovascular status during exercise is also transmitted to the CNS. Ergoreceptors are sensors in skeletal muscle with small, myelinated, and unmyelinated afferent fibers that are stimulated by the metabolic by-products of exercise.[50] In addition to peripheral sensors, there is a "central command" generated in the cortex during exercise that affects signal processing in cardiovascular and respiratory brainstem centers.[7] Normally, exercise increases blood pressure and heart rate; without inputs from ergoreceptors, exercise-related hypertension would stimulate baroreflexes, thereby inhibiting sympathetic outflow and increasing vagal efferent activity. Thus, neural pathways activated during exercise reset the gain of baroreflexes, so that perfusion of muscle will increase and higher metabolic demands can be met.

2.1.2 Respiratory Control

Central autonomic pathways also receive complex afferent information regarding respiration. The neural control of respiration is described in detail in Chapter 4. In summary, the control of respiration is based on a series of feedback loops to regulate carbon dioxide (CO_2) and oxygen (O_2) tensions. Peripheral chemoreceptors are located in the carotid bodies and pulmonary parenchyma;[28] changes in CO_2 cause local changes in hydrogen ion concentration that stimulate chemoreceptors.[70] Central CO_2 receptors are located just beneath the surface of the ventral medulla.[59] Other inputs also participate in central respiratory control. Pulmonary receptors and stretch receptors in the respiratory muscles are at least partially responsible for alterations in respiration during changes in posture. Adjustments of respiration during changes in posture may be related to the presence of vestibulo-respiratory connections (see Chapter 7).[116] Medullary neurons involved in the timing of breathing include inspiratory neurons in the dorsal respiratory group (located in the ventrolateral NTS) and both inspiratory and expiratory neurons in the ventral respiratory group (located near the nucleus ambiguus [NA]).[21]

2.2 Central Integration

After afferent information is collected and processed in the NTS, connections are made with secondary neurons in the baroreceptor reflex arc. A detailed description of the anatomy and physiology of these connections is contained in Chapter 3. In summary, projections from the NTS important in blood pressure regulation include those that affect the excitability of neurons in the RVL medulla and DMV. The RVL medulla is the location of slowly conducting bulbospinal neurons that provide excitatory drive to the preganglionic sympathetic neurons.[86] An increase in arterial blood pressure stimulates the NTS, which then sends inhibitory output to the RVL medulla (through inhibitory interneurons located in other regions), and excitatory outflow to the DMV, which contains cell bodies of vagal cardioinhibitory neurons.[75] Baroreflex-mediated responses to increased arterial blood pressure are relatively rapid, in contrast to the relatively slow response with a long central delay that occurs with a decrease in arterial pressure. This rapid response mediated by vagal outflow controls the beat-to-beat variability of heart rate.[55]

The NTS has other extensive afferent and efferent connections.[7] Some of these connections of the NTS mediate motility and secretions of the gastrointestinal tract, through the adjacent dorsal motor nucleus of the vagus nerve. Efferents from the NTS also ascend to higher brainstem structures, including the parabrachial nucleus, the Kölliker-Fuse nucleus, the A5 noradrenergic cell group, as well as to hypothalamic and limbic structures. Furthermore, the NTS contains bulbospinal neurons and interneurons that influence the activity of respiratory motoneurons in the spinal cord.

The NTS also has reciprocal connections with the central nucleus of the amygdala (CEN), which directly innervates the midbrain, Kölliker-Fuse nucleus, NTS, DMV, and NA.[96] Reciprocal connections between the NTS and CEN include both a dorsal and ventral pathway, which are important in the patterning of affective behavior, such as the "flight-or-fight" response. With stimulation of the flight-or-fight reflex, a patterned response occurs, with increased heart rate, widespread vasoconstriction, and withdrawal of vasoconstrictor tone to skeletal muscles.

Neuroendocrinologic pathways play several roles in central processing of autonomic information related to postural control of blood pressure. The peptide arginine vasopressin (AVP) acts as a hormone in the blood and as a neurotransmitter within the CNS. Normal subjects and patients with spinal cord transection show a significant rise in plasma AVP in response to head-up tilt.[75] Modulation of AVP release from the posterior pituitary involves the dorsal noradrenergic bundle (DNAB) and the ventral noradrenergic bundle (VNAB) which innervate the supraoptic and paraventricular nuclei.[91,99] Projections from the locus coeruleus including A6-noradrenergic cell groups form the DNAB, whereas fibers from A1 and A2 cells in the lateral tegmental region comprise the VNAB.[13,85] Neurons around the A1 region inhibit sympathetic discharge;[14] destruction of the A1 area results in fulminating hypertension and augmented AVP release.[15,47] Hormonal and cardiovascular effects of AVP include vasoconstriction and renal water retention; centrally, AVP has a direct sensitizing effect on the baroreceptor reflex.[107]

2.3 Efferent Connections

Preganglionic sympathetic and parasympathetic nerve fibers constitute the efferent autonomic outflow to the periphery. Preganglionic sympathetic cell bodies are located in the intermediolateral cell column (ILCC) in segments T1-L2 of the spinal cord. These neurons exhibit a low level of spontaneous discharge, largely maintained by tonic excitatory inputs mediated by L-glutamate that originate in the RVL medulla.[41] The excitability and firing pattern of preganglionic neurons is also modulated by neuropeptides and monoamines contributed by multiple descending pathways. In the rat, the paraventricular nucleus of the hypothalamus contributes vasopressin, oxytocin, met-enkephalin, substance P, and other neuropeptides; the RVL medulla contributes epinephrine, neuropeptide-Y (NPY), and substance P; the raphe and ventromedial medulla contribute serotonin, thyrotropin-releasing hormone, substance P, enkephalin, and somatostatin.[10,12]

An excellent review of the gross anatomy of the sympathetic nervous system was published by Harati.[43] Briefly, peripheral processes of preganglionic sympathetic fibers enter the ventral roots of thoracic and lumbar spinal nerves via white rami communicantes, enter the sympathetic trunk, and terminate in paravertebral and prevertebral sympathetic ganglia. Those axons that terminate in paravertebral ganglia either end in the first ganglion entered or pass up or down in the sympathetic trunk, giving off collaterals and terminating in ganglia above or below. Preganglionic axons primarily utilize acetylcholine, which acts on ganglion cells to produce fast excitation via nicotinic receptors and slow modulation via muscarinic receptors. The paravertebral sympathetic trunks consist of two ganglionated cords located along the anterolateral aspects of the vertebral column from the base of the skull to the coccyx. The cervical portion consists of three ganglia (superior, middle, and inferior cervical ganglia) formed by the fusion of the eight segmented cervical ganglia. The superior cervical ganglion gives rise to postganglionic fibers that pass via gray rami to the lower four cranial nerves, the C1-C3 spinal nerves, pharynx, external and internal carotid arteries, and superior cardiac nerves.

Postganglionic sympathetic fibers passing to the external and internal carotid arteries form plexuses, with fibers supplying the levator palpebrae (dilator muscle of the iris), Mullers muscle, blood vessels, sweat glands, and lacrimal glands of the face.[24] Fibers from the superior cervical ganglion form the superior cervical cardiac nerve that passes to the cardiac plexus. The middle cervical ganglion, when present, supplies the gray rami of C5-C6 and the middle cardiac nerve. The inferior cervical ganglion supplies gray rami of C7-T1 and forms the inferior cardiac nerve.[43]

The inferior cervical ganglion frequently fuses with the first thoracic ganglion to form the stellate ganglion. The thoracic, lumbar, and sacral ganglia are segmentally arranged with 11-12 thoracic, 3-4 lumbar, 4-5 sacral, and one fused coccygeal ganglia. Postganglionic sympathetic fibers exit via gray rami to enter all spinal nerves. Unmyelinated postganglionic sympathetic fibers innervate blood vessels, arrector pili, and body wall glands.[43] Postganglionic sympathetic fibers that travel in spinal nerves innervate arterioles which form the resistance bed, which is important in regulating systemic peripheral resistance.[113]

Some sympathetic preganglionic fibers do not synapse in paravertebral ganglia, but pass through and emerge as splanchnic nerves that synapse in prevertebral ganglia. Prevertebral ganglia (celiac, aorticorenal, phrenic, superior mesenteric, and inferior mesenteric) are irregular masses that surround visceral aortic branches. Innervation of the mesenteric capacitance bed is via preganglionic sympathetic fibers traveling in the greater splanchnic nerve, which then synapse in the celiac ganglion, which in turn sends postganglionic sympathetic fibers to capacitance vessels. The greater splanchnic nerve innervates the mesenteric capacitance bed, which is of great importance in vasoconstrictive reflexes in the adjustment of arterial blood pressure.[113]

A large number of neuropeptides is expressed by sympathetic autonomic ganglia, and the organization of ganglion cells expressing different peptides is complex. Furthermore, neurons projecting to different regions and targets are segregated within autonomic ganglia, and form distinct subpopulations. The majority of ganglion cells are noradrenergic. In paravertebral ganglion neurons, NE is mainly co-localized with NPY or somatostatin.[34] NE/NPY neurons have vasoconstrictor function.[62] Galanin, together with NPY, can be released from cardiac sympathetic nerves and decrease cardiovagal activity via presynaptic inhibition of acetylcholine release by parasympathetic terminals.[69] Paravertebral cholinergic neurons serving sudomotor function express vasoactive intestinal polypeptide (VIP). VIP-positive sudomotor neurons may also express calcitonin gene-related peptide (CGRP), substance P, and neurotensin.[34] Almost all prevertebral neurons are noradrenergic, with NE/NPY positive neurons innervating intestinal blood vessels and NE/somatostatin neurons innervating secretomotor neurons in the submucosal ganglia, and NE neurons without coexisting neuropeptides innervating the myenteric ganglia.[10]

Parasympathetic nervous system efferent outflow arises from the brainstem and sacral spinal cord. Cranial parasympathetic efferents arise from brainstem visceral cell groups and course with cranial nerves III, VII, IX, and X. Preganglionic parasympathetic fibers arise from these nuclei and terminate in cranial autonomic ganglia near the structure innervated. The following nuclei in the brainstem contain cell bodies of preganglionic parasympathetic neurons: Edinger-Wesphal nucleus (CN III), superior salivatory nucleus (CN VII), inferior salivatory nucleus (CN IX), nucleus ambiguus and dorsal motor nucleus of vagus (CN X).[43]

The ciliary ganglia receive preganglionic parasympathetic fibers from the Edinger-Wesphal nucleus and send postganglionic fibers via short ciliary nerves to the iris sphincter and ciliary body.[24] Pterygopalatine and submandibular ganglia receive fibers from the superior salivatory nucleus via the intermediate portion of the facial nerve. Preganglionic fibers pass via the major petrossal nerve to the pterygopalantine ganglion; postganglionic fibers then pass to the lacrimal glands, blood vessels, and glands of the mucus membranes of the nose and palate. Preganglionic fibers pass via the chorda tympani to the submandibular ganglion, from which postganglionic fibers pass to the submandibular and sublingual glands. The otic ganglia receive preganglionic fibers from the inferior salivatory nucleus via the glossopharyngeal and the minor petrossal nerve and send postganglionic fibers to the parotid gland.

Ganglia associated with the vagus nerve are located in terminal ganglia in walls of thoracic and abdominal viscera. Preganglionic parasympathetic fibers originate in the NA and DMV. In the thorax, preganglionic fibers enter pulmonary, cardiac, and esophageal plexuses to be distributed to terminal ganglia in the heart and bronchi; short postganglionic parasympathetic fibers then pass to nearby target organs. Vagal innervation of the cardiac ventricles is largely concentrated in the conducting tissue. Most of atrial conducting cells are acetylcholinesterase (ACh-E) positive and some also contain somatostatin. VIP and enkephalin-like immunoreactivity are also present in the axons of the atrial myocardium.[84] The short postganglionic parasympathetic fibers associated with bronchial smooth muscle and glands contain VIP, which relaxes smooth muscle and is a potent vasodilator of pulmonary arteries.[43]

In the abdomen, preganglionic parasympathetic fibers pass through the celiac plexus to end in terminal ganglia of the gastrointestinal tract and kidneys. Vagal postganglionic neurons that innervate the lower esophageal sphincter also contain VIP, which functions in smooth muscle relaxation.[84] In sacral regions, preganglionic parasympathetic fibers arise from preganglionic visceral neurons in segments S2,3,4, with postganglionic parasympathetic fibers arising from terminal ganglia in the walls of pelvic viscera.

3 DISORDERS OF THE AUTONOMIC NERVOUS SYSTEM

3.1 Anatomic Classification

Disorders of the autonomic nervous system may be divided into pathologic conditions that may affect afferent, central, efferent, or any combination of these portions of the baroreflex arc. In many disorders there is evidence of dysfunction of several parts of the autonomic nervous system. Localization of autonomic dysfunction may be based on physiologic and pharmacologic characteristics. Two disorders, pure autonomic failure (PAF, also known as idiopathic orthostatic hypotension) and multiple system atrophy (MSA, also known as Shy-Drager syndrome) have been used as models to explain physiologic and pharmacologic characteristics associated with postganglionic versus preganglionic localization, respectively. These two disorders will be discussed in great detail in the next section.

3.1.1 Mixed Localization

One of the most common pathologic conditions that may be complicated by autonomic dysfunction is diabetes mellitus. Diabetes mellitus is a well-described metabolic disorder which may be accompanied by afferent, central, and efferent autonomic dysfunction.

Patients with diabetes commonly develop afferent vagal dysfunction, documented by the presence of vagal denervation with loss of normal heart rate variation.[111] Patients with diabetes of long duration may additionally develop sympathetic denervation, and ultimately a fixed heart rate. Alcoholism is another disease which may present with multiple sites of autonomic dysfunction. Physiologic evidence of impaired baroreflexes occurs in severe alcoholics with peripheral neuropathy and central nervous system damage.[32] Pathologic changes in alcoholism include vagal degeneration;[40] deposition of PAS-positive amorphous material in sympathetic ganglion cells occurs in alcoholics with orthostatic hypotension.[61] Tabes dorsalis, a disorder associated with long-standing neurosyphillis, is a disease associated with widespread loss of sensory afferents and orthostatic hypotension (Table 1). A variety of autonomic manifestations occur in tabes dorsalis, including pupillary changes, lightening pains, visceral crises, urinary incontinence, sensory complaints, and Charcot joints (neurogenic arthropathy).[2]

3.1.2 Primarily Central Localization

Some conditions that primarily affect the central nervous system may be attended by the presence of autonomic dysfunction (Table 1). Shapiro's syndrome is a rare disease associated with spontaneous hypothermia, polydipsia, polyuria, hyponatremia, and autonomic paroxysms characterized by hypertension, tachycardia, and episodic hyperhydrosis. Dysgenesis of the midline telencephalic structures with agenesis of the corpus callosum is one manifestation of Shapiro's syndrome, and it is frequently associated with abnormalities in the septal region, cingulate gyrus, and other structures.[110] Approximately 20 cases have been reported in the literature. Clinical manifestations include hypothermia that occurs in brief episodes or lasts for months; spontaneous remissions may occur and last for months to years. Profuse and inappropriate diaphoresis may persist for several hours, and may be accompanied by major declines in body temperature and impaired neurologic function.[57] Three- to sixfold increases in basal supine plasma NE levels have been reported in patients with Shapiro's syndrome.[68,90] Although the mechanisms underlying the hypothermia in Shapiro's syndrome have not been completely defined, the structural abnormalities in or near the thermoregulatory center may result in dysfunction of central monoaminergic tone.[4] Treatment modalities include the use of clonidine,[90] cyproheptadine,[57] and oxybutynin.[56]

Central autonomic dysfunction also occurs in Wernicke's encephalopathy. This disorder should be suspected in alcoholic, malnourished, and malignancy patients (including children) that present with hypothermia and hypotension. Pathologic changes in Wernicke's encephalopathy include symmetrical lesions in the paraventricular regions of the hypothalamus and thalamus, the periaqueductal regions of the midbrain, and the floor of the fourth ventricle, particularly in the region of the DMV and vestibular nuclei.[1] Other conditions that affect the perihypothalamic regions can disturb autonomic function. Structural lesions near the hypothalamus cause loss of temperature and blood pressure regulation. A mass obstructing at the foramen of Munroe may be responsible for a constellation of symptoms including hyperthermia, labile blood pressure, vasodilation, diaphoresis, pupillary change, and shivering.[11] Space occupying or compressive lesions in the posterior fossa can cause autonomic dysfunction characterized by either orthostatic hypotension and/or hypertension. Orthostatic and episodic hypertension have been described in patients with ectatic basilar arteries compressing the brainstem.[48] Tumors of the brainstem or cerebellum may present with paroxysmal hypertension or orthostatic hypotension, creating diagnostic confusion.[11] Medications can also cause dysfunction of central autonomic regulation. When autonomic hyperactivity is associated with hyperthermia and muscle rigidity, neuroleptic malignant syndrome should be considered.[33]

TABLE 1

Differential Diagnosis of Autonomic Dysfunction

Afferent lesions

1. Diabetes mellitus
2. Alcoholism
3. Holmes-Adie syndrome
4. Tabes dorsalis

Central lesions

1. Multiple system atrophy (Shy-Drager syndrome)
2. Idiopathic Parkinson's disease
3. Familial dysautonomia
4. Wernicke's encephalopathy
5. Shapiro's syndrome
5. Obstructive hydrocephalus
6. Posterior fossa infarcts
7. Posterior fossa tumors
8. Brainstem arterial-venous malformation
9. Brainstem vascular compression syndromes
10. Spinal cord lesions

Efferent lesion

1. Pure autonomic failure (idiopathic orthostatic hypotension)
2. Metabolic/neuropathic conditions
   a. Diabetes
   b. Amyloidosis
   c. Uremia
   d. Porphria-variegate, acute intermittent
   e. Tangier's disease
   f. Adrenomyeloneuropathy
   g. B12 deficiency
   h. Holmes-Adie syndrome
   i. Ross syndrome
3. Toxins
   a. Alcohol
   b. Organic solvents (*N*-hexane, methyl-butyl ketone)
   c. Box jelly fish stings (Southwest Asia, Australia)
   d. Botulism, type B
   e. Vincristine sulfate
   f. Perhexiline maleate
   g. Vacor
4. Infectious/postinfectious
   a. Guillain-Barré syndrome
   b. Infectious mononucleosis
   c. Rubella
   d. Herpes simplex
   e. HIV
   f. Hansen's disease
5. Neoplastic/paraneoplastic

3.1.3 Predominantly Peripheral Localization

Metabolic disorders commonly impair function of the more efferent portions of the autonomic nervous system in association with peripheral neuropathy (Table 1). The porphyrias, including the acute intermittent[100] and variegate forms,[97] are associated with paroxysmal dysautonomia which may include both hypotension and hypertension. Another condition associated with dysautonomia in association with peripheral neuropathy is amyloidosis. Amyloidosis is a disease of unknown etiology characterized by the deposition of amorphous, hyaline material in mesenchymal tissue. Autonomic ganglia, spinal nerve roots, and peripheral nerves demonstrate the greatest degree of involvement. In addition to peripheral neuropathy, autonomic symptoms such as impotence, orthostasis, dyshidrosis, trophic changes, gastrointestinal dysfunction, and hoarseness may occur in amyloidosis.[71]

Acute dysautonomia may occur in association with viral infections, including rubella,[98] infectious mononucleosis,[114] Guillain-Barré syndrome,[45] human immunodeficiency virus,[15] and herpes simplex.[72] Dysfunction of the autonomic nervous system occurs as a paraneoplastic symptom in many malignancies, such as in Hodgkin's disease.[109] Numerous toxins have been associated with autonomic dysfunction (Table 1), but one of the more unusual includes the venom of the box jelly fish.[19]

Some rare disorders may affect more peripheral portions of the autonomic nervous system without known cause. Ross syndrome is an idiopathic condition characterized by a triad of tonic pupils, loss of muscle stretch reflexes, and dysautonomia. Autonomic symptomatology may include orthostatic hypotension and segmental hypohydrosis. Disturbances in sweating may occasionally precede or follow the development of a tonic pupil. Reported sudomotor manifestations include hemifacial loss of sweating and flushing,[31] segmental hypohydrosis with eczema,[44] and progressive anhydrosis and hyperthermia.[31] Similar to the Holmes-Adie syndrome, the tonic pupils associated with Ross syndrome demonstrate pharmacologic supersensitivity to cholinergic drugs consistent with postganglionic parasympathetic denervation.[89] The absence of muscle stretch reflexes in Ross syndrome indicates that this disorder is not restricted to postganglionic parasympathetic fibers. Pathologic studies of Ross syndrome show evidence of degeneration of dorsal root ganglia and dorsal columns in the lumbosacral region.[106] A presumed postganglionic sympathetic defect is responsible for the segmental sweating disturbance seen in Ross syndrome.[31,89]

3.1.4 *Transient Dysfunction*

Temporary derangement of the autonomic nervous system may occur with motion sickness. Readers are referred to Chapter 5 for a detailed description of the neural substrate of vomiting, and to Chapter 9, which discusses motion sickness in detail. Briefly, vomiting is a frequent accompaniment of vertigo which results from dysfunction or disease of the vestibular system, but can also occur as an isolated phenomenon related to posture, clinically referred to as "posturally evoked vomiting (PEV)."[30] Vestibular inputs are involved in producing motion sickness, with the central structures related to vomiting located in the caudal brainstem, perhaps in the region of NTS.[42] Motion sickness is often accompanied by pallor and sweating, which are the result of increased sympathetic outflow to blood vessels and to sweat glands.[67] However, a recent study in normal volunteers that assessed sympathetic outflow during caloric vestibular stimulation failed to show any significant change in heart rate, blood pressure, muscle sympathetic nervous activity, or plasma NE levels.[23] In contrast, low-frequency oscillations of heart rate variability in space shuttle astronauts were reliably associated with episodes of space motion sickness.[104] Thus, the degree of activation of the sympathetic nervous system during motion sickness is unclear (but see Chapters 9 and 10).

3.2 Physiologic Classification: Preganglionic vs. Postganglionic Localization

Physiologic and pharmacologic characteristics distinguish two well known syndromes characterized predominantly by the presence of autonomic dysfunction. These two disorders, pure autonomic failure (PAF, also known as idiopathic orthostatic hypotension) and multiple system atrophy (MSA, also known as Shy-Drager syndrome) have been used as models to explain physiologic and pharmacologic characteristics associated with postganglionic versus preganglionic localization, respectively. Numerous clinical studies in these two neurologic disorders have elucidated important clinical, pathological, and neuroradiological features of pre- and postganglionic autonomic dysfunction.

3.2.1 Postganglionic Localization

PAF is a slowly progressive disorder, characterized by the presence of orthostatic hypotension without evidence of other neurologic disease. PAF was first described in three patients by Bradbury and Eggleston in 1925.[16] PAF usually has onset in the fourth through sixth decade with equal involvement of men and women.[80] Early symptoms of autonomic dysfunction in PAF include impotence and bladder dysfunction which may precede orthostatic hypotension by nearly a decade. Gastrointestinal symptomatology in PAF frequently includes prominent early severe constipation alternating with diarrhea. Orthostatic hypotension develops in all PAF patients, with symptoms of weakness, dimming of vision, and headache. Prominent occipital-nuccal headache may be a presenting feature of the orthostatic hypotension that occurs in PAF.[88] Symptoms are worse in the morning, after meals, after exercise, and in hot weather. Supine hypertension develops in nearly all patients with PAF. Loss of thermoregulatory sweating with evidence of postganglionic sudomotor failure occurs in most patients with PAF, further complicating heat intolerance.[9,22] Dry mouth, dry eyes, and pupillary abnormalities frequently complicate this disease. Pupillary abnormalities may include alternating anisocoria and Horner's syndrome. The disease is slowly progressive, and duration of disease commonly exceeds ten years. A limited number of pathological studies have examined PAF patients; one study has demonstrated Lewy bodies and dystrophic neurites in NPY-positive ganglion neurons.[66] No etiology has been identified for PAF.

It is possible that sub-clinical abnormalities in vestibular function contribute to PAF, as they do to panic disorder and agoraphobia (see Chapter 11). The connections of the vestibular system with brainstem regions that regulate blood pressure are appropriate for participation of vestibular inputs in the correction of orthostatic hypotension (see Chapter 6). The possible link between vestibular dysfunction and PAF is yet to be thoroughly examined, however.

3.2.2 Preganglionic Localization

The clinical features of multiple system atrophy (MSA) contrast sharply with those of PAF. Multiple system atrophy patients usually develop neurologic features within several years of developing orthostatic hypotension. Less commonly, neurologic symptoms in MSA precede the appearance of autonomic insufficiency. Men are affected twice as frequently as women, and the disease usually begins after age 50.[80] Respiratory abnormalities and sleep apnea often develop in MSA.[39] Laryngeal function may be affected by unilateral or bilateral vocal cord paralysis,[6] and acute severe respiratory stridor may necessitate tracheostomy.[64] Corticospinal tract findings with increased tendon reflexes and extensor plantar responses are often present. Vertigo, with abnormal vestibulo-autonomic reflexes in addition to orthostatic hypotension occur in MSA patients. Two clinical presentations of MSA can be distinguished: (1) predominant striatonigral degeneration (SND) subtype with bradykinesia, and rigidity out of proportion to resting tremor, or (2) an olivopontocerebellar atrophy (OPCA) subtype with truncal and limb ataxia, intention tremor, and slurred speech.[80] Most patients exhibit an overlap of the two symptom constellations. Emotional lability and intellectual deterioration may occur in the late stages of disease, but severe dementia is uncommon in MSA patients. Patients generally succumb seven to ten years after the onset of neurologic symptoms; aspiration, cardiac arythymias, and sleep apnea contribute to their demise.[78]

Neuropathologic changes occurring in MSA involve several CNS regions that play a role in autonomic function. Cell loss is described in many locations, including the hypothalamus, cerebellum, locus coeruleus, DMV, ILCC, and Onuf's nucleus.[95,54] Biochemical correlates of pathologic features in MSA include a decrease in the concentration of substance P in the thoracic region of the spinal cord,[3] and a severe depletion of enkephalin-reactive fibers in the paravertebral ganglia, consistent with loss of preganglionic neurons.[66]

Neuroradiologic features may be useful in differentiating MSA from other autonomic disorders that share clinical characteristics, particularly early in the clinical course. Patients with MSA may have an abnormal decrease in signal intensity in the posterolateral putamen, predominantly on T2 weighted images.[37] A more recent study shows that proton magnetic resonance spectroscopy may be useful in differentiating multiple system atrophy from idiopathic Parkinson's disease.[25]

3.3 Neuropharmacologic Classification

Neuropharmacologic characteristics may also distinguish patients with PAF and MSA.[117] In normal subjects, the resting supine plasma NE level of 200 to 300 pg/ml doubles upon standing. This increment in NE is a reflection of increased sympathetic activity required to maintain blood pressure in the upright position. Patients with PAF have very low supine plasma NE levels (often less than 100 pg/ml) in contrast to the normal or elevated levels seen in patients with MSA. Neither PAF or MSA patients adequately increase levels appropriately upon standing, consistent with a lesion in the baroreflex arc preventing appropriate activation of postganglionic sympathetic neurons.[80] The basal NE levels have localizing value. The normal supine resting levels in MSA suggest that the postganglionic neurons function normally under resting conditions. In PAF, the low supine plasma NE levels resemble the findings in patients with diabetic neuropathy, and suggest that the primary sympathetic dysfunction involves peripheral noradrenergic neurons.[117]

Additional neurochemical characteristics may be useful in distinguishing PAF and MSA. Pressor responses to vasoactive drugs play a role in the diagnosis and treatment of these disorders. Decentralization can be distinguished from denervation on the basis of blood pressure responses to drugs including norepinephrine[82] and isoproterenol.[8] Patients with PAF have a shift to the left of the dose-response curve, suggesting denervation supersensitivity, whereas those with MSA have a greater than normal slope of the dose-response curve, indicating deficient baroreflex modulation.

Characterization of the neurohormonal responses to orthostatic hypotension may be helpful in distinguishing these and other disorders. In normal subjects, AVP is released from the posterior hypothalamus into the systemic circulation in response to baroreceptor activation. In patients with PAF, upright tilt induces marked hypotension and a pronounced increase in plasma AVP levels, illustrating intact afferent and central pathways. However in patients with MSA, tilt induced hypotension does not appropriately increase AVP levels.[52] Food ingestion does not lower BP in normal persons but can cause postprandial hypotension in the various dysautonomias secondary to splanchnic dilatation and insulin release.[18] Postural hypotension after a meal is enhanced in both MSA and PAF, but not in idiopathic Parkinson's disease.[103]

4 TREATMENT OF AUTONOMIC DYSFUNCTION

4.1 Practical Methods

Practical considerations are a cornerstone in the management of the severe orthostatic hypotension that may occur in these disorders. Patients frequently monitor their blood pressure in a haphazard manner and need rational guidelines for blood pressure monitoring. It must be stressed that supine elevation of blood pressure can usually be treated by moving to an upright posture, or if necessary, by walking about for several minutes. Encouraging salt and water intake is crucial in optimizing fluid status. Next, compressive garments, such as Jobst stocking suits, may be of some use in some patients if they or a caregiver are able to place and remove them. Sleeping with head elevated to 30° (reverse Trendelenburg position) will help to decrease some of the nocturnal diuresis that occurs, improving morning standing pressures and decreasing supine hypertension. Physical maneuvers such as leg crossing and squatting improves orthostatic tolerance via augmentation of venous return.[108] Patients should avoid over-the-counter medications since these frequently contain vasoactive substances. Avoiding heat and alcohol and maintaining good bowel and bladder hygiene will minimize some of the complications that can occur in these disorders.

4.2 Volume Expansion

Medications can be used to augment circulating blood volume in orthostatic hypotension. Fludrocortisone causes fluid retention[5] and may also sensitize vascular receptors to pressor amines.[92] Potassium supplementation is required for patients who use fludrocortisone on a chronic basis. Fluid status can be further modified by the use of agents that cause sodium retention and inhibit circulating prostaglandins, such as indomethacin, aspirin, and ibuprofen.[53,105] Reduction in urinary excretion of prostaglandin synthesis does correlate with improvement in blood pressure.[83] Expansion of red cell mass using erythropoetin to augment circulating blood volume has been used in a limited number of patients,[38,76] but the safety and efficacy of long-term use is yet to be determined.

4.3 Vasoactive Agents

Vasoactive agents have been extensively studied in the treatment of orthostatic hypotension. Caffeine may be useful in attenuating falls in blood pressure after eating by blocking adenosine-mediated splanchnic vasodilatation.[73] Dihydroergotamine (DHE) selectively constricts capacitance vessels and reduces peripheral pooling of blood.[49] A variety of DHE preparations have been used including intramuscular, oral,[36] and intranasal aerosol formulations.[93] Other vasoactive agents such as ephedrine, an indirect sympathomimetic agent, may have initial good pressor effects, but long-term use may be complicated by decreased effectiveness over time.[26] Midrodrine, a newer direct alpha-adrenergic agonist has been successfully used in the treatment of orthostatic hypotension,[38] but its long-term use may be complicated by supine hypertension.

Beta blockers, such as propanolol,[17] are useful in some patients at low doses to block beta mediated vasodilation. Clonidine, a calcium channel blocker, has been found useful in some patients with orthostatic hypotension.[87] Dopaminergic blockade via the use of metoclopramide[101] and domperidone[60] improves standing blood pressures in patients with orthostatic hypotension, independent of any effect on aldosterone and plasma renin activity.

Yohimbine, a selective alpha-2 adrenoreceptor antagonist enhances sympathetic nervous system discharge, increasing systolic and diastolic blood pressure and heart rate in autonomic failure patients.[74] Side effects of yohimbine include nausea, vomiting, confusion, and hallucinations and may be dose limiting. Pharmacologic manipulations such as blocking NE uptake (tricyclic antidepressants), inhibition of NE metabolism via monoamine oxidase inhibition (MAOI), and a combination of MAOI with tyramine (to release NE stores) have been disappointing.[81] The use of these agents may be complicated by the occurrence of significant hypertension and hypotension.

4.4 Neurohormonal Replacement

Neurohormonal methods to augment standing blood pressures have had variable success. The vasopressin analogue desmopressin, when administered intramuscularly in the evening, decreases nocturnal polyuria, decreases overnight weight loss, raises supine pressures, and diminishes early morning orthostasis.[65] However, care must be taken to monitor for the occurrence of supine hypertension with the use of desmopressin. More recently, the somatostatin analogue octreotide has been used in the treatment of orthostatic hypotension. Octreotide increases blood pressure via alteration in splanchnic blood flow,[20,77] but only limited case reports are available supporting the use of this agent. Further study to demonstrate long-term efficacy of this somatostatin analogue is required.

5 SUMMARY

Autonomic nervous system reflexes allow the body to make adjustments for numerous physiologic demands, such as those that occur during changes in posture and exercise. Vestibular afferents transmit critical information to the CNS regarding the position of the body in space. This input, along with baroreceptor and chemoreceptor information, is integrated in brainstem nuclei, including the NTS, DMV, and other nearby structures. From these integration stations, processed information is transmitted to both higher and lower levels as part of many efferent reflex arcs.

Many diseases can cause dysfunction of the autonomic nervous system. The symptomatology resulting from this dysfunction is based on which component (afferent, central, or efferent) of the autonomic nervous system is involved. Prominent dizziness, orthostatic hypotension, vertigo, and other neurovegetative symptoms commonly occur with these disorders. Understanding the anatomy and physiology of the autonomic nervous system and the pathophysiology of these clinical disorders will facilitate improved treatments of autonomic syndromes.

REFERENCES

1. Adams, R. D. and Victor, M., Diseases of the nervous system due to nutritional deficiency, in *Principles of Neurology*, Adams, R. D. and Victor, M., Eds., McGraw-Hill, New York, 1989, 820.
2. Adams, R. D. and Victor, M., Nonviral infections of the nervous system, in *Principles of Neurology*, Adams, R. D. and Victor, M., Eds., McGraw-Hill, New York, 1989, 554.
3. Anand, P., Bannister, R., McGregor, G. P., Ghatei, M. A., Mulderry, P. K., and Bloom, S. R., Marked depletion of dorsal spinal cord substance P and calcitonin gene-related peptide with intact skin flare responses in multiple system atrophy, *J. Neurol. Neurosurg. Psychiatr.*, 51, 192, 1988.
4. Arroyo, H. A., DiBlasi, M., and Grinszpan, G. J., A syndrome of hyperhydrosis, hypothermia, and bradycardia possibly due to central monoaminergic dysfunction, *Neurology*, 40, 556, 1990.

5. Bannister, R., Ardill, L., and Fentem, P., An assessment of various methods of treatment of idiopathic orthostatic hypotension, *Q. J. Med.*, 28, 377, 1969.
6. Bannister, R., Gibson, W., Michaels, L., and Oppenheimer, D. R., Laryngeal abductor paralysis in multiple system atrophy, *Brain*, 104, 351, 1981.
7. Barron, K. D. and Chokroverty, S., Anatomy of the autonomic nervous system: brain and brainstem, in *Clinical Autonomic Disorders*, Low, P. A., Ed., Little, Brown, Rochester, MN, 1993, 3.
8. Baser, S. M., Brown, R. T., Curras, M. T., Baucom, C. E., Hooper, D. R., and Polinsky, R. J., Beta-receptor sensitivity in autonomic failure, *Neurology*, 41, 1107, 1991.
9. Baser, S. M., Meer, J., Polinsky, R. J., and Hallet, M., Sudomotor function in autonomic failure, *Neurology*, 41, 1564, 1991.
10. Bennaroch, E., Neuropeptides in the sympathetic nervous system: presence, plasticity, modulation, and implications, *Ann. Neurol.*, 36, 6, 1994.
11. Benarroch, E. E., Central autonomic disorders, in *Clinical Autonomic Disorders*, Low, P. A., Ed., Little, Brown, Rochester, MN, 1993, 463.
12. Benarroch, E. E., Aimone, L. D., and Yaksh, T. L., Segmental analysis of neuropeptide concentrations in the human spinal cord, *Neurology*, 40, 137, 1990.
13. Blessing, W. W. and Reis, D. J., Inhibitory cardiovascular function of neurons in the causal ventrolateral medulla of the rabbit: relationship to the area containing A1 containing noradrenergic cells, *Brain. Res.*, 253, 161, 1982.
14. Blessing, W. W., Sved, A. F., and Reis, D. J., Destruction of noradrenergic neurons in the rabbit brainstem elevates plasma vasopressin causing hypertension, *Science*, 217, 661, 1982.
15. Blessing, W. W., West, W. T., and Chalmers, J., Hypertension, bradycardia, and pulmonary edema in the conscious rabbit after brainstem lesions coinciding with the A1 group of catecholamine neurons, *Circ. Res.*, 49, 949, 1981.
16. Bradbury, S. and Eggleston C., Postural hypotension: a report of three cases, *Am. Heart J.*, 1, 73, 1925.
17. Brevetti, G., Chiariello, M., Giudice, P., DeMichele, G., Mansi, D., and Campanella, G., Effective treatment of orthostatic hypotension by propanolol in the Shy-Drager syndrome, *Am. Heart J.*, 40, 938, 1981.
18. Brown, R. T., Polinsky, R. J., Lee, G. K., and Deeter, J. A., Insulin-induced hypotension and neurogenic orthostatic hypotension, *Neurology*, 36, 1402, 1986.
19. Chand, R. P. and Selliah, K., Reversible parasympathetic dysautonomia following stinging attributed to the box jelly fish (chironex fleckeri), *Aust. N. Z. J. Med.*, 14, 673, 1984.
20. Chaudhuri, K. R., Thomaides, T., Pavit, T. D., and Mathias, C. J., Octreotide prevents alcohol-induced hypotension and mesenteric vasodilatation in primary autonomic failure, *Clin. Auton. Res.*, 4, 77, 1994.
21. Chokroverty, S., Sleep apnea and respiratory disturbances in multiple system atrophy with autonomic failure, in *Autonomic Failure: A Textbook of Clinical Disorders of the Autonomic Nervous System*, Bannister, R., Ed., Oxford University Press, London, 1983, 432.
22. Cohen, J., Low, P., Fealey, R., Sheps, S., and Jiang, N. S., Somatic and autonomic function in progressive autonomic failure and multiple system atrophy, *Ann. Neurol.*, 22, 692, 1987.
23. Costa, F., Lavin, P., Robertson, D., and Biaggioni, I., Effect of neurovestibular stimulation on autonomic cardiovascular regulation, *Clin. Auton. Res.*, 4, 214, 1994.
24. Cross, S. A., Autonomic innervation of the eye, in *Clinical Autonomic Disorders*, Low, P. A., Ed., Little, Brown, Rochester, MN, 1993, 133.
25. Davie, C. A., Wenning, G. K., Barker, G. J., Tofts, P. S., Kendall, B. E., Quinn, N., McDonald, W. I., Marsden, C. D., and Miller, D. H., Differentiation of multiple system atrophy from idiopathic Parkinson's disease using proton magnetic resonance spectroscopy, *Ann. Neurol.*, 37, 204, 1995.
26. Davies, B., Bannister, R., and Sever, P., Pressor amines and monoamineoxidase inhibitors for the treatment of postural hypotension in autonomic failure: limitations and hazards, *Lancet*, 1, 172, 1978.
27. Davies, I. B., Bannister, R., Hensby, C., and Sever, P.S., The pressor actions of noradrenaline and angiotensin II in chronic autonomic failure treated with indomethacin, *Br. J. Clin. Pharmacol.*, 10, 223, 1980.
28. Dempsey, J. A. and Foster, T. N., Mediation of ventilatory adaptations, *Physiol. Rev.*, 62, 262, 1982.
29. Doba, N. and Reis, D. J., Role of the cerebellum and vestibular apparatus in regulation of orthostatic reflexes of the cat, *Circ. Res.*, 34, 9, 1974.
30. Drachman, D. A., Diamond, E. R., and Hart, C. W., Posturally-evoked vomiting: association with posterior fossa lesions, *Ann. Otol.*, 86, 97, 1977.
31. Drummond, P. D. and Edis, R. H., Loss of facial sweating and flushing in Holmes-Adie syndrome, *Neurology*, 40, 847, 1990.
32. Duncan, G., Johnson, R. H., Lambie, D. G., and Whiteside, E. A., Evidence of vagal neuropathy in chronic alcoholics, *Lancet*, 2, 1053, 1980.

33. Ebadi, M., Pfeiffer, R. F., and Murrin, L. C., Pathogenesis and treatment of neuroleptic malignant syndrome, *Gen. Pharmacol.*, 21, 367, 1990.
34. Elfvin, L. G., Lindh, B., and Hökfelt, T., The chemical neuroanatomy of sympathetic ganglia, *Annu. Rev. Neurosci.*, 16, 471, 1993.
35. Eyzaguirre, C. and Lewin, J., Effect of different oxygen tensions on the carotid body in vitro, *J. Physiol.*, 159, 238, 1961.
36. Fouad, F. M., Tarazi, R. C., and Bravo, E. L., Dihydroergotamine in idiopathic orthostatic hypotension: short term intramuscular and long-term oral therapy, *Clin. Pharmacol. Ther.*, 30, 782, 1981.
37. Fulham, M. J., Dubinsky, R. M., Polinsky, R. J., Brooks, R. A., Brown, R. T., Curras, M. T., Baser, S. M., Hallet, M., and DiChiro, G., Computerized tomography, magnetic resonance imaging, positron emission tomography with 18 F-flurodeoxyglucose in multiple system and pure autonomic failure, *Clin. Auton. Res.*, 1, 27, 1991.
38. Grubb, B. P., Lachant, N. A., and Kosinski, D., Erythropoetin as therapy for severe refractory orthostatic hypotension, *Clin. Auton. Res.*, 4, 212, 1994.
39. Guilleminault, C., Briskin, J. G., Greenfield, M. S., and Silvestri, R., The impact of autonomic nervous system dysfunction on breathing during sleep, *Sleep*, 4, 263, 1981.
40. Guo, Y.-P., McLeod, J. G., and Baverstock, J., Pathologic changes in the vagus nerve in diabetics and chronic alcoholics, *J. Neurol. Neurosurg. Psychiatr.*, 50, 1449, 1987.
41. Guyenet, P. G., Haselton, J. R., and Sun, M. K., Sympathoexcitatory neurons of the rostroventrolateral medulla and the origin of the sympathetic vasomotor tone, *Prog. Brain. Res.*, 81, 105, 1989.
42. Hatcher, R. A. and Weiss, S., Studies on vomiting, *J. Pharmacol. Exp. Ther.*, 22, 139, 1923.
43. Harati, Y., Anatomy of the spinal and peripheral autonomic nervous system, in *Clinical Autonomic Disorders*, Low, P. A., Ed., Little, Brown, Rochester, MN, 1993, 17.
44. Heath, P. D., Moss, C., and Cartlidge, N. E. F., Ross syndrome and skin changes, *Neurology*, 32, 1041, 1982.
45. Hodson, A. K., Hurwitz, B. J., and Albrecht, R., Dysautonomia in Guillain-Barre syndrome with dorsal root ganglioneuropathy, Wallerian degeneration, and fatal myocarditis, *Ann. Neurol.*, 15, 88, 1984.
46. Hornbein, T. F., Griffo, Z. J., and Roos, A., Quantitation of chemoreceptor activity: interrelation of hypoxia and hypercapnia, *J. Neurophysiol.*, 24, 561, 1961.
47. Imaizumi, T., Granata, A. R., Benarroch, E. E., Sved, A. F., and Reis, D. J., Contributions of arginine vasopressin and the sympathetic nervous system to fulminating hypertension after destruction of the neurons of the caudal ventrolateral medulla in the rat, *J. Hypertens.*, 3, 491, 1985.
48. Jannetta, P. J., Segal, R., and Wolfson, S. K., Neurogenic hypertension: etiology and surgical treatment. 1. Observations in 53 patients, *Ann. Surg.*, 201, 391, 1985.
49. Jennings, G., Esler, M., and Holmes, R., Treatment of orthostatic hypotension with dihydroergotamine, *Br. Heart J.*, 2, 307, 1979.
50. Kaufman, M. P. and Rybicki, K. J., Discharge properties of group III and IV muscle afferents: their responses to mechanical and metabolic stimuli, *Circ. Res.*, 61 (Suppl. I), I60,1987.
51. Kaufmann, H., Brannan, T., Krakof, L., Yahr, M. D., and Mandeli, J., Treatment of orthostatic hypotension due to autonomic failure with a peripheral alpha-adrenergic agonist (midodrine), *Neurology*, 38, 951, 1988.
52. Kaufmann, H., Oribe, E., Miller, M., Knott, P., Whilshire-Clement, B. A., and Yahr, M. D., Hypotension-induced vasopressin release distinguishes between pure autonomic failure and multiple system atrophy with autonomic failure, *Neurology*, 42, 590, 1992.
53. Kochar, M. S. and Itskovitz, H. D., Treatment of idiopathic orthostatic hypotension (Shy-Drager syndrome) with indomethicin, *Lancet*, 8072, 1011, 1978.
54. Kume, A., Takahashi, A., Hashizume, Y., and Asai, J., A histometrical and comparative study on Purkinje cell loss and olivary nucleus cell loss in multiple system atrophy, *J. Neurol. Sci.*, 101, 178, 1991.
55. Levy, M. N., Sympathetic-parasympathetic interactions in the heart, *Circ. Res.*, 29, 437, 1971.
56. LeWitt, P., Hyperhydrosis and hypothermia responsive to oxybutynin, *Neurology*, 38, 506, 1988.
57. LeWitt, P. A., Newman, R. P., Greenberg, H. S., Rocher, L. L., Calne, D. B., and Ehrenkranz, J. R., Episodic hyperhidrosis, hypothermia, and agenesis of the corpus callosum, *Neurology*, 33, 1122, 1983.
58. Lin-Greenberg, A. and Taneja-Uppal, N., Dysautonomia and infection with the human immunodeficiency virus, *Ann. Intern. Med.*, 106, 167,1987.
59. Long, S. E. and Duffin, J., The medullary respiratory neurons: a review, *Can. J. Physiol. Pharmacol.*, 62, 161, 1984.
60. Lopes de Faria, S. R., Zanella, M. T., Andriolo, A., Ribeiro, A. B., and Chacra, A. R., Peripheral dopaminergic blockade for the treatment of diabetic orthostatic hypotension, *Clin. Pharmacol. Ther.*, 44, 670,1988.

61. Low, P. A., Walsh, J. C., Huang, C. Y., and McLeod, J. G., The sympathetic nervous system in alcoholic neuropathy: a clinical and pathological study, *Brain*, 98, 357, 1975.
62. Lundberg, J. M., Terenius, L., and Hökfelt, T., Neuropeptide Y (NPY)-like immunoreactivity in peripheral noradrenergic neurons and effects of NPY on somatic function, *Acta Physiol. Scand.*, 116, 477,1982.
63. Mancia, G. and Mark, A. L., Arterial baroreflexes in humans, in *Handbook of Physiology, Sec. 2: The Cardiovascular System. Vol. III: Peripheral Circulation and Organ Blood Flow, Part 2,* Shepherd, J. T. and Abboud, F. M., Eds., American Physiological Society, Bethesda, MD, 1983, 755.
64. Martinovits, G., Leventon, G., Goldhammer, Y., and Sadeh, M., Vocal cord paralysis as a presenting sign in the Shy-Drager syndrome, *J. Laryngol. Otol.*, 102, 280, 1988.
65. Mathias, C. J., Fosbraey, P., Da Costa, D. F., Thornley, A., and Bannister, R., The effect of desmopressin on nocturnal polyuria, overnight weight loss and morning postural hypotension in patients with autonomic failure, *Br. Med. J.*, 293, 353, 1986.
66. Matthews, M. R., Autonomic ganglia in multiple system atrophy and pure autonomic failure, in *Autonomic Failure: A Textbook of Clinical Disorders of the Autonomic Nervous System*, Bannister, R. and Mathias, C. J., Eds., Oxford Medical, Oxford, 1992, 593.
67. Money, K. E., Motion sickness, *Physiol. Rev.*, 50, 1,1970.
68. Mooradian, A. D., Morley, G. K., McGeachie, R., Lundgren, S., and Morley, J. E., Spontaneous periodic hypothermia, *Neurology*, 34, 79,1984.
69. Moriarty, M., Gibbins, I. L., Potter, E. K., and McCloskey, D. I., Comparison of the inhibitory roles of neuropeptide Y and galanin on cardiac vagal action in the dog, *Neurosci. Lett.*, 139, 275, 1992.
70. Mueller, R. A., Lundberg, D. B. A., Breese, G. R., Hedner, J., Hedner, T., and Jonason, J., The neuropharmacology of respiratory control, *Pharmacol. Rev.*, 34, 255, 1982.
71. Munsat, T. L. and Poussaint, A. F., Clinical manifestations and diagnosis of amyloid polyneuropathy: report of three cases, *Neurology*, 12, 413, 1962.
72. Neville, B. G. R. and Sladen, G. E., Acute autonomic neuropathy following primary herpes simplex infection, *J. Neurol. Neurosurg. Psychiatr.*, 47, 648, 1984.
73. Onrot, J., Golberg, M. R., Biaggioni, I., Hollister, A. S., Kincaid, D., and Robertson, D., Hemodynamic and humoral effects of caffeine in autonomic failure. Therapeutic implications for post-prandial hypotension, *N. Engl. J. Med.*, 313, 549, 1985.
74. Onrot, J., Goldberg, M. R., Biaggioni, I., Wiley, R. G., Hollister, A. S., and Robertson, D., Oral yohimbine in human autonomic failure, *Neurology*, 37, 215, 1987.
75. Palkovitz, M. and Zaborsky, L., Neuroanatomy of central cardiovascular control: nucleus tractus solitarii; afferent and efferent neuronal connections in relation to the baroreceptor reflex arc, in *Progress in Brain Research: Hypertension and Brain Mechanisms*, Dejong, W., Provost, A, P., and Shapiro, A. P., Eds., Elsevier, Amsterdam, 1977, 9.
76. Perera, R., Isola, L., and Kaufman, H., Erythropoetin therapy in primary autonomic failure, *Clin. Auton. Res.*, 4, 81, 1994.
77. Piazza, S., Furlan, R., Bevilacqua, M., Turiel, M., Crema, C., Porta, A., Lombardi, F., and Malliani, A., Short-term effects of octreotide in pure autonomic failure: changes in haemodynamics and neural cardiovascular control, *Clin. Auton. Res.*, 4, 202, 1994.
78. Polinsky, R. J., Shy-Drager syndrome, in *Parkinson's Disease and Movement Disorders,* Jankovic, J. and Tolosa, E., Eds., Urban and Schwarzenberg, Baltimore, 153.
79. Polinsky, R. J., Central nervous system control of blood pressure, in *Mild Hypertension: Current Controversies and New Approaches,* Weber, M. and Mathias, C., Eds., Steinkopff-Verlag, Darmstadt, 1984, 11.
80. Polinsky, R. J., Multiple system atrophy: clinical aspects, pathophysiology, and treatment, *Neurol. Clin.*, 2, 487, 1984.
81. Polinsky, R. J., Clinical autonomic pharmacology, *Neurol. Clin.*, 8, 77, 1990.
82. Polinsky, R. J., Kopin, I. J, Ebert, M. H., and Weise, V., Pharmacologic distinction of different orthostatic hypotension syndromes, *Neurology*, 31, 1, 1981.
83. Polinsky, R. J., Kopin, I. J., and Gullner, H. G., Effect of indomethacin inhibition of prostaglandin synthesis on blood pressure and plasma norepinephrine levels in patients with orthostatic hypotension, *Neurology*, 31 (Suppl.), 63, 1981.
84. Rechardt, L., Aalto-Setala, K., Purjeranta, M., Pelto-Huikko, M., and Kyosola, K., Peptidergic innervation of the human atrial myocardium: an electron microscopal and immunocytochemical study, *J. Auton. Nerv. Syst.*, 17, 21, 1986.
85. Reid, R. L., Central and peripheral autonomic control mechanisms, in *Autonomic Failure: A Textbook of Clinical Disorders of the Autonomic Nervous System*, Bannister, R., Ed., Oxford University Press, London, 1983, 17.

86. Reis, D. J., Ross, C., Granata, A. R., and Ruggiero, D. A., Role of the C1 area of the rostroventrolateral medulla in cardiovascular control, in *Brain Peptides and Catecholamines in Cardiovascular Control,* Buckley, J. P. and Ferrario, C., Eds., Raven Press, New York, 1987, 1.
87. Robertson, D., Goldberg, M. R., Hollister, A. S., Wade, D., and Robertson, R. M., Clonidine raises blood pressure in severe idiopathic orthostatic hypotension, *Am. J. Med.,* 74, 193, 1983.
88. Robertson, D., Kincaid, D. W., Haile, R. M., and Robertson, R. M., The head and neck discomfort of autonomic failure: an unrecognized aetiology of headache, *Clin. Auton. Res.,* 4, 99, 1994.
89. Ross, A. T., Progressive selective sudomotor degeneration, *Neurology,* 8, 809, 1958.
90. Sanfield, J. A., Linares, O. A., Cahalan, D. D., Forrester, J. M., Halter, J. B., and Rosen, S. G., Altered norepinephrine metabolism in Shapiro's syndrome, *Arch. Neurol.,* 46, 53, 1989.
91. Sawchenko, P. E. and Swanson, L. W., The organisation of noradrenergic pathways from the brainstem to the paraventricular and supraoptic nuclei in the rat, *Brain. Res. Rev.,* 4, 275, 1982.
92. Schmid, P. G., Eckstein, J. W., and Abboud, F. M., Effect of 9 a-fluorohydrocortisone on forearm vascular responses to norepinephrine, *Circulation,* 34, 620, 1966.
93. Siminoski, K., Lukinuk, C., and Armstrong, P. W., Treatment of postural hypotension with aerosol ergotamine, *C. M. A. J.,* 139, 52, 1988.
94. Spector, R. H. and Bachman, D. L., Bilateral Adie's tonic pupil with anhydrosis and hyperthermia, *Arch. Neurol.,* 41, 342, 1984.
95. Spokes, E., Bannister, R., and Oppenheimer, D., Multiple system atrophy with autonomic failure: clinical, histological, and neurochemical observations on four cases, *J. Neurol. Sci.,* 43, 59, 1979.
96. Spyer, K. M., Central nervous system control of the cardiovascular system, in *Autonomic Failure: A Textbook of Clinical Disorders of the Autonomic Nervous System,* Bannister, R., Ed., Oxford University Press, Oxford, 1989, 56.
97. Stewart, P. M. and Hensley, W. J., An acute attack of variegate porphyria complicated by severe autonomic neuropathy, *Aust. N. Z. J. Med.,* 11, 82, 1981.
98. Summers, Q. and Harris, A., Autonomic neuropathy after rubella infection, *Med. J. Aust.,* 147, 353, 1987.
99. Swanson, L. W., Sawchenko, P. E., Berod, A., Hartman, B. K., Helle, K. E., and Van Orden, D. E., An immunohistochemical study of the organisation of catecholaminergic cells and terminal fields in the paraventricular and suproptic nuclei of the hypothalamus, *J. Comp. Neurol.,* 196, 271, 1981.
100. Taddien, L. and Watson, C. J., The clinical porphyrias, *Semin. Hematol.,* 5, 335, 1968.
101. Takeda, R., Takeda, Y., Koshida, H., Ikeda, M., Yasuhara, S., Nagai, K., Morise, T., and Miyamori, I., Effect of metoclopramide on orthostatic hypotension in diabetes mellitus: failure to demonstrate the role of vasoconstrictive hormones, *Exp. Clin. Endocrinol.,* 91, 197, 1988.
102. Tang, P. C. and Gernandt, B. E., Autonomic responses to vestibular stimulation, *Exp. Neurol.,* 24, 558, 1969.
103. Thomaides, M. D., Bleasdale-Barr, K., Chaudhuri, K. R., Pavitt, D., Marsden, C. D., and Mathias, C. J., Cardiovascular and hormonal responses to liquid food challenge in idiopathic Parkinson's disease, multiple system atrophy and pure autonomic failure, *Neurology,* 43, 900, 1993.
104. Toscano, W. B. and Cowings, P. S., Heart rate variability during early adaptation to space, *Clin. Auton. Res.,* 4, 216, 1994.
105. Tsuda, Y., Kimura, K., Yoneda, S., Asai, T., Handa, N., Tanouchi, J., Inoue, M., and Abe, H., Hemodynamics in Shy-Drager syndrome and treatment with indomethacin, *Eur. Neurol.,* 22, 42, 1983.
106. Ulrich, J., Morphological basis of Adies's syndrome, *Eur. Neurol.,* 19, 390, 1980.
107. Unger T., Rohmeiss, P., Demmert, G., Ganton, D., Lanf, R. E., and Luft, F., Opposing cardiovascular effects of brain and plasma AVP: role of V1 and V2 AVP receptors, in *Brain Peptides and Catecholamines in Cardiovascular Control,* Buckley, J. P. and Ferrario, C., Eds., Raven Press, New York, 1987, 393.
108. van Lieshout, J. J., Ten Harkel, A. D., and Wieling, W., Physical manoeuvers for combating orthostatic dizziness in autonomic failure, *Lancet,* 339, 897, 1992.
109. van Lieshout, J. J., Wieling, W., Van Montfrans, G.A., Settels, J. J., Speelman, J. D., Endert, E., and Karemaker, J. M., Acute dysautonomia associated with Hodgkin's disease, *J. Neurol. Neurosurg. Psychiatr.,* 49, 830,1986.
110. Walker, B. R., Anderson, J. A., and Edwards, C. R., Clonidine therapy for Shapiro's syndrome, *Q. J. Med.,* 82, 235,1992.
111. Watkins, P. J. and Mackay, J. D., Cardiac denervation in diabetic neuropathy, *Ann. Intern. Med.,* 92, 304, 1980.
112. Widdicombe, J. G., Pulmonary and respiratory tract receptors, *J. Exp. Biol.,* 100, 41, 1982.
113. Wieling, W. and van Lieshout, J. J., Maintenance of postural normotension in humans, in *Clinical Autonomic Disorders,* Low, P. A., Ed., Little, Brown, Rochester, MN, 1993, 69.

114. Yahr, M. D. and Frontera, A. T., Acute autonomic neuropathy: its occurrence in infectious mononucleosis, *Arch. Neurol.*, 32, 132, 1975.
115. Yates, B. J., Goto, T., and Bolton, P. S., Response of neurons in the rostral ventrolateral medulla of the cat to natural vestibular stimulation, *Brain Res.*, 601, 255, 1993.
116. Yates, B. J., Jakus, J., and Miller, A. D., Vestibular effects on respiratory outflow in the decerebrate cat, *Brain Res.*, 629, 209, 1993.
117. Ziegler, M., Lake, C., and Kopin, I., The sympathetic nervous system defect in primary orthostatic hypotension, *N. Engl. J. Med.*, 296, 293, 1977.

INDEX